Particle Packing Method for Recycled Aggregate Concrete

Subhasis Pradhan · Shailendra Kumar ·
Sudhirkumar V. Barai

Particle Packing Method for Recycled Aggregate Concrete

Springer

Subhasis Pradhan
Birla Institute of Technology and Science
Pilani, Rajasthan, India

Sudhirkumar V. Barai
Birla Institute of Technology and Science
Pilani, Rajasthan, India

Shailendra Kumar
Department of Civil Engineering
Guru Ghasidas Vishwavidayalaya
Bilaspur, India

ISBN 978-981-99-7518-1 ISBN 978-981-99-7516-7 (eBook)
https://doi.org/10.1007/978-981-99-7516-7

This Springer imprint is published by the registered company Springer Nature Singapore Pte Ltd.
The registered company address is: 152 Beach Road, #21-01/04 Gateway East, Singapore 189721, Singapore

The man who moves a mountain
begins by carrying away small stones.

—Confucius

Dedicated to all structural designers and architecture who commit to create buildings as ecosystems!

Preface

Concrete is an attested building material because of its versatility and cost-effectiveness. Its production and use have escalated by about 12 times since the Second World War and in the present scenario, the yearly per capita concrete production is about 4.8 tonnes. The continued increase in concrete production causes the quick depletion of suitable aggregates, as it forms about 70% of the concrete volume and also produces huge amount of construction and demolition (C&D) waste. The use of waste concrete generated from C&D waste as a source of aggregate is a sustainable approach to reduce the huge dependency of concrete industries on the non-renewable resources as well as to reduce solid waste management issues. This book discusses the systematic research and major achievements by the authors while utilizing recycled coarse aggregate (RCA) for structural concrete preparation.

This book comprises 11 chapters, which are partitioned under 5 major parts. First part of the book discusses the material processing and characterization of all the raw materials utilized in the research. Second part presents the concrete mix proportioning and mixing methods employed for the preparation of recycled aggregate concrete (RAC) and natural aggregate concrete (NAC). The short-term mechanical performance of the prepared concrete mixes is discussed at multi-scale level in the third part. Also, in this part, the fracture behaviour and fracture parameters of RAC are critically analysed. Sustainability of the RAC is quantified through Life Cycle Assessment (LCA), which is explained in the fourth part. Finally, in the fifth part of the book, the performance of structural members (beam, column, and slab) is discussed critically.

The entire research is focussed to substitute the natural coarse aggregate (NCA) completely by commercially available RCA (yielded by crushing the waste concrete from C&D debris) and assess the performance of resulting RAC and structural members. The performance of fresh and hardened RAC is adversely affected owing to the inferior properties of RCA (mainly due to the presence of adhered mortar layer, micro-cracks, and old interfacial transition zone (ITZ)). In this context, Particle Packing Method (PPM) of mix design approach is proposed, which is based on the idea of void minimization through efficient packing of aggregates mixture consisting of different sizes of aggregates. Additionally, the Two-Stage Mixing Approach

(TSMA) is adopted. An improved short-term macro-mechanical property (compressive strength, tensile strength, and modulus of elasticity) substantiates the 100% use of RCA in RAC while employing PPM mix design and TSMA mixing methods.

The three-point bending (TPB) test is performed using three different sizes of single edge notched beam. The fracture energy is evaluated by using the load-CMOD curve obtained from the TPB test. The fracture energy and fracture toughness parameters of RAC are compared to NAC. The impact of mix design methods and types of coarse aggregates on fracture properties are analysed critically.

A multi-scale microstructural characterization is executed by using thermogravimetric analysis (TGA), nanoindentation technique, image analysis of back-scattered electrons (BSE) images, and X-ray microtomographic (XRT) images. The influence different types of aggregate and mix design approach on degree of hydration is investigated using the TGA test of hydration products. A method is proposed by combining two earlier methods to estimate degree of hydration from the recorded chemically bound water. The study on the influence of aggregate types and mix design methods on degree of hydration is presented. The relationship between compressive strength and degree of hydration is discussed. The nanoindentation technique and image analysis of BSE images are employed to determine the thickness, voids content, and unhydrated cement content of ITZ. The image analysis of XRT images facilitates the voids content in concrete specimens. An expression is proposed to predict the compressive strength of concrete from its microstructural characteristics and its correlation with the experimental results is substantiated.

A systematic analysis of the influence of RCA and PPM mix design method on life cycle assessment (LCA) of concrete as compared to those concrete prepared using NCA and conventional mix design method is conducted. Considering the Indian scenario a LCA based on cradle-to-gate theory is conducted and the environmental impacts are measured using CML baseline method with the help of SimaPro and Ecoinvent 3.1 database. The primary data regarding the preparation of NCA and RCA are collected from the respective production facilities. Transport activities are the second largest contributor in each category of environmental impacts after the influence of cement. From the sensitivity analysis, the maximum collection distance of C&D waste is optimized for different supply distance of processed RCA to obtain comparable environmental impact with natural aggregate concrete prepared using conventional mix design method in Indian context. The maximum possible supply distance of RCA is determined for different impact categories with specific collection distance of C&D waste.

The performance of reinforced RAC structural members (beam, column, and slab) is studied with the 100% use of RCA. For this, the concrete is prepared using PPM and TSMA due to the improved macro-mechanical properties, fracture behaviour, microstructural characteristics, and lesser environmental impacts. The flexure and shear behaviour of RAC beams is studied through experimental investigation and compared with convention concrete beams. Expressions are proposed to predict the ultimate strength and diagonal cracking strength of RAC beams with and without transverse reinforcement, respectively, by operating the database prepared by compiling the existing and present test results of RAC beams. Axial compressive strength of RAC columns is discussed in comparison with the NAC columns. The influence of the spacing of lateral ties is critically analysed. The existing expressions in codes for the estimation of the axial load carrying capacity are checked for their applicability of RAC columns. The experimental results on the punching shear strength and ultimate deflection of RAC slabs are compared with the NAC slabs. The initial cracking load and strain energy absorption by RAC slabs are analysed. Available codal expressions and Critical Shear Crack Theory for the prediction of the punching shear capacity of RAC slabs are verified.

The authors hope that this book will help the readers to understand the behaviour and performance of RAC. It can be an useful reference book for the graduate students, researchers, and concrete technologists to comprehend the research conducted on RAC and explore further in this domain. It can serve as a handy reference material for the practice engineers to employ PPM mix design method for preparing RAC and its subsequent structural applications. Also, this book will further accentuate the need of sustainable construction practice and the role of RCA yielded from C&D waste in this regard.

Pilani, India — Subhasis Pradhan
Bilaspur, India — Shailendra Kumar
Pilani, India — Sudhirkumar V. Barai

Acknowledgments

Sustainable development is no longer an option, rather it is a necessity. A sustainable development satisfies the needs of the present generation while keeping the ability of the future generations to accomplish their needs. The three pillars of sustainability, i.e. economy, society, and environment can be maintained by following the three important principles of sustainability, i.e. recycle, reuse, and reduce (the 3Rs). Being motivated from this principle, an effort has been made to shape the research conducted by the authors on utilization of recycled aggregate from construction and demolition waste and present in the form of a book. This could be made possible today just because of the direct and indirect support of many well-wishers, who helped to stay determined, drove to march fast, and gave strength to overcome the impediments in achieving the final objective.

At the outset, Subhasis expresses his gratitude to the co-authors Prof. Sudhirkumar V. Barai and Prof. Shailendra Kumar for their intense involvement, guidance, support, inspiration, and constant encouragement to explore new things and overcome the failures and proceed towards the goal during his Ph.D. program. Their comprehensive knowledge on the subject, grit, determination, hunger for creative ideas, and thought-provoking discussion helped to tackle new challenges in the subject. It has been a privilege for him to be associated with such wonderful persons in his research journey in the beautiful campus of IIT Kharagpur. He is grateful to Prof. Brajesh Kumar Dubey for his help and support in carrying out the Life Cycle Assessment (LCA) studies.

Subhasis expresses his sincere thanks to the Civil Engineering Department of Birla Institute of Technology & Science, Pilani, Pilani Campus, India for providing an amiable ambiance leading to successful completion of the book. This book could not have been completed without the full support and cooperation of the author's family members. He is grateful to his father Bhaskar Pradhan and mother Sabita Pradhan, who lent their unconditional love and emotional support and implanted the confidence to fight against all odds and helped to overcome the problems. He would like to express a word of admiration to his lovely wife Suchismita Subadini for her unprejudiced and diligent encouragement, patience, and understanding while enduring the completion of the book. He would like to thank his sister Bindi, brother-in-law Dibya, relatives, and friends for their unconditional love, affection, and care.

Shailendra is thankful to Guru Ghasidas Vishwavidyalaya, Bilaspur, India, for providing a conducive environment for the research and extending the support to avail the laboratory facilities for the Ph.D. research of Subhasis. Further, he would like to express a word of appreciation to his lovely wife, Neelam Prabha, for her support and encouragement and his sons Somil and Shankhin for bearing with the hardship during the preparation of the manuscript of the book.

Sudhir is grateful to Indian Institute of Technology, Kharagpur, India and Birla Institute of Technology & Science, Pilani, Pilani Campus, India, for providing stimulating environment for working on this book. Behind every accomplishment, the family plays an important role in sacrificing the precious time of togetherness. He would like to thank his wife—Parama—and lovely daughters—Sristi and Shailey—for all their warmth and support and helping him to sail through during difficult times.

Authors express their sincere thanks to Ministry of Human Resource and Development (MHRD), India for the financial support provided for the project on "Sustainable and Cost Effective Housing using Recycled Aggregate Based Concrete" under the mega project on Future of Cities. The entire project could smoothly run because of the support extended by IL&FS Environmental Infrastructure and Services Ltd. Plant (New Delhi) for providing recycled aggregate. The laboratory facilities provided by Indian Institute of Technology Kharagpur, West Bengal, India and Guru Ghasidas Vishwavidyalaya, Bilaspur, Chhattisgarh, India to carry out the experiments are gratefully acknowledged.

Above all, we would like to express profound gratitude and appreciation to the Almighty.

Pilani, India — Subhasis Pradhan
Bilaspur, India — Shailendra Kumar
Pilani, India — Sudhirkumar V. Barai

Contents

About the Authors

Dr. Subhasis Pradhan is presently working as an Assistant Professor in the Department of Civil Engineering at Birla Institute of Technology & Science, Pilani, India. He graduated from Veer Surendra Sai University of Technology, India in 2011. He completed his M.Tech. in Structural Engineering from Indian Institute of Technology (IIT) Guwahati in 2014 and subsequently joined IIT Kharagpur for his Ph.D. research. His research on "Performance of Recycled Aggregate Concrete and Structural Members: Particle Packing Method of Mix Design Approach" during Ph.D. earned the acknowledgement from the research community. In January 2020, he joined Nanyang Technological University (NTU) as a Postdoctoral Research Fellow. During Postdoc he worked on alternative binders, such as LC^3 binder and one-part geopolymer while using marine clay, which is one of the major construction wastes in Singapore. His Ph.D. and postdoc research works are published in journals of international repute. Moreover, a Singapore patent is filed based on his research on LC^3 binder using marine clay. His main research interests are in concrete technology, sustainable construction materials, alternative binders, geopolymer, CO_2 sequestration in concrete, microstructural characterization, fracture behavior of concrete, Life Cycle Assessment and behavior of reinforced concrete members (beam, column and slab).

Dr. Shailendra Kumar is a Professor of Civil Engineering Department, School of Engineering & Technology, Guru Ghasidas Vishwavidyalaya, India. Before this, he was a faculty member at National Institute of Technology (NIT) Jamshedpur. He pursued his undergraduate degree from NIT Jamshedpur, post graduate degree from NIT Rourkela and Ph.D. from Indian Institute of Technology (IIT) Kharagpur. His main research areas are concrete fracture mechanics, fibrereinforced concrete, alternate construction materials, etc. He published more than 75 research papers in national/ international journals/conferences and authored 02 books on Concrete Fracture Models (published by Springer) and Applications and Simplified Testing Methods of Double-K Concrete Fracture Model. He has executed 02 sponsored research projects under MHRD/UGC and supervised many M.Tech. and Ph.D. thesis works. He has been one of the members of the Technical Committee of "RILEM

TC265-TDK" who developed the theoretical and experimental standardization of Double-K Fracture Model of concrete which have been published through 04 research papers in Materials & Structures in 2021.

Dr. Sudhirkumar V. Barai is a Director and Senior Professor of Civil Engineering at Birla Institute of Technology & Science Pilani, India. He is also a Professor in the Department of Civil Engineering, Indian Institute of Technology (IIT) Kharagpur availing lien for the current position since January 2020. In 1995, he obtained Ph.D (Eng.) from Indian Institute of Science (IISc) Bangalore. He received the degrees of B.E. (Civil) and M.E. (Civil) with specialization in structural engineering in 1987 and 1989, respectively, from the Faculty of Engineering, MS University of Baroda, India. He was Erskine Visiting Fellow at University of Canterbury, Christchurch, New Zealand during May-June 2008. He was also visiting scientist at National University of Singapore during May-July 2003. He was recipient of BOYSCAST fellowship and visited Department of Civil and Environmental Engineering, Carnegie Mellon University, Pittsburgh, USA during May-November 2000. He was a post-doctoral fellow at Department of Solid Mechanics, Materials and Structures, Tel Aviv University, Israel during February 1997-July 1998. His areas of research are sustainable construction materials, structural health monitoring and computational intelligence in engineering. He has published more than 250+ papers in leading international journals and conferences in his research fields. He has co—authored four books—(1) Concrete Fracture Models and Applications (2) Shear Strengthening of T—beam with GFRP: A Systematic Approach and (3) Systematic Approach of Characterisation and Behaviour of Recycled Aggregate Concrete, (4) Stability and Failure of High Performance Composite Structures, which have been published by Springer.

Nomenclature

Abbreviations

ACI	American Concrete Institute
ADP (fossil fuels)	Abiotic depletion due to fossil fuels
ADP	Abiotic depletion
AP	Acidification potential
ASTM	American standard testing of materials
BIS	Bureau of Indian Standard
BS	British Standard
BSE	Back-scattered electrons
C&D	Construction and demolition
CH	Calcium hydroxide
CMOD	Crack mouth opening displacement
CSA	Canadian Standards Association
CSH	Calcium silicate hydrate
DM	Double mixing method
EP	Eutrophication potential
FT-IR	Fourier Transformed Infrared Spectroscopy
GWP	Global warming potential
INR	International Organization for Standardization currency code for the Indian rupee
IS	Indian Standard
ISO	International Organization for Standardization
ITZ	Interfacial Transition Zone
JSCE	Japan Society of Civil Engineers
LCA	Life cycle assessment
LCI	Life cycle inventory
LCIA	Life cycle impact assessment
NAC IS	NAC prepared using IS:10262 mix design method
NAC PPM	NAC prepared using PPM mix design approach
NAC	Natural aggregate concrete

NZS	New Zealand Standard
ODP	Ozone layer depletion
OPC	Ordinary Portland Cement
PD	Packing density
POCP	Tropospheric ozone photochemical oxidants
PPM	Particle packing method
RA	Recycled aggregate
RAC IS	RAC prepared using IS:10262 mix design method
RAC PPM	RAC prepared using PPM mix design approach
RAC	Recycled aggregate concrete
RC	Reinforced concrete
RCA	Recycled coarse aggregate
RFA	Recycled fine aggregate
SEM	Scanning electron microscope
SEN	Single edge notch
TM	Triple mixing method
TPB	Three-point bending
TSMA	Two-stage mixing approach
VC	Voids content

Symbols

f_c	Compressive strength
f_t	Split tensile strength
f_r	Flexural tensile strength
E	Modulus of elasticity
r	Replacement ratio of RCA
G_F	Fracture energy
K_{Ic}^{ini}	Initial fracture toughness
K_{Ic}^{un}	Unstable fracture toughness
K_{Ic}^{c}	Cohesive fracture toughness
P	Load
P_{ini}	Initial cracking load
P_u	Peak load
CTOD	Crack tip opening displacement
CTOD_c	Critical crack tip opening displacement
β_n	Brittleness index
$l_{ch,mod}$	Modified characteristic length
d_a	Maximum size of the aggregate
W_B	Chemically bound water
$W_{B\infty}$	Chemically bound water at infinite time
α	Degree of hydration

E_r	Reduced elastic modulus
H	Hardness
ν	Poisson's ratio
ν_i	Poisson's ratio of the indenter
E_i	Modulus of elasticity of indenter
p_i	Porosity of the ITZ
t_i	ITZ thickness
p	Porosity of the specimen
f_{ca}	Coarse aggregate fraction
A_{cr}	Aggregate crushing value
ρ	Longitudinal reinforcement content
ρ_t	Tensile reinforcement content
ρ_c	Compression reinforcement content
a/d	Shear span-to-depth ratio
V_{cr}	Cracking load
V_y	Yield load
V_u	Ultimate load
δ_{cr}	Displacement corresponds to V_{cr}
δ_y	Displacement corresponds to V_y
δ_u	Displacement corresponds to V_u
f_y	Yield strength of longitudinal reinforcement
ρ_w	Transverse reinforcement content
f_{yw}	Yield strength of transverse reinforcement
v_c	Diagonal cracking strength of RC beam
v_u	Ultimate strength of RC beam
$v_{c,NAC}$	Diagonal cracking strength of NAC beams
$v_{c,RAC}$	Diagonal cracking strength of RAC beams
$v_{u,NAC}$	Ultimate strength of NAC beams
$v_{u,RAC}$	Ultimate strength of RAC beams
ρ_t	Transverse reinforcement content in column
s	Spacing of lateral ties in column
$V_{R,cr}$	Initial cracking load
$V_{R,u}$	Ultimate punching load
$V_{Rd,c}$	Punching shear strength
b_0	Control perimeter
ψ	Rotation of the slab

List of Figures

List of Tables

Chapter 1
Background on Techniques for Sustainable Use of Recycled Aggregate and Application of Particle Packing Method

1.1 Overview on Sustainable Development

Concrete is an attested building material because of its versatility and cost-effectiveness. Thus, it is the second most consumed material in the world after water (ISO/TC 71 2005; Weil et al. 2006; Knoeri et al. 2013). Its production and use is escalated by about 12 times since the Second World War and in present scenario the yearly per capita concrete production is about 4.8 tonnes (Monkman and MacDonald 2017). The rapid socio-economic development leads to the unavoidable process of industrialization and urbanization, which is the inception of this situation. Concrete is a composite material, which typically consists of coarse aggregates, fine aggregates, and a binder matrix formed from the chemical reactions of cement and water. Approximately, 70% to 80% of its total volume is occupied by aggregates (Mindess et al. 2003; Alexander and Mindess 2005), which is extracted from nature. The uncontrolled mining of qualified aggregates for construction industry causes the rapid depletion of the natural resources. In this incessant process, a balance must be maintained between development of the society and preserving the integrity of the environment. Moreover, large quantities of construction and demolition (C&D) wastes are generated by the construction activities and it comprises approximately 30%–35% of global waste (Construction materials recycling association 2005; Fischer and Werge 2009; Marzouk and Azab 2014). In the recent past, the waste concrete of C&D wastes has been explored as a source of aggregates by the researchers. The initial research on the recycled aggregate (RA) extracted from waste concrete showed less satisfactory results considering its mechanical performance in concrete. However, the sustainability aspects of it was imaginable, which fueled for future research to have deeper understanding and improvement of this special concrete. Gradually different techniques were suggested with a background of comprehensive research to improve the quality of recycled aggregate concrete (RAC), which made recycled aggregate a viable alternative to conventional aggregate. This boosts the market size of global C&D waste, which will be about USD 34.4 billion by 2026 as per Construc-

S. Pradhan et al., *Particle Packing Method for Recycled Aggregate Concrete*,
https://doi.org/10.1007/978-981-99-7516-7_1

tion & Demolition Waste Market by type (Sand, Soil & Gravel, Concrete, Bricks & Masonry, Wood, Metal), Source (Residential, Commercial, Industrial, Municipal), & Region (APAC, North America, Europe, MEA, & South America)–Global forecast to 2026 (2021).

1.2 RAC in Sustainable Construction

Sustainable development is no longer an option, rather it is a necessity. A sustainable development satisfies the needs of the present generation while keeping the ability of the future generations to accomplish their needs. The three pillars of sustainability, i.e. economy, society, and environment can be maintained by following the three important principles of sustainability, i.e. recycle, reuse, and reduce (the 3Rs). The ever-increasing demand of new constructions urges for the usage of sustainable construction materials to reduce the associated carbon footprint. The life cycle of a concrete structure starts with the extraction of raw materials for aggregates, binder, water, and reinforcement from nature. Usually after the service life of a structure it is demolished and the waste may be dumped in the landfill. In this whole cycle, it not only depletes the natural resources but also causes damage to the environment. Owing to the huge demand for suitable aggregates, sustainable alternatives are continuously explored. The RA attracts the concrete technologists as it satisfies the initial criteria of being sustainable and the reasons are presented in Fig. 1.1. However, researchers are still engaged to establish the appropriate use of this material in concrete with satisfactory performance.

The RA can be in the form of crushed concrete, crushed brick, or broken glass pieces. The scope of this book is limited to the discussion on RA yielded from waste concrete. The source of waste concrete can be from demolished building, concrete road beds, unused concrete in ready-mix concrete plant, rejected precast

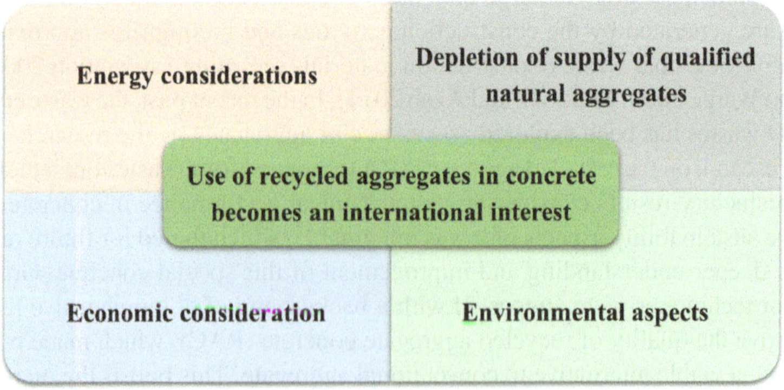

Fig. 1.1 Reasons for RA as an alternative to natural aggregates (Authors')

concrete member, and laboratory-tested concrete specimens. Based on their size the extracted RA can be categorized as recycled coarse aggregate (RCA) or recycled fine aggregate (RFA). The natural aggregate can be partially or completely replaced by RA to prepare the RAC. As the RA is yielded from the C&D waste, the burden on natural resources and environmental issues due to the solid waste generation and landfill can be minimized simultaneously. All these aspects should be accounted to define its sustainability aspects. Life cycle assessment (LCA) is a standard protocol to quantitatively assess the environmental impact of a product or system according to ISO (ISO 14040 2006; ISO 14044 2006). In order to compare the sustainability advantages of RAC with respect to the conventional concrete, the extraction of RA from C&D waste as well as the preparation of RAC using RA is to be accounted in the study.

1.3 Problems in RAC

1.3.1 General Problems

The engineering properties of RAC are not consistent along with its sustainability aspects. Fundamentally, RA is categorized separately from the natural aggregate on the basis of the presence of attached mortar layer. Natural aggregate and adhered hardened mortar are the components of RA. The adhered mortar layer and its quantity are the primary reasons for the inferior physical and mechanical properties of RA. Apart from these, the type of parent aggregate, strength of the parent concrete, age and exposure condition of parent concrete, number of crushing stages, and the presence of micro-cracks also affect the properties of RA. Consequently, the performance of RAC is also influenced by the incorporation of RA. Moreover, the quantity of replacement of RA is an important factor to dictate the performance of RAC. The variability in macro-mechanical performance of RAC is very significant due to the large variation in the inherent characteristics of RA. The degree of hydration, voids content, and interfacial transition zone (ITZ) in RAC are different from the conventional concrete. These parameters influence the micro-mechanical properties, which is also reflected in its macro-mechanical performance. Consequently, the behaviour and performance of reinforced RAC structural members are adversely affected due to the effect of RA and resulting RAC.

1.3.2 Problems in India

India being a developing country invests around 10% of the GDP on construction activities (Construction market in India 2015). The required aggregates with necessary quality are mined from the nature. A downside to this, approximately 25 to

30 million tonnes of C&D waste are generated in India and only 5% of which is processed to extract RA (Central Pollution Control Board, Ministry of Environment, Forests & Climate Change, India 2017). The awareness as well as the facilities to utilize the waste concrete sorted from C&D waste is fundamentally dissimilar in India as compared to the other developed countries, which requires reliable scientific support for its confident application to prepare structural concrete. Moreover, IS: 383 (2016) suggests for the replacement of RA up to 100% for lean concrete and recycled coarse aggregate (RCA) up to 100%, 20%, and 25% for lean concrete, plain concrete, and reinforced concrete, respectively. The prevalent use of RCA in reinforced concrete with higher replacement content is possible by preparing RAC with satisfactory performance with respect to natural aggregate concrete (NAC). Hence, necessary methods must be employed, which should be simple and effective to encourage its large-scale application, while replacing completely the NCA with RCA. Moreover, a synergistic effort among various government departments, research groups, designers, construction engineers, and certifying authorities is required.

1.4 Solutions to Address the Problems in RAC

The major problem in RAC arises due to the mortar layer adhered to the RA. This affects the fresh as well as hardened concrete properties. Moreover, the long-term performance of concrete and behaviour of reinforced concrete members are adversely affected because of the inferior quality of this fundamental entity. Consequently, extensive researches are conducted to deal with the attached mortar-related issues in different ways. Direct techniques are developed to remove the attached mortar and strengthen the attached mortar, which help in improving the RA characteristics and thus the performance of RAC. Moreover, some indirect approaches are developed, such as modified mix design methods and modified mixing methods to improve the RAC performance.

1.4.1 Treatment Methods to Remove the Attached Mortar

The characteristics of RA can be improved by the removal of attached mortar through chemical, mechanical, or thermal treatment methods. Pre-soaking method proposed by Katz (2004) treats the RCA by water washing, which partially removes the attached mortar and dust particles. The engineering properties of RCA are improved by ultrasonic bath treatment, and hence the compressive strength of RAC is improved by 7%.

A modified mechanical grinding method proposed by Dimitriou et al. (2018) in combination with pre-soaking method to remove the attached mortar. However, the formation of micro-cracks is unavoidable in this method (Xiao 2018). Thus, the partial removal of attached mortar and micro-cracks formation will have contrasting influence on the characteristics of RCA as well as RAC.

Bru et al. (2014) proposed a combined microwave heat treatment and impact crushing process to remove the attached mortar in RCA. As this method follows a selective heating approach, the degradation of RCA at higher heat due to mineralogical transformation is less likely to happen. However, the feasibility of the method to produce large quantity of RCA is questionable (Mistri et al. 2020).

The chemical treatment of RCA by immersing it in acidic solution to loosen the bond between aggregate and mortar was explored by several researchers (Tam et al. 2007; Kumar and Minocha 2018; Ismail and Ramli 2013; Katkhuda and Shatarat 2017; Kim et al. 2018). In this context, different acids, such as hydrochloric acid, sulfuric acid, acetic acid, and phosphoric acids, are used in suitable concentrations. Kumar and Minocha (2018) proposed a combined thermal and chemical treatment method, in which the RA is exposed at 300°C and 0.7 M hydrochloric acid solution to detach the attached mortar. However, the chemical treatment or thermo-chemical methods will not only contaminate the RCA but also create safety issues while employing in large-scale production.

1.4.2 Treatment Methods to Strengthen the Attached Mortar

The mortar adhered on the surface of RA is porous in nature. The methods to strength the attached mortar target to reduce the porosity by using different fillers. The fillers can be pozzolana slurry (Li 2009; Li et al. 2017), emulsion polymer (Kou and Poon 2010), or calcium carbonate ($CaCO_3$) precipitants from accelerated carbonation (Pu et al. 2021; Liu et al. 2021) or bacterial deposition (Singh et al. 2018) processes. These techniques strengthen the attached mortar as well as improve the old ITZ between the parent aggregate and adhered mortar (Mistri et al. 2020). In addition to these methods, coating of cement slurry (Martirena et al. 2017) and emulsion of sodium silicate solution (Pan et al. 2017) are used to surface coat the RA and reduce its porosity with resulting products. Although these methods improve the characteristics of RA and consequently the performance of RAC, the commercial application of these laboratory-tested methods seems less suitable.

1.4.3 Mix Design Methods for RAC

The various treatment methods focus on removal or strengthening of the major culprit in RA, i.e. the adhered mortar. However, the feasibility of these methods in mass application is not considered. To address this issue in addition to the inherent problems associated with RA, modified mix design methods were explored. Direct weight replacement (DWR) method (Dhir et al. 1999), equivalent mortar volume (EMV) method (Fathifazl et al. 2009), and direct volume replacement (DVR)

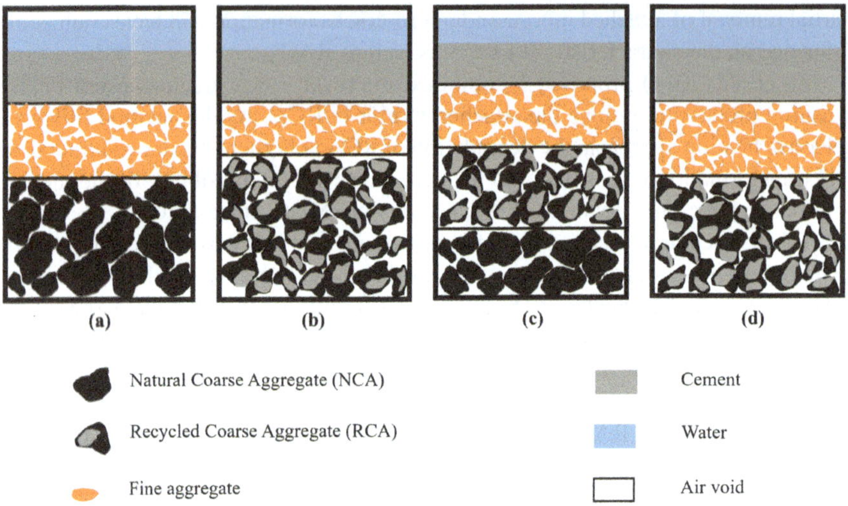

Fig. 1.2 Different mix design methods: **a** Conventional method, **b** DWR method, **c** EMV method, and **d** DVR method (Authors')

method (Knaack and Kurama 2013) are three different aggregate replacement methods employed by different researchers for the proportioning of RAC. The comparative illustration of required materials in these three mix design methods with respect to the conventional mix design method for NAC is represented in Fig. 1.2. Recently, the Particle Packing Method (PPM) is employed by Pradhan et al. (2017) to design the mix for RAC. This method minimizes the negative effects of RA in RAC by reducing the porosity through a compact concrete mixture.

Direct Weight Replacement (DWR) Method

In DWR method, the total weight of coarse aggregate (mixture of NCA and RCA), cement, and water content is kept constant for any replacement percentage of RCA (Fig. 1.2b). This method can be employed for any replacement level of RCA. The density of RCA is lower than NCA, and thus the volume of RCA is higher than NCA for the same weight. To compensate the increase in volume due to RCA replacement, fine aggregate content is reduced in DWR method-based RAC concrete. As the fine aggregate content is lower with respect to the conventional mix design method, the fresh mortar content will also be lower. Consequently, the total amount of mortar, i.e. fresh and old adhered mortar will be controlled to limited extent. The compressive strength performance with increasing RCA replacement level shows insignificant difference with respect to the NAC (Dhir et al. 1999; Knaack and Kurama 2013).

Equivalent Mortar Volume (EMV) Method

Fathifazl et al. (2009) proposed the EMV method and treated RCA as a two-phase material rather than a single coarse aggregate, i.e. the adhered mortar and parent aggregate. The determination of attached mortar content in RCA is essential in this method as the same amount of fresh mortar content will be adjusted in the designed RAC (Fig. 1.2c). Thus, the content of attached mortar along with the fresh mortar of RAC is kept same as the fresh mortar content of NAC. Fathifazl et al. (2009) exposed the RCA in five freeze–thaw cycles after immersing it in sodium sulphate solution for 24 hours to determine the content of attached mortar. The difficulties in removing the attached mortar completely are discussed in earlier section. Thus, the 100% replacement of RCA will not be theoretically possible in EMV method (Pradhan et al. 2017). Marginal improvement in the mechanical performance was reported for EMV method-based RAC (Fathifazl et al. 2009; Knaack and Kurama 2013).

Direct Volume Replacement (DVR) Method

The direct volume replacement (DVR) method considers RCA as a single-phase coarse aggregate. In this method, the volume of replacing RCA is same as the volume of NCA. Hence, the fine aggregate, cement, and water contents are unaffected as the total volume of coarse aggregate remains constant. The existing codes for mix design of conventional concrete can be followed in this regard, as RCA is not considered as a different type of coarse aggregate as compared to NCA. This method does not help to improve the mechanical performance of RAC and significant reduction in performance is observed as the replacement level of RCA increases (Knaack and Kurama 2013).

The mix proportions of RAC were designed by employing the aforementioned three methods by Knaack and Kurama (2013) and compared the workability, compressive strength, and modulus of elasticity. The test results concluded that DVR method provides better workability condition as compared to DWR and EMV methods of mix design. In EMV method, workability reduces significantly and for replacement ratio more than 20% needs changes in mix proportions and high dose of water reducing admixture. However, the compressive strength is not influenced significantly by any of these mix design methods at any replacement ratio.

Particle Packing Method (PPM)

Researchers in a variety of technological sectors, including ceramics, powder metallurgy, asphalt, and concrete technology, recognize the benefits of enhanced particle packing and the use of the Particle Packing Method (PPM) in this context. It has, however, been utilized in concrete mix proportioning very seldom. Using a mixture of different sized particles (coarse and fine aggregates in concrete), PPM aims to achieve the highest packing density (PD) achievable. By inserting smaller particles into the spaces between larger ones, this technique allows for the optimization of varied sized particles. As a result, voids are minimized while the PD of the particle combination increases. The porosity of the aggregate mixture will be reduced by using this PPM technique in concrete mix proportioning. As a result, the quantity of paste (a binder and water mixture) needed can be reduced. Using this technique, a concrete mix that is both economically and environmentally sustainable can be designed.

For PPM-based mix proportioning, the best combination of the various aggregate sizes can be found experimentally or via theoretical models. In an experimental procedure, smaller aggregates are mixed in with larger aggregates. The maximum achievable PD for the combination of coarse and fine aggregate and the associated weight percentage of the chosen size of coarse aggregates and fine aggregate are computed. PD and voids content (VC) are then estimated using Eqs. 1.4.1 and 1.4.2, respectively, after the bulk density is determined in this context.

$$\mathrm{PD} = \sum \frac{\text{Bulk density} \times \text{Weight fraction}}{\text{Specific gravity}} \tag{1.4.1}$$

$$\mathrm{VC} = 1 - \sum \frac{\text{Bulk density}}{\text{Specific gravity}} \tag{1.4.2}$$

The developed theoretical packing models can be divided into discrete and continuous models. In the particle distribution system, the continuous model assumes the existence of aggregates in every plausible size. According to the foundational research done by Feret (1892) and Fuller and Thompson (1907), the aggregate can be continuously graded to improve the qualities of concrete. In order to reach the highest density, Fuller and Thompson (1907) devised the aggregate gradation curve. Later, by assuming the smallest particle to be infinitesimally small, Andreasen and Andersen (1930) provided an equation (Eq. 1.4.3) for particle packing. By taking into account the impact of the smallest sized particle on aggregate gradation, Funk and Dinger (2013) proposed a modified Andreasen equation (Eq. 1.4.4).

$$CPFT = \left(\frac{d}{D}\right)^n \times 100 \tag{1.4.3}$$

$$CPFT = \left(\frac{d^q - d_0^q}{D^q - d_0^q}\right) \times 100 \tag{1.4.4}$$

where $CPFT$ = cumulative percent finer than, d = particle size, d_0 = minimum size of the particle, D = maximum size of the particle, n = exponent of the equation $(0.45 - 0.70)$, and q = distribution coefficient (0.25–0.37).

A number of discrete particle packing models, such as Furnas model, Aim model, Modified Toufar model, Strovall model, Dewar model, Linear Packing Model (LPM), Solid Suspension Model (SSM), Compressible Packing Model (CPM), and Three-Parameter model are available to estimate packing density (Jones et al. 2002). These models can be divided into three categories: multimodal packing models, ternary packing models, and binary packing models. In order to attain the highest density possible, the fundamental premise of these discrete particle packing models takes into account the contribution of each size class of particle in the mixture of particles. The binary packing models put out by Furnas (1929) and Aim and Le Goff (1968) have shown that mixing two different sized particles can minimize voids. Without compromising the packing of the larger ones, the smaller sized particles can be used to fill in the spaces between the larger ones. These two models, however, are not appropriate for use in concrete mix proportioning since their theories only take into account spherical-shaped particles and do not take interaction effects into account. Later, Powers (1968) took into account the interaction effect linked to the packing of the aggregates, such as wall effect and loosening effect, and came to the conclusion that the void ratio could be minimized for a specific combination of two different sizes of aggregates.

By combining the coarse, medium, and fine aggregates, Toufar et al. (1976) expanded the idea of the binary packing model to create a ternary mixture. The final addition to the combination of coarse and medium-sized aggregates in this ternary packing technique is the fine aggregate. With a binary mixing of coarse and medium-sized aggregates, it is represented as a binary packing model. Later, Goltermann et al. (1997) added the characteristic diameter and eigenpacking degree of the aggregates to get beyond the Toufar model's constraints resulting from the assumption of spherical and mono-sized particles. Characteristic diameter is the diameter for which the cumulative probability of the Rosin–Rammler distribution is 0.37 (size associated with 63% of the material passing).

By taking into account the log mean size of each material (coarse aggregate, fine aggregate, and cementitious material) included in the mixture, a method to estimate the void ratio for a specific combination of the aggregates mixture was proposed by Anderson and Dewar (2003). A multi-grain mixture's packing density is taken into account by the multi-component mixture model (linear packing model) put forth by Stovall et al. (1986). This packing density is depicted as a function of the fractional solid volume of each size of grain present in the mixture. By adapting the Linear Packing Model (LPM), De Larrard (De Larrard 1989, 1999; De Larrard and Sedran 1994, 2002) introduced Solid Suspension Model (SSM) and Compressible Packing Model (CPM) for the design of concrete mix. The SSM is based on an analogy of Mooney's viscosity model (Mooney 1951), where a portion of the water introduced to the concrete mix helps to fill in the spaces between the aggregates and binder mixture and the remaining portion aids in workability. For the same water content, the mixture's reduced voids aid in making concrete more workable. By introducing a virtual packing factor, De Larrard et al. were able to account for the random packing of particles in SSM, which is distinct from ideal packing. Later, De Larrard and Sedran (1994) introduced a concrete mix design approach based on SSM.

The CPM was derived based on the concept of LPM (De Larrard 1989, 2009; De Larrard and Sedran 1994, 2002). Apart from the interaction effects (wall effect and loosening effect), De Larrard also accounted the compaction effort associated with the process of packing of particle mixture in CPM. In this regard, a compaction index (K) was introduced for the estimation of the actual packing density (α_t) from virtual packing density (β). Kwan et al. (Kwan and Wong 2008; Kwan et al. 2012, 2013; Wong and Kwan 2005, 2008; Wong 2013) proposed a three-parameter model, where another interaction effect known as the wedging effect is accounted along with wall effect and loosening effect. The trapped incomplete layers of small particles between the coarse particles cause the wedging effect that is seen. Like the wall effect or the loosening effect, it is independent of the relative dominance of coarse or fine particles. Compared to previous two-parameter models, the three-parameter model displays a better association with the experimental results. The use of it for aggregates with irregular shapes, however, needs more research.

Glavind et al. (Glavind et al. 1993; Glavind and Pedersen 1999; Glavind 2009) shown that it is always preferable to obtain minimal voids for concrete mix design by packing the aggregates as densely as possible, which lowers the binder content. In addition to boosting the strength and durability (creep and shrinkage) performance of concrete, this is advantageous economically because it reduces the amount of required binder. Rao and Krishnamoothy (1993) noted that the proportioning of various-sized aggregates can be done using the theoretical Fuller grading to produce the least amount of void content, and he developed an empirical equation by fitting this trend to figure out the ratios of coarse and fine aggregates. By filling the vacant space with cement paste, Jacobsen and Arntsen (2008) conducted studies at various aggregate packing densities and discovered that there is a decrease in packing density as a result of interaction, aggregate shape, and particle size distribution. To achieve the highest packing density possible, Kolonko et al. (2010) conducted a systematic search of the particle size distribution. In this procedure, the larger size spheres serve as the inner boundary of the container while the smaller size particles serve as filler material in between the larger ones. With the development of the Compaction-Interaction Packing Model, which can take into account aggregates and fillers when calculating packing density, higher packing densities were achieved (Fennis et al. 2009, 2012a, b). Therefore, less water needs to be added to voids when packing has a higher density.

To optimize the components of the concrete mix, Kwan (2013) proposed a three-tiered technique. The packing densities of cementitious materials, aggregates less than 1.2 mm, and aggregates bigger than 1.2 mm, respectively, are used to determine the water demand, paste demand, and mortar demand sequentially in successive stages. Wet packing method (Fung et al. 2009; Kwan et al. 2012; Wong 2013; Li and Kwan 2014) can further improve cementitious material packing density. Kwan et al. (2012) suggested that wet packing should be used since new mortar and concrete are both wet. It was found that the proper blending of fine and coarse aggregate increases packing density under wet conditions compared to dry conditions, and that less fine aggregate is needed to get maximum packing density.

1.4.4 Mixing Methods for RAC

The characteristics of RCA can be improved by treating it with appropriate coating. This idea was employed by researchers while modifying the normal mixing approach (Fig. 1.3). The methods, such as double mixing (DM) method (Otsuki et al. 2003), two-stage mixing approach (TSMA) (Tam et al. 2005; Tam and Tam 2008), and Triple Mixing (TM) method (Kong et al. 2010), were developed to enhance the performance of RAC. In essence, water is added to each of these mixing techniques in stages. However, each stage of these mixing processes uses a different volume of water, and each stage's mixing time is also different. A coating of cement or mineral admixture is formed on the surface of RCA with the help of the addition of water in several processes. This enhances the strength of both the old and new ITZ and lowers RCA porosity.

Otsuki et al. (2003) added half of the water to the dry mix of coarse and fine aggregate in DM method (Fig. 1.4). This method improved the ITZ quality, strength and durability of RAC. Similarly, in TSMA, two equal halves of the total water are added at two stages of the mixing process as shown in Fig. 1.5. The TSMA helped in developing a stronger ITZ by effectively filling the voids and cracks present in RCA (Tam et al. 2005). The increment in compressive strength of RAC was up to 21% by using TSMA, which was also effective in enhancing durability. Kong et al. (2010) proposed the triple mixing method (TM) to improve the quality of ITZ. In this method, mineral admixture, cement, and mixture of water and superplasticizer are added at three different phases (Fig. 1.6). The TM contributed significantly in improving workability, compressive strength, and flexural strength in comparison to DM method (Li 2009).

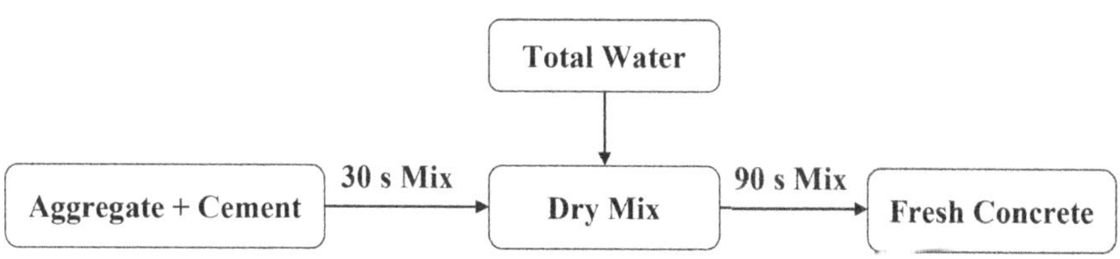

Fig. 1.3 Schematic diagram of normal mixing approach (Authors')

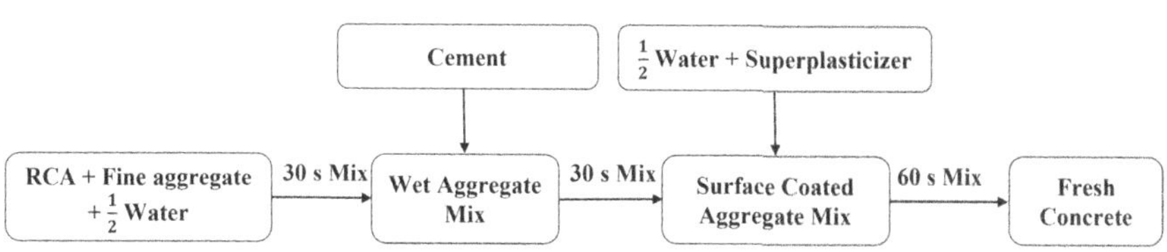

Fig. 1.4 Schematic diagram of double mixing method (Otsuki et al. 2003)

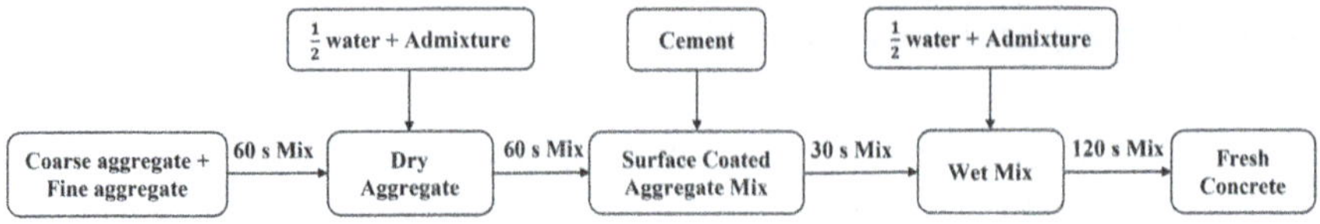

Fig. 1.5 Schematic diagram of two-stage mixing approach (Tam et al. 2005)

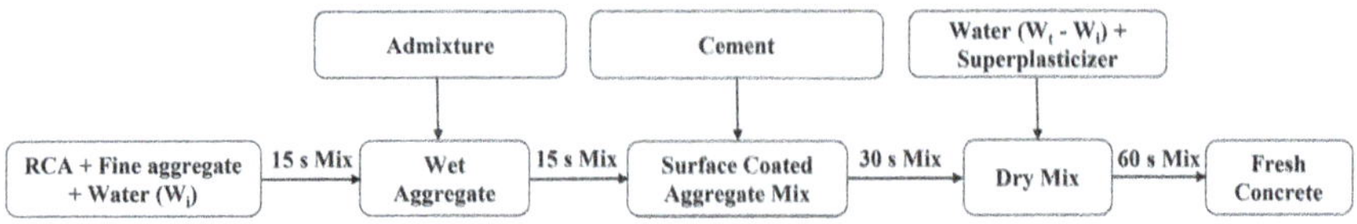

Fig. 1.6 Schematic diagram of triple mixing method (Kong et al. 2010)

1.5 Organization of the Book

This book meticulously examines how to create sustainable RAC using RCA, as well as how to apply it to structural elements. It is divided into five main sections: material processing and characterization (Part A); proportioning and mixing of the concrete mix (Part B); multi-scale performance evaluation of the concrete; sustainability evaluation of the recycled aggregate concrete; and structural applications (Part E). With the exception of this chapter, the book has ten chapters and is divided into five parts. A brief outline of each part and the associated chapter(s) is shown in Fig. 1.7.

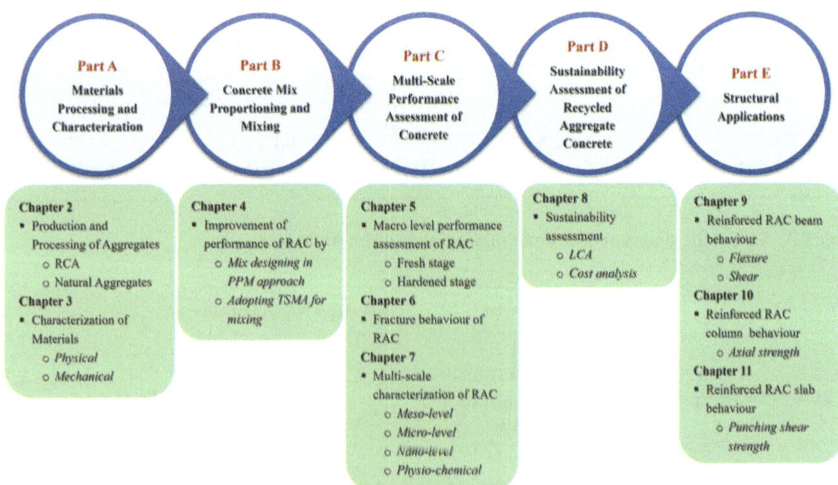

Fig. 1.7 Organization of the book at a glance (Authors')

Chapter "Background on Techniques for Sustainable Use of Recycled Aggregate and Application of Particle Packing Method" provides an introduction on the problems in practicing the confident application of RAC and discusses briefly on possible solutions.

Chapter "Production and Processing of Aggregates" discusses the recycling processes involved in commercially available RA. The methods used from collection of C&D waste to extraction of RA are discussed. The statistics regarding the availability of C&D waste processed to extract RA and applied in RAC preparation for different countries are presented.

Chapter "Characterization of Materials" presents the physical and mechanical characteristics of NCA and RCA. The procedures and codes used in this regard are emphasized.

Chapter "Particle Packing Method of Mix Proportioning and Modified Mixing Approach" describes the procedure of determining the mix proportions of aggregates and paste content in PPM approach. It is compared with the proportioning of conventional mix design approach. In addition to this, the TSMA and normal mixing methods are highlighted.

Chapter "Macro-level Performance Assessment of Concrete: Conventional Approach" analyses the compressive strength, tensile strength, and modulus of elasticity test results of RAC prepared using PPM mix design approach and normal mixing method. The results are compared with the RAC mixes designed using conventional approach. Moreover, the experimental results of NAC mixes prepared using both PPM and conventional method are compared to analyse the influence of RCA as well as PPM in RAC performance.

Chapter "Macro-level Performance Assessment of Concrete: Experimental Fracture Analysis" is dedicated to analyse the influence of RCA and PPM mix design approach on the fracture properties of concrete by conducting the three-point bending (TPB) test on single edge notched (SEN) beam specimens. The influence of size effect is also presented by executing the TPB test for SEN beams of different dimensions.

Chapter "Performance Assessment of Concrete: Meso-, Micro-, Nano-level, and Physio-chemical Analysis" examines the characteristics of RAC at meso-, mico- and nano-level. In this regard, the results of X-ray µ-CT image analysis, scanning electron microscopy (SEM) image analysis, and nanoindentation are analysed systematically. Moreover, the degree of hydration is measured using thermogravimetric analysis (TGA). The relationship between these multi-scale results is employed to establish a correlation with the compressive strength of RAC.

Chapter "Life Cycle Assessment and Cost Analysis" compares the environmental impacts of the concrete mixes prepared by using NCA, RCA, PPM mix design approach and conventional mix design method. In this context, the standard Life Cycle Assessment (LCA) is executed. Moreover, the sensitivity analysis is discussed for different transport scenarios in raw material collection and supply of processed aggregate.

Chapter "Structural Applications: Beam" contains the comprehensive study on the flexure and shear behaviour of reinforced RAC beams. The influence of 100% RCA and different longitudinal reinforcement contents is analysed for the concrete mixes prepared by using PPM mix design approach. The suitability of existing design standards is analysed critically by using a database prepared by collecting all the test results existed of RAC beams and appropriate recommendations are presented.

Chapter "Structural Applications: Column" discusses the effect of 100% RCA and spacing of lateral ties on the axial load carry capacity of RAC columns prepared by using PPM mix design approach. The capacity of RAC columns is compared with NAC columns. The modifications needed in the existing design standards are recommended by comparing critically with experimental results. Based on the proposed constitutive model for RAC, the axial strength of reinforced RAC columns is analysed using numerical simulation in ABAQUS platform.

Chapter "Structural Applications: Slab" assesses the punching shear capacity of PPM mix-designed reinforced concrete slabs and analyses the suitability of prevailing design provisions for RAC slabs. The numerical simulation of punching shear behaviour of reinforced slabs is presented.

1.6 Closure

The rationale for using the RA removed from C&D waste is explored. The difficulties of using RCA for the preparation of structural concrete are underlined. The advantages and disadvantages of various treatment methods to dissolve and fortify the adhering mortar are explored. To provide a clear image of how to employ the best strategy for RAC preparation, various mix design methods and modified mixing procedures are provided in a methodical manner. This chapter highlights the crucial justifications for using the PPM mix design technique and TSMA for RAC.

References

Aim RB, Le Goff P (1968) Effet de paroi dans les empilements désordonnés de sphères et application à la porosité de mélanges binaires. Powder Technol 1(5):281–290

Alexander M, Mindess S (2005) Aggregates in concrete. Taylor & Francis

Anderson R, Dewar JD (2003) Manual of ready-mixed concrete. CRC Press

Andreasen AHM, Andersen J (1930) Relation between grain size and interstitial space in products of unconsolidated granules. Kolloid-Zeitschrift 50(3):217–228

Bru K, Touzé S, Bourgeois F, Lippiatt N, Ménard Y (2014) Assessment of a microwave-assisted recycling process for the recovery of high-quality aggregates from concrete waste. Int J Min Process 126:90–98

Central Pollution Control Board, Ministry of Environment, Forests & Climate Change, India (2017). Guidelines on Environmental Management of Construction & Demolition (C&D) Wastes. Technical report

Construction & Demolition Waste Market by type (Sand, Soil & Gravel, Concrete, Bricks & Masonry, Wood, Metal), Source (Residential, Commercial, Industrial, Municipal), & Region (APAC, North America, Europe, MEA, & South America)–Global forecast to 2026 (2021). Technical report
Construction market in India (2015). A report by dmg events India. Technical report
Construction materials recycling association (2005) Technical report. Illinois, Chicago
De Larrard F (1989) Ultrafine particles for the making of very high strength concretes. Cement Concr Res 19(2):161–172
De Larrard F (1999) Concrete mixture proportioning: a scientific approach. CRC Press
De Larrard F (2009) Concrete optimisation with regard to packing density and rheology. 3rd RILEM international symposium on rheology of cement suspensions such as fresh concrete. France
De Larrard F, Sedran T (1994) Optimization of ultra-high-performance concrete by the use of a packing model. Cement Concr Res 24(6):997–1009
De Larrard F, Sedran T (2002) Mixture-proportioning of high-performance concrete. Cement Concr Res 32(11):1699–1704
Dhir RK, Limbachiya MC, Leelawat T (1999) Suitability of recycled concrete aggregate for use in BS 5328 designated mixes. Proceed Inst Civil Eng: Struct Build 134(3):257–274
Dimitriou G, Savva P, Petrou MF (2018) Enhancing mechanical and durability properties of recycled aggregate concrete. Constr Build Mater 158:228–235
Fathifazl G, Abbas A, Razaqpur AG, Isgor OB, Fournier B, Foo S (2009) New mixture proportioning method for concrete made with coarse recycled concrete aggregate. J Mater Civil Eng 21(10):601–611
Fennis SAAM, Walraven JC, Den Uijl JA (2009) The use of particle packing models to design ecological concrete. Heron 54(2–3):183–202
Fennis SAAM, Walraven JC, Den Uijl JA (2012a) Compaction-interaction packing model: regarding the effect of fillers in concrete mixture design. Mater Struct 463–478
Fennis SAAM, Walraven JC, Den Uijl JA (2012b) Defined-performance design of ecological concrete. Mater Struct 639–650
Feret R (1892) Sur la Compacité des mortiers hydrauliques. Dunod, Vve C
Fischer C, Werge M (2009) EU as a recycling society-present recycling levels of municpal waste and construction & demolition waste in the EU. Technical report
Fuller W, Thompson S (1907) The laws of proportioning concrete. Am Soc Civil Eng 33:223–298
Fung WWS, Kwan AKH, Wong HHC (2009) Wet packing of crushed rock fine aggregate. Mater Struct 42(5):631–643
Funk JE, Dinger D (2013) Predictive process control of crowded particulate suspensions: applied to ceramic manufacturing. Springer Science & Business Media
Furnas CC (1929) Flow of gases through beds of broken solids, vol 300. US Government Printing Office
Glavind M (2009) Sustainability of cement, concrete and cement replacement materials in construction. Sustainability of construction materials. Wood Head Publishing in Materials. Great Abington, Cambridge, UK, pp 120–147
Glavind M, Olsen GS, Munch Petersen C (1993) Packing calculation and concrete mix design
Glavind M, Pedersen EJ (1999) Packing calculations applied for concrete mix design. 1–10
Goltermann P, Johansen V, Palbøl L (1997) Packing of aggregates: an alternative tool to determine the optimal aggregate mix. ACI Mater J 94(5):435–443
IS: 383, (2016) Coarse and fine aggregate for concrete-specification. Bureau of Indian Standards, New Delhi, India
Ismail S, Ramli M (2013) Engineering properties of treated recycled concrete aggregate (RCA) for structural applications. Constr Build Mater 44:464–476
ISO 14040, (2006) Environmental management-life cycle assessment-principles and framework. ISO, Geneva
ISO 14044, (2006) Environmental management-life cycle assessments-requirements and guidelines. ISO, Geneva

ISO/TC 71 (2005) Business plan. Concrete, reinforced concrete and prestressed concrete. International Organization for Standarization
Jacobsen S, Arntsen B (2008) Aggregate packing and -void saturation in mortar and concrete proportioning. Mater Struct 41(4):703–716
Jones R, Zheng L, Newlands M (2002) Comparison of particle packing models for proportioning concrete constituents for minimum voids ratio. Mater Struct 35(6):301–309
Katkhuda H, Shatarat N (2017) Improving the mechanical properties of recycled concrete aggregate using chopped basalt fibers and acid treatment. Constr Build Mater 140:328–335
Katz A (2004) Treatments for the improvement of recycled aggregate. J Mater Civil Eng 16(6):597–603
Kim Y, Hanif A, Kazmi SM, Munir MJ, Park C (2018) Properties enhancement of recycled aggregate concrete through pretreatment of coarse aggregates-Comparative assessment of assorted techniques. J Clean Prod 191:339–349
Knaack AM, Kurama YC (2013) Design of concrete mixtures with recycled concrete aggregates. ACI Mater J 110(5):483–492
Knoeri C, Sanyé-Mengual E, Althaus HJ (2013) Comparative LCA of recycled and conventional concrete for structural applications. Int J Life Cycle Assess 18(5):909–918
Kolonko M, Raschdorf S, Wäsch D (2010) A hierarchical approach to simulate the packing density of particle mixtures on a computer. Granular Matter 12(6):629–643
Kong D, Lei T, Zheng J, Ma C, Jiang J, Jiang J (2010) Effect and mechanism of surface-coating pozzalanics materials around aggregate on properties and ITZ microstructure of recycled aggregate concrete. Constr Build Mater 24(5):701–708
Kou SC, Poon CS (2010) Properties of concrete prepared with PVA-impregnated recycled concrete aggregates. Cement Concr Compos 32(8):649–654
Kumar GS, Minocha AK (2018) Studies on thermo-chemical treatment of recycled concrete fine aggregates for use in concrete. J Mater Cycles Waste Manag 20(1):469–480
Kwan AKH, Chan KW, Wong V (2013) A 3-parameter particle packing model incorporating the wedging effect. Powder Technol 237:172–179
Kwan AKH, Li LG, Fung WWS (2012) Wet packing of blended fine and coarse aggregate. Mater Struct 45(6):817–828
Kwan AKH, Wong HHC (2008) Packing density of cementitious materials: part 2-packing and flow of OPC + PFA + CSF. Mater Struct 41(4):773–784
Li LG, Kwan AKH (2014) Packing density of concrete mix under dry and wet conditions. Powder Technol 253:514–521
Li W, Long C, Tam VW, Poon CS, Hui Duan W (2017) Effects of nano-particles on failure process and microstructural properties of recycled aggregate concrete. Constr Build Mater 142:42–50
Li X (2009) Recycling and reuse of waste concrete in China. Part II. Structural behaviour of recycled aggregate concrete and engineering applications. Resour Conser Recycl 53:107–112
Liu S, Shen P, Xuan D, Li L, Sojobi A, Zhan B, Poon CS (2021) A comparison of liquid-solid and gas-solid accelerated carbonation for enhancement of recycled concrete aggregate. Cement Concr Compos 118(11):103988
Martirena F, Castaño T, Alujas A, Orozco-Morales R, Martinez L, Linsel S (2017) Improving quality of coarse recycled aggregates through cement coating. J Sustain Cement-Based Mater 6(1):69–84
Marzouk M, Azab S (2014) Environmental and economic impact assessment of construction and demolition waste disposal using system dynamics. Resour Conser Recycl 82:41–49
Mindess S, Young JF, Darwin D (2003) Concrete. Pearson Education Limited
Mistri A, Bhattacharyya SK, Dhami N, Mukherjee A, Barai SV (2020) A review on different treatment methods for enhancing the properties of recycled aggregates for sustainable construction materials. Constr Build Mater 233:117894
Monkman S, MacDonald M (2017) On carbon dioxide utilization as a means to improve the sustainability of ready-mixed concrete. J Clean Prod 167:365–375
Mooney M (1951) The viscosity of a concentrated suspension of spherical particles. J Colloid Sci 6(2):162–170

Otsuki N, Miyazato S-I, Yodsudjai W (2003) Influence of recycled aggregate on interfacial transition zone, strength, chloride penetration and carbonation of concrete. J Mater Civil Eng 15(5):443–451
Pan X, Shi Z, Shi C, Ling TC, Li N (2017) A review on concrete surface treatment Part I: Types and mechanisms. Constr Build Mater 132:578–590
Powers TC (1968) The properties of fresh concrete. John Wiley and Sons
Pradhan S, Kumar S, Barai SV (2017) Recycled aggregate concrete: particle packing method (PPM) of mix design approach. Const Build Mater 152:269–284
Pu Y, Li L, Wang Q, Shi X, Luan C, Zhang G, Fu L, El-Fatah Abomohra A (2021) Accelerated carbonation technology for enhanced treatment of recycled concrete aggregates: A state-of-the-art review. Constr Build Mater 282:122671
Rao VVLK, Krishnamoothy S (1993) Aggregate mixtures for least-void content for use in polymer concrete. Cement Concr Aggr 15(2):97–107
Singh LP, Bisht V, Aswathy MS, Chaurasia L, Gupta S (2018) Studies on performance enhancement of recycled aggregate by incorporating bio and nano materials. Constr Build Mater 181:217–226
Stovall T, De Larrard F, Buil M (1986) Linear packing density model of grain mixtures. Powder Technol 48(1):1–12
Tam VWY, Gao XF, Tam CM (2005) Microstructural analysis of recycled aggregate concrete produced from two-stage mixing approach. Cement Concr Res 35(6):1195–1203
Tam VWY, Tam CM (2008) Diversifying two-stage mixing approach (TSMA) for recycled aggregate concrete: TSMAs and TSMAsc. Constr Build Mater 22(10):2068–2077
Tam VWY, Tam CM, Wang Y (2007) Optimization on proportion for recycled aggregate in concrete using two-stage mixing approach. Constr Build Mater 21(10):1928–1939
Toufar W, Born M, Klose E (1976) Contribution of optimisation of components of different density in polydispersed particles systems. Freiberger Book A 558:29–44
Weil M, Jeske U, Schebek L (2006) Closed-loop recycling of construction and demolition waste in Germany in view of stricter environmental threshold values. Waste Manag Res 24(3):197–206
Wong HHC, Kwan AKH (2005) Packing density:a key concept for mix design of high performance concrete. In: Materials science and technology in engineering conference (MaSTEC). Hong Kong, pp 1–15
Wong HHC, Kwan AKH (2008) Packing density of cementitious materials: part 1-measurement using a wet packing method. Mater Struct 41(4):689–701
Wong V (2013) Applying theories of particle packing and rheology to concrete for sustainable development. Org Technol Manag Const Int J 5(2):844–851
Xiao J (2018) Recycled aggregate concrete structures. Springer-Verlag, Heidelberger

Part I
Materials Processing and Characterization

Chapter 2
Production and Processing of Aggregates

2.1 Introduction

In India, the extensive use of reinforced concrete and prestressed concrete structures happened in the second half of the nineteenth century. At the time of construction, the reuse of building waste after their service life was not thought about. The destruction of structures due to natural disasters is inevitable. Moreover, replacement of old structures with new and higher capacity structures is essential in order to satisfy the need of the growing population and traffic. Uncontrolled demolition of old structures is a problem owing to the improper waste management. At the same time, the waste concrete from ready-mix plant, laboratory-tested specimens, and rejected precast concrete members aggravate this issue.

Gradual depletion of qualified natural resources for new construction draws the interest towards C&D waste in search for suitable alternative materials. Waste concrete of C&D waste is exploited as the source of aggregates to replace the conventional aggregates. A combination of appropriate techniques is essential in the recycling process to extract good quality recycled aggregates with the expense of less energy. The various techniques for the demolition of existing structures are explained in detail by Rao et al. (2019). The recycling strategy and techniques practised differently in different countries (Xiao 2018). The equipment usually used in recycling process are jaw crusher, impact crusher, gyratory crusher, magnate separator, and sieve (Rao et al. 2019). The procedure followed in India is explained in the subsequent section.

2.2 Extraction of Recycled Coarse Aggregate

Recycling of waste concrete involves several manual as well as mechanical steps to generate reusable RCA. The C&D waste from different sites, such as dismantled structures, ready-mix concrete batching plant, and concrete testing laboratories,

S. Pradhan et al., *Particle Packing Method for Recycled Aggregate Concrete*,
https://doi.org/10.1007/978-981-99-7516-7_2

is collected and transported to the recycling plant. Manual screening and sorting of waste concrete from C&D debris is the first step of the recycling process. Subsequently, the waste concrete goes through different crushing processes to acquire desirable grading of recycled aggregate. Firstly, the large concrete blocks are broken into pieces using impact crusher and also done manually by hammer. The smaller concrete blocks are then carried by the poclain machine to the hopper with apron feeder. It is transported to the jaw crusher through conveyor belt. The crushed concrete blocks coming out of jaw crusher are transported towards the vibrating sieve through conveyor belt. In between, the ferrous materials are removed by using the magnetic separator and using blower the dust particles are separated. Thus, the crushed concrete reaching at vibrating sieve is free of ferrous materials and fine dust. The vibrating sieve is equipped with multi-sized sieves, which enables to separate the RCA of different sizes and RFA efficiently. The flowchart of the complete recycling process is presented in Fig. 2.1.

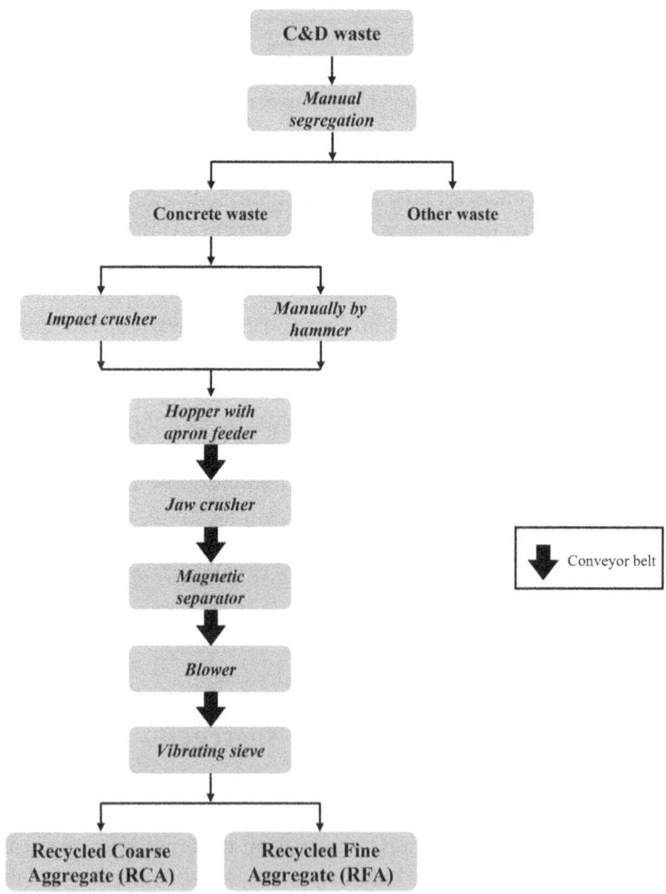

Fig. 2.1 Recycled aggregates production process (Authors')

In the processed recycled aggregates, the parent aggregates are partially or completely covered with mortar. Moreover, in accordance to the study of Mistri et al. (2019), a substantial quantity of the total recycled aggregates is also consisting of only cement mortar (without any parent aggregate), especially in case of smaller sized coarse aggregates (Mistri et al. 2019). This attached mortar is the primary reason for poor physical and mechanical properties of recycled aggregate. Consequently, it adversely affects the performance of prepared concrete mix. Researchers have developed methods like ultrasonic treatment method (Katz 2003), pre-soaking treatment (Tam et al. 2007), chemical treatment, freeze–thaw method (Razaqpur et al. 2007), thermal treatment (de Juan and Gutiérrez 2009), microwave heating method (Akbarnezhad et al. 2011), autogenous cleaning process (Pepe et al. 2014), heating and rubbing method, and mechanical grinding method for removing adhered mortar to obtain a higher quality of recycled aggregates. Additionally, coating of pozzolana slurry (Li 2009; Li et al. 2017), emulsion polymer (Kou and Poon 2010), or calcium carbonate ($CaCO_3$) precipitants from accelerated carbonation (Pu et al. 2021; Liu et al. 2021) or bacterial deposition (Singh et al. 2018) processes is implemented by the researchers to enhance the physical and mechanical characteristics of recycled aggregates. The various treatment methods to improve the quality of recycled aggregates are discussed in chapter "Background on Techniques for Sustainable Use of Recycled Aggregate and Application of Particle Packing Method".

2.3 Production and Processing of Natural Coarse Aggregates

Naturally available rocks are the sources of conventional aggregates. The petrological classification includes basalt group, granite group, limestone group, quartzite group, schist group, flint group, gritstone group, porphyry group, gabbro group, and hornfels group of rocks. During mining of rocks, explosive is used. After expulsion of the rock, it is carried to the crusher. Initially, the large rocks are broken into pieces using the impact crusher. These rock pieces are fed to the jaw crusher to bring into aggregate form. The vibrating sieve is used to separate the aggregates of different sizes. The detailed processing procedure of natural coarse aggregate is shown in Fig. 2.2.

2.4 Closure

Waste generated from the demolition of existing structure is the major source of C&D waste. Simply dumping this waste in landfill will create serious impact on social, economic, and environment life cycle. A detailed process to extract valuable and reusable recycled aggregate from waste concrete is presented in this chapter. Additional crushing stages and treatment procedure can be followed to extract cleaner

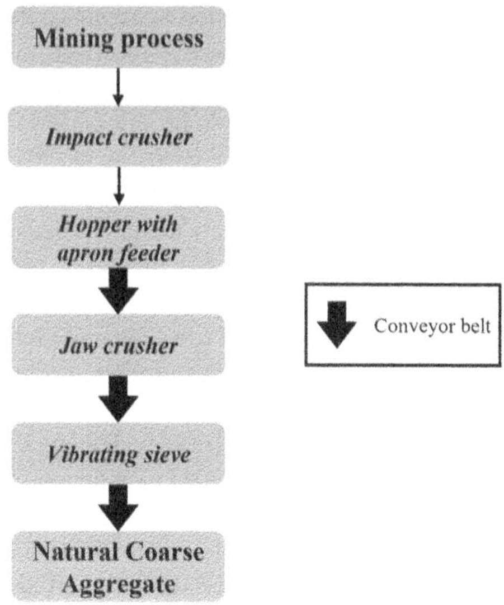

Fig. 2.2 Natural coarse aggregate production process (Authors')

and better recycled aggregates with lesser attached mortar content. The processing steps of natural coarse aggregate from naturally available rocks are explained. The source and processing procedure and number of steps involved for the preparation of natural coarse aggregate are different to that of recycled coarse aggregate.

References

Akbarnezhad A, Ong KCG, Zhang MH, Tam CT, Foo TWJ (2011) Microwave-assisted beneficiation of recycled concrete aggregates. Constr Build Mater 25(8):3469–3479

de Juan MS, Gutiérrez PA (2009) Study on the influence of attached mortar content on the properties of recycled concrete aggregate. Constr Build Mater 23(2):872–877

Katz A (2003) Properties of concrete made with recycled aggregate from partially hydrated old concrete. Cement Concr Res 33(5):703–711

Kou SC, Poon CS (2010) Properties of concrete prepared with PVA-impregnated recycled concrete aggregates. Cement Concr Compos 32(8):649–654

Li W, Long C, Tam VW, Poon CS, Hui Duan W (2017) Effects of nano-particles on failure process and microstructural properties of recycled aggregate concrete. Const Build Mater 142:42–50

Li X (2009) Recycling and reuse of waste concrete in China. Part II. Structural behaviour of recycled aggregate concrete and engineering applications. Resour Conserv Recycl 53:107–112

Liu S, Shen P, Xuan D, Li L, Sojobi A, Zhan B, Poon CS (2021) A comparison of liquid-solid and gas-solid accelerated carbonation for enhancement of recycled concrete aggregate. Cement Concr Compos 118(2020):103988

Mistri A, Bhattacharyya SK, Dhami N, Mukherjee A, Barai SV (2019) Petrographic investigation on recycled coarse aggregate and identification the reason behind the inferior performance. Constr Build Mater 221:399–408

Pepe M, Toledo Filho RD, Koenders EA, Martinelli E (2014) Alternative processing procedures for recycled aggregates in structural concrete. Const Build Mater 69:124–132

Pu Y, Li L, Wang Q, Shi X, Luan C, Zhang G, Fu L, El-Fatah Abomohra A (2021) Accelerated carbonation technology for enhanced treatment of recycled concrete aggregates: A state-of-the-art review. Constr Build Mater 282:122671

Rao MC, Bhattacharyya SK, Barai SV (2019) Systematic approach of characterisation and behaviour of recycled aggregate concrete. Springer

Razaqpur AG, Abbas A, Fournier B, Fathifazl G, Isgor OB, Foo S (2007) Proposed method for determining the residual mortar content of recycled concrete aggregates. J ASTM Int 5(1):1–12

Singh LP, Bisht V, Aswathy MS, Chaurasia L, Gupta S (2018) Studies on performance enhancement of recycled aggregate by incorporating bio and nano materials. Constr Build Mater 181:217–226

Tam VWY, Tam CM, Le KN (2007) Removal of cement mortar remains from recycled aggregate using pre-soaking approaches. Resour Conserv Recycl 50(1):82–101

Xiao J (2018) Recycled aggregate concrete structures. Springer-Verlag, Heidelberger

Chapter 3
Characterization of Materials

3.1 Introduction

Concrete is a composite material and a heterogeneous mixture of cement, coarse aggregate, fine aggregate, water, and chemical admixture. The behaviour and performance of concrete in the fresh as well as hardened stages largely depend on the physical, mechanical, and chemical properties of its constituents. Hence, it is important to determine the characteristics of the raw materials prior to the designing of concrete mix for their intended application. This book aims to compare the performance of concrete, prepared using two different types of coarse aggregates (natural coarse aggregate (NCA) and recycled coarse aggregate (RCA)). The subsequent sections exclusively discuss the differences in physical and mechanical properties between NCA and RCA.

3.2 Coarse Aggregates

About 70–80% of the concrete volume is occupied by aggregates. The inclusion of aggregates provides considerable advantages to make the concrete not only economical but also strong, stiff, stable, and durable. The conventional source of aggregates is natural rocks. However, waste concrete present in C&D waste can also be a viable source of aggregates. Accordingly, based on their origin, these aggregates can be classified as NCA and RCA. The physical, chemical, and mechanical properties of RCA are inferior to that of NCA, primarily due to the adhered mortar layer. The key parameters for the quality assessment of these coarse aggregates are appended in the following sections.

S. Pradhan et al., *Particle Packing Method for Recycled Aggregate Concrete*,
https://doi.org/10.1007/978-981-99-7516-7_3

3.2.1 Physical Properties Coarse Aggregate

The physical properties of the aggregates are assessed by using the gradation curve, surface texture, shape, bulk density, specific gravity, and water absorption parameters. RCA obtained by the crushing of waste concrete is partially or fully covered with a layer of mortar. The quantity of attached mortar to recycled aggregate increases as its size decreases (de Juan and Gutiérrez 2009; Florea and Brouwers 2013). Moreover, the quantity and quality of adhered porous mortar layer to the parent aggregate influence the physical properties of RCA.

Particle Size Distribution

Bairagi et al. (1990, 1993) observed relatively higher fineness modulus of recycled aggregate. However, the grading curves of NCA and RCA do not exhibit significant difference. The experimental investigation conducted by Zega et al. (2010) also observed similar particle size distribution for both NCA and RCA irrespective of the type of aggregate and w/c ratio used in the parent concrete. Katz (2003, 2004) extracted recycled aggregate from the laboratory-tested specimens, which were partially hydrated. It was observed that, irrespective of the age of the parent concrete, the gradation of RCA and NCA was similar. Chakradhara Rao et al. (2011) studied the physical properties of RCA obtained from three different sources and observed relatively finer than NCA. However, the gradation curves of RCA from all the sources were within the specified limits. The RCA processed in the laboratory by Mukharjee (2014) was not well graded, and hence a mixture of 65:35 was prepared using 20–10 mm and 10–4.75 mm aggregates. The study conducted by Salem and Burdette (1998) reported higher angularity and rough and porous texture of RCA owing to the attached mortar layer. The multiple crushing steps and use of impact crusher involved in the processing of recycled aggregate resulted in more angular and less flaky and elongated RCA (Bairagi et al. 1993; Ho et al. 2013). Pradhan et al. (2017) conducted the sieve analysis of NCA and commercially available RCA (multiple sources of parent concrete) in accordance to the procedure specified in IS: 383 (2016) to obtain the particle size distribution. The study revealed that the RCA is relatively finer than the NCA as the fineness modulus of RCA is 1.25% lower than NCA. The particles finer than 4.75 mm in RCA is 2.5%. However, the gradation curve (Fig. 3.1) shows that the gradation of RCA is within the lower and upper bounds specified in IS: 383 (2016).

Flakiness and Elongation Indices

IS: 2386 (Part I) (1963) explains the characteristics for flakiness and elongation indices of coarse aggregate, which define the shape of the coarse aggregate. The thickness of the aggregate less than 0.6 times of the mean size of the sieve to which

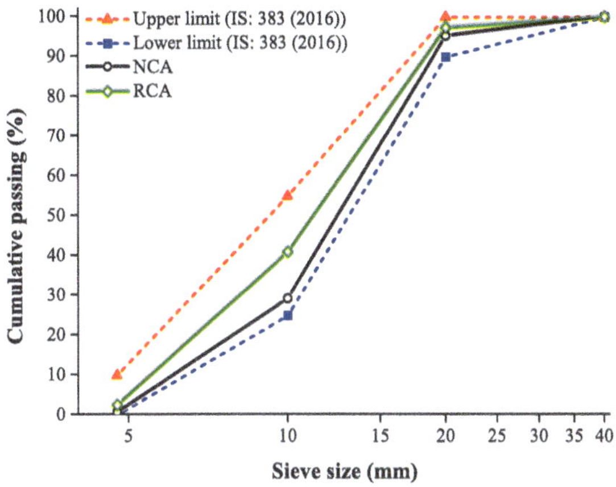

Fig. 3.1 Particle size distribution of NCA and RCA (Authors')

Table 3.1 Physical properties of NCA, RCA, and sand (Pradhan et al. 2017)

Aggregate type	Size of the aggregate (mm)	Specific gravity	Water absorption (%)	Bulk density (kg/m^3)		Flakiness index(%)	Elongation index(%)
				Loose	Compacted		
NCA	20–12.5	2.93	1.59	1447	1648	19.14	49.42
	12.5–10	2.88	1.67	1436	1599		
	10–4.75	2.80	1.06	1415	1594		
RCA	20–12.5	2.54	3.63	1304	1374	15.17	18.24
	12.5–10	2.53	3.40	1231	1315		
	10–6.3	2.43	4.20	1192	1272		
	6.3–4.75	2.29	5.09	1147	1288		
Sand	–	2.66	0.24	1444	1578	–	–

the aggregate particle belongs to can be called as flaky. Similarly, when the longest dimension of the aggregate particle is greater than 1.8 times of the mean sieve size to which it belongs to can be termed as elongated. The calculated flakiness and elongation indices of NCA and RCA are reported in Table 3.1. The values indicate the shape of the RCA is partly rounded.

Bulk Density

Bulk density of aggregate is defined as the mass of the aggregate per unit volume, which is determined at the loose as well as compacted state. It gives an idea about the packing capacity of the aggregate. The bulk density and specific gravity of attached mortar are very less as compared to the natural aggregate, because of its porous nature (Topcu and Şengel 2004; Evangelista and de Brito 2007; Yang et al. 2008; Fathifazl et al. 2011; Limbachiya et al. 2012; McNeil and Kang 2013). Owing to the same reason, the bulk density and specific gravity of RCA were lower than NCA as reported in the earlier studies (Sagoe-Crentsil et al. 2001; Shayan and Xu 2003; Domingo-Cabo et al. 2009; Ulloa et al. 2013). The density of RCA in saturated surface dry (SSD) condition varied between 2290 and 2510 kg/m^3 as observed by Chakradhara Rao (2010). The study by de Juan and Gutiérrez (2009) reported that, bulk density and SSD density of RCA reduce with the increase in attached mortar content. According to the procedure stated in IS: 2386 (Part III) (1963), the loose and compacted bulk densities of both NCA and RCA were determined by Pradhan et al. (2017) and presented in Table 3.1. The loose and compacted bulk densities of RCA were lower than that of NCA, which is attributed to the attached lighter and porous mortar layer to RCA.

Specific Gravity

Specific gravity is the ratio of the weight of a certain volume of aggregate to the weight of water occupying same volume. The procedure stated in IS: 2386 (Part III) (1963) was executed to determine the specific gravity of both NCA and RCA and shown in Table 3.1. For each size of RCA, the specific gravity is lower than that of NCA. This is due to the same reason explained for lower bulk density. It is observed that, for both NCA and RCA the specific gravity decreases as the size of the aggregate decreases, which is due to the increase in surface area with the reduction in the size of the aggregate.

Water Absorption

Water absorption of coarse aggregate is an important parameter as it can influence the workability as well as the hydration process of the concrete. The attached mortar layer to RCA is porous in nature and increases its surface area, which in turn increases its water absorption capacity (Hansen and Narud 1983). The water absorption of RCA was observed to be two to five times of NCA (Gómez-Soberón 2002; Shayan and Xu 2003; Etxeberria et al. 2007; Xiao et al. 2015). Moreover, Poon et al. (2004) experienced the water absorption of RCA up to 15%. Padmini et al. (2009) observed an increase in water absorption of RCA with the increase in compressive strength of parent concrete, which is due to the higher content of attached mortar in RCA processed from the concrete with higher strength. Chakradhara Rao (2010) and Mukharjee (2014) observed a maximum of 3.9% and 4.6% water absorption of RCA. Thomas et al. (2018) observed that the water absorption of RCA increases as the number of recycling cycle increases owing to the increment in adhered mortar content. The method explained in IS: 2386 (Part III) (1963) was followed by Pradhan et al. (2017) to determine the water absorption of both NCA and RCA and the observed values are reported in Table 3.1. As expected, the water absorption of RCA was observed to be higher than NCA owing to the attached porous mortar layer.

3.2.2 Mechanical Properties of Coarse Aggregate

Crushing value, impact value, ten percent fines value, and Los Angeles abrasion value are the representation of the mechanical characteristics of coarse aggregate. The presence of old ITZ, which is the weak link between old mortar and parent aggregate the disintegration of attached mortar is inevitable. Hence, the strength of parent concrete influences the mechanical properties of RCA. Shayan and Xu (2003) reported the crushing value of 24% for RCA and Chakradhara Rao (2010) and Mukharjee (2014) obtained the crushing value about 31%–37%. The crushing value of RCA was observed to be more than two times of NCA (Xiao et al. 2015). Poon et al. (2004) and Chakradhara Rao et al. (2011) observed lower ten percent fines value for RCA irrespective of the strength of the parent concrete, whereas the impact value was observed to be higher for RCA (López-Gayarre et al. 2009; Limbachiya 2010; Mukharjee and Barai 2014). Zhang et al. (2017) correlated the crushing value of RCA and macro-mechanical properties of concrete and a decreasing trend is reported for the macro-mechanical properties with the increase in crushing value. The reduction in Los Angeles abrasion value was observed with the increase in compressive strength of parent concrete (Hansen and Narud 1983; de Juan and Gutiérrez 2009). However, Zega et al. (2010) observed that the w/c of the parent concrete has less significant influence on the Los Angeles abrasion value.

The mechanical properties of coarse aggregate, namely, Los Angeles abrasion value, impact value, and crushing value were assessed by Pradhan et al. (2017) for both NCA and RCA in accordance to the procedure specified in IS: 2386 (Part IV) (1963) and presented in Table 3.2. Los Angeles abrasion value is the measure of toughness and abrasion characteristics of coarse aggregate. RCA exhibited higher Los Angeles abrasion value with respect to NCA, which is attributed to the old ITZ and loose mortar layer adhered to RCA. However, it is less than the maximum value of 50% specified in IS: 383 (2016). Impact value is a measure of the resistance of the aggregate to sudden impact. The impact resistance of RCA is lower than NCA (Table 3.2) owing to the same reason explained for Los Angeles value. However, it is within the maximum limit of impact value of 45% given in IS: 383 (2016). Crushing value is the relative measure of the resistance offered by the aggregate to crushing while a compressive load is applied gradually. Owing to the similar reason for higher Los Abrasion value and impact value of RCA, the crushing value is also observed to be higher. Nevertheless, the value is within the upper limit of 45% stated in IS: 383 (2016).

Table 3.2 Mechanical properties of NCA and RCA (Pradhan et al. 2017)

Aggregate type	Properties		
	Los Angeles abrasion value (%)	Impact value (%)	Crushing value (%)
NCA	14.34	13.87	17.66
RCA	34.08	24.23	23.32

3.3 Closure

The characterization of the raw materials is essential to verify their suitability in concrete preparation and further application in structural member. The physical as well as mechanical properties of RCA are inferior to NCA. The water absorption of RCA is significantly higher (two to five times) than NCA. Moreover, the specific gravity and bulk density of RCA are lower than NCA. The partly rounded shape of RCA can be confirmed from the observed flakiness and elongation indices. However, the estimated properties of RCA by Pradhan et al. (2017) are within the permissible limits specified by BIS and suitable for concrete preparation.

References

Bairagi N, Ravande K, Pareek V (1993) Behaviour of concrete with different proportions of natural and recycled aggregates. Resour Conserv Recycl 9:109–126

Bairagi NK, Vidyadhara HS, Ravande K (1990) Mix design procedure for recycled aggregate concrete. Constr Build Mater 4(4):188–193

Chakradhara Rao M (2010) Characterisation and behaviour of recycled aggregate concrete. PhD thesis, IIT Kharagpur

Chakradhara Rao M, Bhattacharyya SK, Barai SV (2011) Influence of field recycled coarse aggregate on properties of concrete. Mater Struct 44:205–220

De Juan MS, Gutiérrez PA (2009) Study on the influence of attached mortar content on the properties of recycled concrete aggregate. Constr Build Mater 23(2):872–877

Domingo-Cabo A, Lázaro C, López-Gayarre F, Serrano-López MA, Serna P, Castaño-Tabares JO (2009) Creep and shrinkage of recycled aggregate concrete. Constr Build Mater 23:2545–2553

Etxeberria M, Marí AR, Vázquez E (2007) Recycled aggregate concrete as structural material. Mater Struct 40(5):529–541

Evangelista L, de Brito J (2007) Mechanical behaviour of concrete made with fine recycled concrete aggregates. Cement Concr Compos 29(5):397–401

Fathifazl G, Ghani Razaqpur A, Burkan Isgor O, Abbas A, Fournier B, Foo S (2011) Creep and drying shrinkage characteristics of concrete produced with coarse recycled concrete aggregate. Cement Concr Compos 33(10):1026–1037

Florea MVA, Brouwers HJH (2013) Properties of various size fractions of crushed concrete related to process conditions and re-use. Cement Concr Res 52:11–21

Gómez-Soberón JM (2002) Porosity of recycled concrete with substitution of recycled concrete aggregate. Cement Concr Res 32(8):1301–1311

Hansen TC, Narud H (1983) Strength of recycled concrete made from crushed concrete coarse aggregate. Concr Int 5:79–83

Ho NY, Lee YPK, Lim WF, Zayed T, Chew KC, Low GL, Ting SK (2013) Efficient utilization of recycled concrete aggregate in structural concrete. J Mater Civil Eng 25(3):318–327

IS: 2386 (Part I) (1963) Method of Test for aggregate for concrete. Part I-Particle size and shape, Bureau of Indian Standards, New Delhi, India

IS: 2386 (Part III) (1963) Method of Test for aggregate for concrete. Part III–Specific gravity, density, voids, absorption and bulking. Bureau of Indian Standards, New Delhi, India

IS: 2386 (Part IV) (1963) Method of Test for aggregate for concrete. Part IV-Mechanical properties, Bureau of Indian Standards, New Delhi, India

IS: 383 (2016) Coarse and fine aggregate for concrete-specification. Bureau of Indian Standards, New Delhi, India

Katz A (2003) Properties of concrete made with recycled aggregate from partially hydrated old concrete. Cement Concr Res 33(5):703–711

Katz A (2004) Treatments for the improvement of recycled aggregate. J Mater Civil Eng 16(6):597–603

Limbachiya M, Meddah MS, Ouchagour Y (2012) Use of recycled concrete aggregate in fly-ash concrete. Constr Build Mater 27:439–449

Limbachiya MC (2010) Recycled aggregates: Production, properties and value-added sustainable applications. J Wuhan Univ Technol Mater Sci Edn 25(6):1011–1016

López-Gayarre F, Serna P, Domingo-Cabo A, Serrano-López MA, López-Colina C (2009) Influence of recycled aggregate quality and proportioning criteria on recycled concrete properties. Waste Manag 29(12):3022–3028

McNeil K, Kang TH-K (2013) Recycled concrete aggregates: a review. Int J Concr Struct Mater 7(1):61–69

Mukharjee BB (2014) Behaviour of concrete incorporating recycled aggregates and nanosilica. PhD Thesis, IIT Kharagpur

Mukharjee BB, Barai SV (2014) Influence of incorporation of nano-silica and recycled aggregates on compressive strength and microstructure of concrete. Constr Build Mater 71:570–578

Padmini AK, Ramamurthy K, Mathews MS (2009) Influence of parent concrete on the properties of recycled aggregate concrete. Constr Build Mater 23(2):829–836

Poon CS, Shui ZH, Lam L (2004) Effect of microstructure of ITZ on compressive strength of concrete prepared with recycled aggregates. Constr Build Mater 18(6):461–468

Pradhan S, Kumar S, Barai SV (2017) Recycled aggregate concrete: particle packing method (PPM) of mix design approach. Constr Build Mater 152:269–284

Sagoe-Crentsil KK, Brown T, Taylor AH (2001) Performance of concrete made with commercially produced coarse recycled concrete aggregate. Cement Concr Res 31:707–712

Salem RM, Burdette EG (1998) Role of chemical and mineral admixtures on physical properties and frost resistance of recycled aggregate concrete. ACI Mater J 95(5):558–563

Shayan A, Xu A (2003) Performance and properties of structural concrete made with recycled concrete aggregate. ACI Mater J 100(5):371–380

Thomas C, de Brito J, Gil V, Sainz-Aja J, Cimentada A (2018) Multiple recycled aggregate properties analysed by X-ray microtomography. Constr Build Mater 166:171–180

Topcu IB, Şengel S (2004) Properties of concretes produced with waste concrete aggregate. Cement Concr Res 34(8):1307–1312

Ulloa VA, García-Taengua E, Pelufo MJ, Domingo A, Serna P (2013) New views on effect of recycled aggregates on concrete compressive strength. ACI Mater J 110(6):687–696

Xiao J, Li L, Shen L, Poon CS (2015) Compressive behaviour of recycled aggregate concrete under impact loading. Cement Concr Res 71:46–55

Yang K-H, Chung H-S, Ashour AF (2008) Influence of type and replacement level of recycled aggregates on concrete properties. ACI Mater J 105(3):289–296

Zega CJ, Villagrán-Zaccardi YA, Di Maio ÁA (2010) Effect of natural coarse aggregate type on the physical and mechanical properties of recycled coarse aggregates. Mater Struct 43:195–202

Zhang Z, Zhang Y, Yan C, Liu Y (2017) Influence of crushing index on properties of recycled aggregates pervious concrete. Constr Build Mater 135:112–118

Part II
Concrete Mix Proportioning and Mixing

Chapter 4
Particle Packing Method of Mix Proportioning and Modified Mixing Approach

4.1 Introduction

One of the main industries responsible for the rapid depletion of natural resources and the production of greenhouse gases in order to obtain the essential raw materials is the construction industry. To ease the strain on natural aggregate mining, aggregates, which make up around 70—80% of the volume of concrete, must be replaced with a sustainable alternative. Additionally, the construction industry and the generation of C&D trash go hand in hand. Natural aggregate can potentially be replaced with recycled aggregate (RA), which is produced from leftover concrete from the C&D waste. RA can be divided into RCA and RFA categories according to size. The use of RCA alone for structural concrete preparation is the exclusive focus of this work.

The inferior physical and mechanical properties of RCA as discussed in chapter "Characterization of Materials" are bound to have adverse effects on the performance of RAC. In order to address this issue, a novel Particle Packing Method (PPM) of mix design approach is proposed by Pradhan et al. (2017). In order to enhance the ITZ of RAC, the established Two-Stage Mixing Approach (TSMA) is used during mixing. In the sections that follow, the PPM approach to concrete mix proportioning is thoroughly covered. Additionally, both the NAC and the RAC are created utilizing the traditional mix design approach for comparison study.

4.2 Particle Packing Method (PPM)

PPM application draws engineers from the disciplines of ceramics and powder metallurgy. Through a tightly packed mixture of particles, this enables the development of superior materials. In practice, the explicit application of PPM for concrete mix proportioning is not very common. The packing of the aggregate mixture and, consequently, the performance of concrete are directly impacted by the size (coarse or

S. Pradhan et al., *Particle Packing Method for Recycled Aggregate Concrete*,
https://doi.org/10.1007/978-981-99-7516-7_4

Fig. 4.1 Schematic diagram of PPM (Authors')

fine), shape (rounded or angular), and type of parent rock of the aggregate. As a result, correct aggregate proportioning is crucial, and PPM is quite helpful in this regard.

The main objective of PPM is to minimize voids by combining particles or aggregates of various sizes. The foundation of this strategy is the packing properties of a single size of aggregates and how they affect the packing of a mixture of aggregates. The goal of PPM is to get the highest packing density feasible by combining various sizes of coarse and fine materials. In this case, the smaller particles are employed to raise the packing density of the mixture by filling in the spaces between the larger particles. The increased packing density lowers the requirement of binder and water. Figure 4.1 illustrates the void minimization process of PPM through multi-sized aggregates. In step-I, only larger RCA (coarse recycled aggregate$_L$) is used, and hence the required cement paste to fill up the remaining spaces is the highest. In step-II, with the use of both larger and smaller RCA (smaller RCA: coarse recycled aggregate$_S$) the cement paste content is lowered. Subsequent addition of fine aggregates to coarse recycled aggregate$_L$ and coarse recycled aggregate$_S$ mixture in step-III further reduces the voids content and required cement paste content.

The appropriate proportion of different sizes of aggregates can be optimized experimentally as well as theoretically. The Compressible Packing Model (CPM) was employed by Amario et al. (2017) for the proportioning of aggregates in RAC mixture. In this context, both fine aggregate and RCA were segregated into three size classes each. Subsequently, Puente de Andrade et al. (2020) used the CPM approach to proportion the RAC mix while replacing the RFA 0%, 25%, and 50%. Both the studies prepared the RAC mix by using the simulated composition of mixture as performed in Bentolab Pro 3 software. The explanation for the paste content determination is not detailed by Amario et al. (2017) and Puente de Andrade et al. (2020). Babalola et al. (2020) developed an algorithm of densified mix design approach (DMDA) based on the framework of PPM. The DMDA was employed to proportion the RAC with ternary binders. Optimum weights for each binder were estimated to determine the quantity of each binder. However, the importance of multi-sized RCA and their packing characteristics in optimum packing density is not highlighted by Babalola et al. (2020). Kirthika et al. (2020) developed a novel mix design method by combining the particle packing density and minimum paste theory for mix proportioning of RAC with RFA. The adjustment in total water content owing to the high water absorption of RFA was necessitated for achieving the saturated surface dry (SSD) condition. This chapter discusses the detailed process of experimental optimization of coarse aggregate and fine aggregate mixture and validation using CPM for both NAC and RAC as conducted by Pradhan et al. (2017). Moreover, the steps for optimum fresh paste content determination are elaborated. Two important steps involved in the PPM mix design approach are

(a) Determination of aggregate fractions and packing density.
(b) Determination of paste content.

4.2.1 Determination of Aggregate Fractions and Packing Density

Experimental Procedure

Traditionally, Fuller's curve is used to describe the particle size distribution of aggregates. A well-distributed aggregate satisfying Fuller's curve does not ensure the maximum packing density. In PPM, the optimal combination of the aggregates mixture satisfies the proper distribution of aggregates as well as maximum packing density. In the experimental optimization process, the coarse aggregates of different sizes required for the study are selected and segregated in different size zones from coarser to finer, such as CA_1, CA_2, CA_3, etc. The compacted bulk density and specific gravity of aggregates are required to determine separately for each size zone. In the hierarchy of aggregates mixture, the mixing of coarse aggregates is carried out in descending order. The bulk density and packing density (PD) of the aggregates mixture increase with the increase in the smaller sized aggregates fraction up to a certain proportion, whereas further increase in smaller aggregates content reduces the bulk density and PD. This behaviour is due to the loosening effect. The PD and voids content (VC) are to be calculated using Eqs. 4.2.1 and 4.2.2, respectively, as stated in IS: 2386 (Part III)(1963) or ASTM C29 (2009).

$$\mathrm{PD} = \sum \frac{\text{Bulk density} \times \text{Weight fraction}}{\text{Specific gravity}} \tag{4.2.1}$$

$$\mathrm{VC} = 1 - \sum \frac{\text{Bulk density}}{\text{Specific gravity}} \tag{4.2.2}$$

The first mixture is to be prepared by blending CA_1 and CA_2 in different proportions by mass, such as 90:10, 80:20, 70:30, etc., and for each mixture proportion of CA_1 and CA_2 the bulk density, PD (using Eq. 4.2.1), and VC (using Eq. 4.2.2) should be determined. The maximum PD is determined for the aggregate mixture of CA_1 and CA_2. The weight fraction of CA_1 and CA_2 aggregate combination yielding maximum PD is kept fixed while preparing the subsequent mixture with CA_3. The combination of CA_1 and CA_2 aggregates is mixed with CA_3 aggregates in different proportions by mass (90:10, 80:20, 70:30, etc.). The bulk density, PD, and VC are calculated for each weight fraction of these aggregate mixture. The maximum PD of the mixture and its corresponding weight fraction were determined. Subsequently, CA_4 is added in different proportions by mass to the combination of CA_1-, CA_2-, and CA_3-sized coarse aggregates and the proportion at which maximum PD obtained is determined. The smaller sized coarse aggregates are added to the mixture one at a time in decreasing sequence, and for each mixture, the maximum PD and its associated weight percentage of each aggregate size are determined. The addition of fine aggregate will next be made in a variety of mass proportions to the mixture of coarse aggregates of various sizes. The weight percentage of the chosen sizes of coarse aggregates and fine aggregates is calculated, together with the highest achievable PD for the combination of the two.

Theoretical Methods

The CPM is used to determine the theoretical packing density of the aggregate mixture. The CPM is selected out of different models as discussed in chapter "Background on Techniques for Sustainable Use of Recycled Aggregate and Application of Particle Packing Method", because it accounts the interaction effects (wall effect and loosening effect) as well as the compaction index associated with the compaction effort. Consideration of compaction effort makes this model more practical for the estimation of packing density of the aggregate mixture. In CPM, the particle size distribution (PSD) is measured for the constituent aggregates and it is expressed as the fraction retained (% by volume) in different size groups. The compaction index (K) introduced by De Larrard is used to calculate actual packing density (α_t) from virtual packing density (β). The β can be defined as the maximum potential packing density of a mixture if the particles would have been placed one by one in such a way that they use the minimum amount of space. The general expression for β of a mixture containing n size classes with class i being dominant is represented in Eq. 4.2.3.

$$\gamma_i = \frac{\beta_i}{1 - \sum_{j=1}^{i-1}\left[1 - \beta_i + b_{ij}\beta_i\left(1 - \frac{1}{\beta_j}\right)\right] y_j + \sum_{j=i+1}^{n}\left[1 - \frac{a_{ij}\beta_i}{\beta_j}\right] y_j} \qquad (4.2.3)$$

where y_j = volumetric proportion of size class i, β_i = virtual packing density of the size class i compacted alone, γ_i = virtual packing density of the mixture when class i is dominant, a_{ij} = interaction coefficient due to the loosening effect, and b_{ij} = interaction coefficient due to the wall effect. The interaction coefficients can be determined from Eqs. 4.2.4 and 4.2.5.

$$a_{ij} = \sqrt{1 - \left(1 - \frac{d_j}{d_i}\right)^{1.02}} \quad (4.2.4)$$

$$b_{ij} = 1 - \left(1 - \frac{d_i}{d_j}\right)^{1.5} \quad (4.2.5)$$

where d_i and d_j are the diameters of the granular classes i and j as defined by sieve sizes. The loosening effect occurs, when the coarse grains are the dominant class and the fine grains are not small enough to fit into the interstitial space between the coarse grains without disturbing its packing arrangement. So, by inserting the fine grains into the interstices of coarse grains, the packing arrangement of coarse particles gets loosened which results in a local reduction in volume of the coarse grains. The wall effect occurs, when the fine grains are the dominant class and the added coarse grains create voids in its interface, which are small enough to be filled by other classes of grains present in the mixture of grains.

The real packing density is lower than the virtual packing density and it depends on the applied compaction energy. A scalar K is introduced in order to determine real packing density, which depends on the applied compaction only. As K tends to infinity, the α_t tends to β. The packing density α_t can be determined indirectly from Eq. 4.2.6.

$$K = \sum_{i=1}^{n} K_i = \sum_{i=1}^{n} \frac{y_i \beta_i}{\frac{1}{\alpha_t} - \frac{1}{\gamma_i}} \quad (4.2.6)$$

Determination of Paste Content

The total packing density (PD) obtained by mixing different sized coarse aggregates and fine aggregate is used to determine the voids content (VC) of the mixture using Eq. 4.2.7.

$$\mathrm{VC} = 1 - \mathrm{PD} \quad (4.2.7)$$

In a concrete mix, the paste (cementitious material and water) not only fills up the voids present in the aggregate mixture but also provides desired workability. Hence, the total paste content in this method is the sum of the voids content, and the excess quantity of paste required to bind the aggregate particles and obtain desired workability of the concrete. The total quantity of paste is determined by multiple trials to achieve the necessary conditions, i.e. the desired workability (without any segregation and bleeding) as well as the compressive strength at the same time. In order to satisfy the necessary conditions, different w/c ratios are employed in the trial process. The cement and water content in the PPM mix design is determined according to the following steps:

(a) Total voids content = VC + Excess paste content × VC
(b) Volume of aggregates = 1 − Total voids content
(c) Total solid volume of aggregates = $\sum \frac{\text{Weight fraction of aggregates}}{\text{Specific gravity}}$
(d) Weight of aggregate = $\frac{\text{Volume of the aggregates}}{\text{Total solid volumes of aggregates}}$ × Weight fraction × 1000
(e) If, $w/c = n$; $w = n \times c$
(f) Total paste content = $c + w = \frac{c}{SG_{\text{cement}}} + \frac{n \times c}{SG_{\text{water}}}$
(g) Cement content = $\frac{\text{Total voids content}}{\text{Total paste content}}$ × 1000
(h) Water content = w/c × Cement content

4.3 Mix Proportion

4.3.1 Aggregate Content

The experimental optimization process was executed by Pradhan et al. (2017) for proportioning of aggregates in the mixture. The study considered three different categories of NCA sizes, i.e. 20–12.5 mm, 12.5–10 mm, and 10–4.75 mm and similarly four different categories of RCA sizes, i.e. 20–12.5 mm, 12.5–10 mm, 10–6.3 mm, and 6.3–4.75 mm. Along with the coarse aggregates the Zone II fine aggregate was mixed to optimize the combination of aggregates for maximum packing density. The compacted bulk density and specific gravity for each size of aggregates were determined and reported in chapter "Characterization of Materials". The bulk density and PD of the aggregate mixture (coarse and fine aggregates) were determined in accordance to the procedure discussed in section "Experimental Procedure". After conducting the whole exercise, maximum bulk density and PD for the NCA and fine aggregate mixture were found to be 1.99 and 0.713, respectively. The proportion of 20–12.5-mm-, 12.5–10-mm-, 10–4.75-mm-sized NCA and fine aggregate to achieve maximum packing density was 38.4:9.6:12:40. Similarly, the maximum bulk density and PD were found to be 1.865 and 0.728, respectively, for aggregate proportion 31.92:13.68:11.4:3:40 of 20–12.5-mm-, 12.5–10-mm-, 10–6.3-mm-, and 6.3–4.75-mm-sized RCA and fine aggregate. The bulk density and PD obtained for different aggregate combinations at different proportions are depicted in Figs. 4.2 and 4.3.

The aggregate proportions obtained experimentally were analysed and compared with the modified Andreasen model (Andreasen and Andersen 1930) using EMMA Mix Analyser software. It was reported that the theoretical curve generated using modified Andreasen equation shows good correlation with the particle size distribution curves obtained in PPM. The particle size distribution curves for both RCA+fine aggregate mixture and NCA+fine aggregate mixtures are shown in Figs. 4.4 and 4.5,

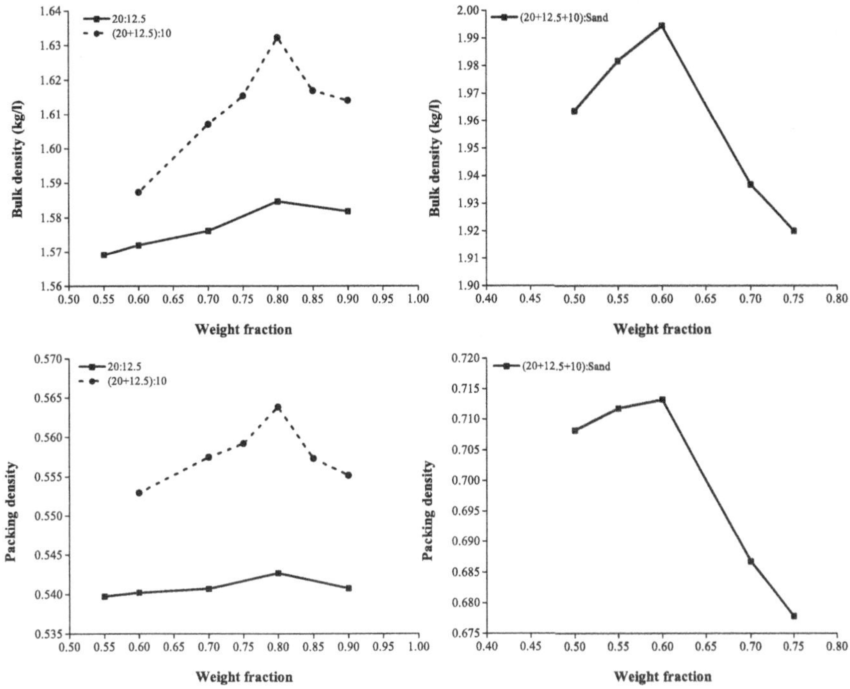

Fig. 4.2 Bulk density and packing density of NCA and sand mixture (Authors')

respectively. Along with the particle size distribution curves of these aggregate mixtures, the respective theoretical curve of the modified Andreasen model (Andreasen and Andersen 1930) is also presented.

The aggregate proportions obtained experimentally in PPM were used to estimate the theoretical packing density using CPM. In CPM, the K for dry rodding procedure to prepare dry granular packing is 4.5 (De Larrard 1999). By using the CPM approach and K value of 4.5, the theoretical packing density was obtained to be 0.67 and 0.66 for RCA+fine aggregate mixture and NCA+fine aggregate mixture, respectively. This shows that the theoretical packing density for both the compositions does not reflect good correlation with the experimental results. Because of the lower value and significant deviation from the experimental results, the obtained theoretical packing density value cannot be suggested for the determination of packing density. But, interestingly by using a higher K value of 6.5, the theoretical packing density was found out to be 0.725 and 0.713 for RCA+fine aggregate mixture and NCA+fine aggregate mixture, respectively. These theoretical packing density values are nearly equal as that of experimental values.

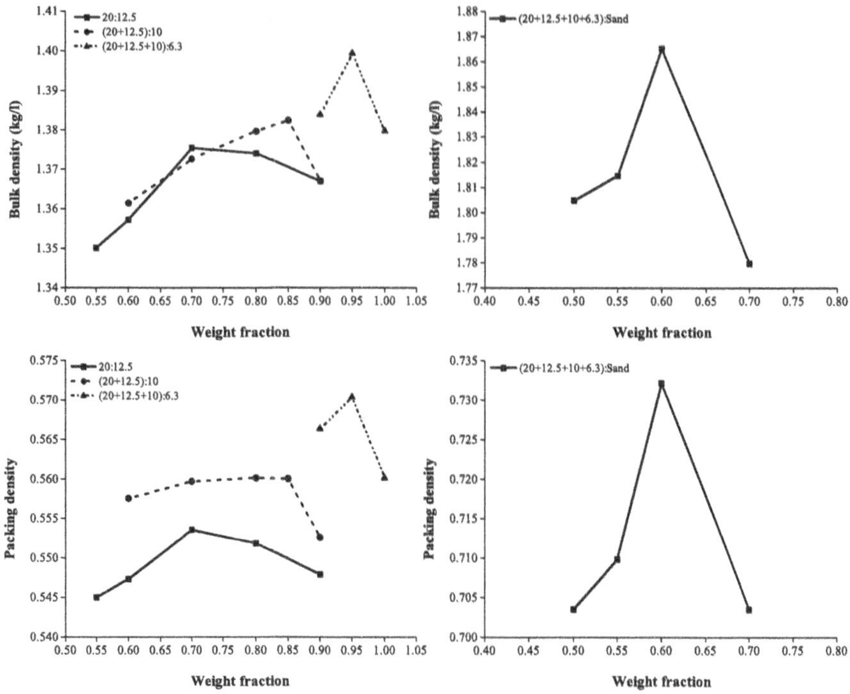

Fig. 4.3 Bulk density and packing density of RCA and sand mixture (Authors')

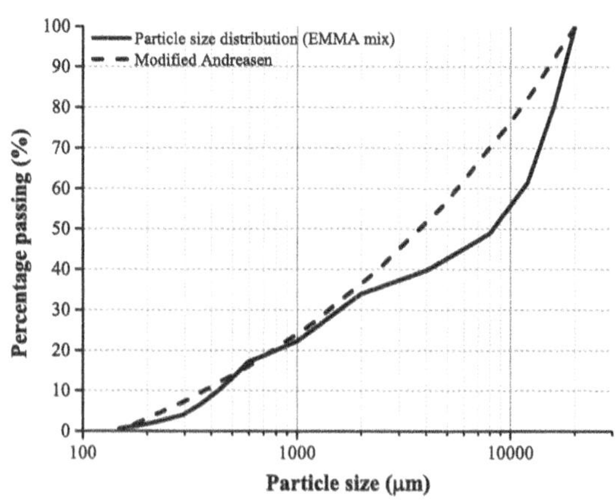

Fig. 4.4 Particle size distribution for NCA and fine aggregate mixture (Pradhan et al. 2017)

4.3.2 *Paste Content*

The total paste content in PPM mix design approach can be determined by employing the steps discussed in section "Determination of Paste Content". As mentioned earlier, the total paste content is the sum of the voids content and the excess quantity of

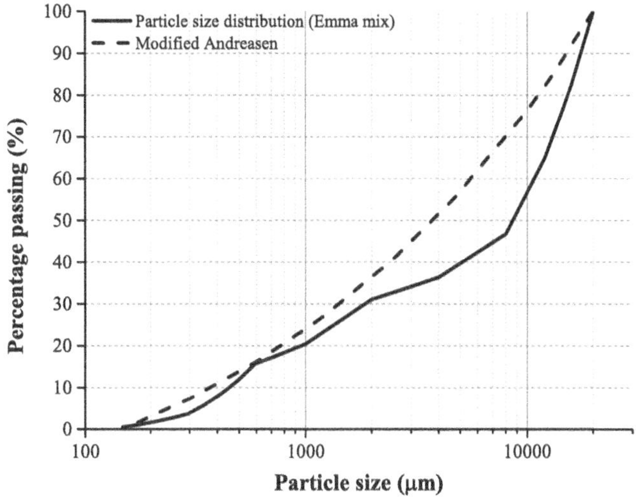

Fig. 4.5 Particle size distribution for RCA and fine aggregate mixture (Pradhan et al. 2017)

paste required to bind the aggregate particles as well as to obtain desired workability of the concrete. Multiple trials were carried out at w/c ratio 0.40, 0.43, 0.45, and 0.47 to achieve the target mean strength for M30 grade (compressive strength of 30 MPa after 28 days of curing) of concrete with desired slump value of 100 mm. In this process, 16% excess paste content was incorporated for each w/c ratio to obtain desired workability as well as strength.

The mixture proportions of both NAC and RAC using conventional approach are presented in Appendix A. Both NAC and RAC were proportioned using conventional as well as PPM mix design approaches. Consequently, four concrete mixes were prepared in total. The quantities of all the constituents of both NAC and RAC for the two different mix design methods are represented in Table 4.1. For all the mixes w/c of 0.45 was considered. As it can be seen in Table 4.1, the requirement of water and cement is on the lower side for both NAC and RAC in the PPM mix design approach. However, 24.9% and 13% more fine aggregates were required for NAC and RAC, respectively, in the PPM mix design approach as compared to conventional method. The total coarse aggregate quantities were almost same in both the mix design methods. From the mix proportions, the PPM mix design approach can be considered as economical than conventional method as it requires a lesser quantity of cement.

Table 4.1 Mix proportions of concrete

Type of concrete	Mix design method	Effective w/c	Water (kg/m^3)	Cement (kg/m^3)	Sand (kg/m^3)	Coarse Aggregate (kg/m^3)			
						20–12.5 mm	12.5–10 mm	10–6.3 mm	6.3–4.75 mm
NAC	Conventional	0.45	215.9	438.1	610.9	755.8	–	485.6	–
NAC	PPM	0.45	203.5	410.0	762.1	666.9	247.7	228.6	–
RAC	Conventional	0.45	234.2	422.2	621.2	671.5	–	416.7	–
RAC	PPM	0.45	225.8	410.0	702.0	560.2	240.1	200.1	52.7

4.4 Mixing Process

The Two-Stage Mixing Approach (TSMA) proposed by Tam et al. (2005) was adopted by Pradhan et al. (2017) for the PPM mix-designed concrete. However, the part of water added in two steps, i.e. to the dry aggregate mixture and surface-coated aggregate mixture, and the mixing time of each step is different to the Tam et al. (2005) proposed TSMA. The nomenclature of this mixing method is basically due to the addition of water in two stages and the steps involved in TSMA are illustrated in Fig. 4.6. In the first phase, $1/3^{\mathrm{rd}}$ water was added to the dry mix of coarse and fine aggregates mixture and subsequently the rest of the water was added to the wet aggregates and cement mixture. The mixing time for each stage is specified in Fig. 4.6. The concrete prepared using conventional mix design method is mixed according to the normal mixing approach. In normal mixing approach, all the water is added at a time to the dry mixture of coarse aggregates, fine aggregates, and cement. The steps and mixing time involved in normal mixing approach are illustrated in Fig. 4.7.

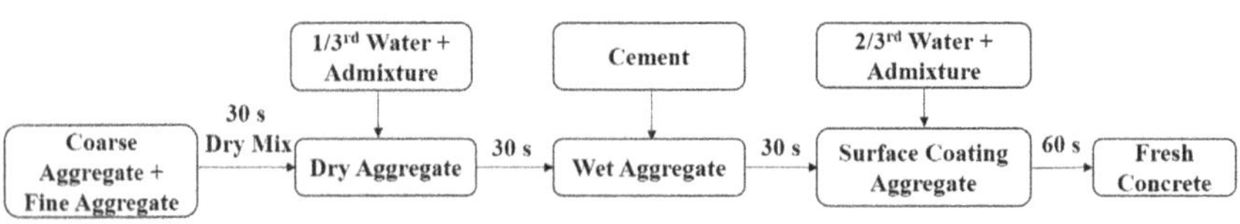

Fig. 4.6 Schematic diagram of modified two-stage mixing approach (Pradhan et al. 2017)

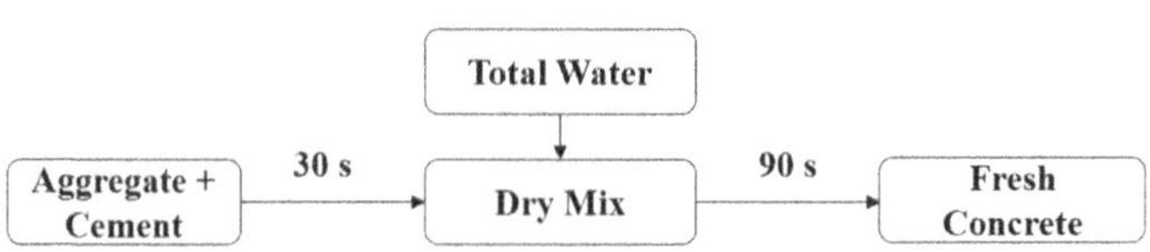

Fig. 4.7 Schematic diagram of normal mixing approach (Authors')

4.5 Closure

The proposed PPM mix design for making concrete was covered in the current chapter. The determination of aggregate proportions and paste content is described in great detail. Utilizing PPM will reduce the amount of voids in concrete, lowering the amount of cement paste needed. For lesser paste content, the overall mortar content (fresh + adhered hardened mortar) will be decreased in RAC, so that the performance of RAC can be improved. For both NCA and RCA mixtures with fine aggregate, the experimental and theoretical PD and voids contents were calculated. In order to examine the effectiveness of PPM mix-designed concrete, both NAC and RAC were made using the traditional mix design approach. For PPM and conventional mix-designed concrete, the modified TSMA and regular mixing techniques, respectively, were employed. These mixing procedures' steps were covered in detail.

References

Amario M, Rangel CS, Pepe M, Toledo Filho RD (2017) Optimization of normal and high strength recycled aggregate concrete mixtures by using packing model. Cement Concr Compos 84:83–92

Andreasen AHM, Andersen J (1930) Relation between grain size and interstitial space in products of unconsolidated granules. Kolloid-Zeitschrift 50(3):217–228

ASTM C29 (2009) Standard test method for bulk density ("unit weight") and voids in aggregate

Babalola OE, Awoyera PO, Tran MT, Le DH, Olalusi OB, Viloria A, Ovallos-Gazabon D (2020) Mechanical and durability properties of recycled aggregate concrete with ternary binder system and optimized mix proportion. J Mater Res Technol 9(3):6521–6532

De Larrard F (1999) Concrete mixture proportioning: a scientific approach. CRC Press

IS: 2386 (Part III) (1963) Method of Test for aggregate for concrete. Part III–Specific gravity, density, voids, absorption and bulking. Bureau of Indian Standards, New Delhi, India

Kirthika SK, Singh SK, Chourasia A (2020) Performance of recycled fine-aggregate concrete using novel mix-proportioning method. J Mater Civil Eng 32(8):04020216

Pradhan S, Kumar S, Barai SV (2017) Recycled aggregate concrete: particle packing method (PPM) of mix design approach. Constr Build Mater 152:269–284

Puente de Andrade G, de Castro Polisseni G, Pepe M, Toledo Filho RD (2020) Design of structural concrete mixtures containing fine recycled concrete aggregate using packing model. Constr Build Mater 252:119091

Tam VWY, Gao XF, Tam CM (2005) Microstructural analysis of recycled aggregate concrete produced from two-stage mixing approach. Cement Concr Res 35(6):1195–1203

Part III
Multi-Scale Performance Assessment of Concrete Mixes

Chapter 5
Macro-level Performance Assessment of Concrete: Conventional Approach

5.1 Introduction

The performance of concrete is significantly impacted by its constituents and their composition. As a result, it is expected that the usage of different coarse aggregate, namely, NCA and RCA, will cause the performance of NAC and RAC to vary. The physical and mechanical properties of RCA that are detailed in chapter "Characterization of Materials" are inferior to NCA. Thus, it is also anticipated that concrete's performance at the macro-level will suffer as a result of aggregate's weaker qualities. In chapter "Particle Packing Method of Mix Proportioning and Modified Mixing Approach", a new mix design approach, i.e. PPM, is suggested and discussed in order to address this problem. The performance of PPM mix-designed concrete in the fresh and hardened stages is covered in this chapter, along with a comparison to mixes made using the standard IS: 10262 (2009) method. The performance of four various concrete mixes in both the fresh and hardened stages is described in the following sections.

5.2 Performance of Concrete Mixes

5.2.1 Fresh Concrete

The adhering porous mortar layer to the natural aggregate is principally responsible for RCA's increased water absorption capability. Fresh concrete's workability suffers as a result of RCA's increased water absorption value (Padmini et al. 2009). Ravindrarajah and Tam (1985) and Hansen and Narud (1983) proposed that RAC needed an additional 5% water to be nearly as workable as NAC, whereas Tabsh and Abdelfatah (2009) added an additional 10% water. However, utilizing RCA taken from the laboratory-tested specimens, Bairagi et al. (1990) observed equal work-

S. Pradhan et al., *Particle Packing Method for Recycled Aggregate Concrete*,
https://doi.org/10.1007/978-981-99-7516-7_5

ability for RAC as NAC. Use of saturated or pre-wetted RCA was recommended by Sagoe-Crentsil et al. (2001) to attain a workability comparable to NAC. When employing oven and air-dried RCA instead of RCA in SSD state, the experimental inquiry by Poon et al. (2004b) indicated higher slump values and quicker slump disappearance of RAC. Such behaviour results from the early availability of a larger free water content. Furthermore, Poon et al. (2007) noted that the slump value increased as the replacement percentage of RCA increased, and the slump value increased as fly ash (25%) was used to replace cement in both RAC and NAC. The two-stage mixing approach employed by Tam et al. (2005) helped in improving the workability of RAC. Yang et al. (2008) found that the greater water absorption capacity of RCA had a minor impact on the usability of fresh RAC. Because smaller sized RCA absorb more water, the reduction in coarse aggregate size led to a decrease in slump value (Padmini et al. 2009). With an increase in the percentage of RCA replacement, Chakradhara Rao (2010) observed a decrease in the workability of RAC.

A target slump of 75–100 mm slump was kept constant for NAC as well as RAC by Pradhan et al. (2017). The workability of the fresh concrete was determined by conducting the slump cone test as specified in IS: 7320 (1974). A metal base plate was used to support the slump cone mould. Three layers of newly produced concrete were poured into the cone, and each layer was uniformly compacted by 25 blows using a tamping rod with a 16 mm diameter. A trowel was used to flatten the top surface. The mould was then gently removed, and the slump value was determined by how much the height of the unsupported concrete had decreased. Without the use of any admixtures, conventional concrete was able to achieve the requisite workability. However, in RAC without adding any admixture, a maximum of only 10 mm slump was observed. In order to achieve the desired slump of 75–100 mm, MasterGlenium ACE 30 (BASF) admixture was added to the RAC mixes at a rate of 0.32% by weight of cement.

5.2.2 *Hardened Concrete*

Compressive Strength

The petrological features of aggregates, their physical and mechanical qualities, w/c, the kind of binder, the presence of mineral admixtures, the mix design procedure, and the mixing technique all affect the compressive strength of concrete (Poon et al. 2004a, b; McNeil and Kang 2013; Behera et al. 2014). Hansen and Narud (1983) investigated the impact of the parent concrete's strength (low, medium, and high) on the strength of RAC. The investigation supported the hypothesis that the compressive strength of RAC depends on the durability and strength of parent concrete. While the w/c of RAC was observed to be the same or less than parent concrete, the compressive strength of RAC was observed to be the same or more than parent concrete. The compressive strength of RAC increases when w/c decreases, similar to NAC (Ravindrarajah and Tam 1985; Padmini et al. 2009). As the strength of parent concrete improves, so does the compressive strength of RAC (Etxeberria et al. 2007b; Kou and Poon 2015). According to Kou et al. (2008); Xiao et al. (2012), the compressive strength of RAC decreased as the replacement level of RCA increased. Up to 30% substitution of RCA, the drop in RAC's compressive strength was found to be less substantial (Xiao et al. 2005; Etxeberria et al. 2007b; Chakradhara Rao et al. 2011; Fonseca et al. 2011; Behera et al. 2014). However, up to 30% reduction in compressive strength of RAC was experienced at 100% replacement of NCA (Hansen and Narud 1983; Ravindrarajah and Tam 1985; Poon et al. 2004b; Xiao et al. 2005; Behera et al. 2014). Bairagi et al. (1993) and Katz (2003) both found that utilizing 100% RCA reduced the compressive strength of RAC by up to 60% and 76%, respectively. One of the causes of this abnormal behaviour is the heterogeneity of the RCA, according to Etxeberria et al. (2007a). Katz (2003) noted that due to the potential for subsequent hydration processes when utilizing unhydrated cement, the age of the parent concrete can also affect the strength of RAC. By employing high-strength parent concrete, Casuccio et al. (2008) observed similar compressive strength of RAC as NAC, but a considerable reduction in compressive strength of RAC was documented when using RCA extracted from medium-strength concrete. Due to the presence of two ITZs, which is considered as the weakest region in concrete, RAC has a reduced compressive strength (Tam et al. 2007; Mukharjee and Barai 2014a; Behera et al. 2014). Although the compressive strength of RAC was comparable with NAC at the early stage (up to 7 days), the rate of increase in compressive strength of RAC was less than NAC (Salem and Burdette 1998; Etxeberria et al. 2007a; Chakradhara Rao 2010). By filling the voids with secondary hydration products and enhancing the compressive strength of the nano-engineered RAC, the introduction of nano-silica enhanced the ITZ quality (Mukharjee and Barai 2014a, b). The amount of attached mortar affects the compressive strength of RAC and is influenced by the parent concrete's strength, the number of crushing stages utilized, and the size of the aggregates (Akbarnezhad et al. 2013). According to Brand et al. (2015), when oven dried, partially saturated, and fully saturated RCA were compared to each other and the compressive strength of RAC, it was found that using partially saturated RCA and a two-stage mixing method improved compressive strength.

Pradhan et al. (2017) evaluated the compressive strength of the hardened concrete by using the cube specimens of 150 mm size in accordance to the procedure given in IS: 516 (1959). The specimens were placed in a 3000 kN capacity compression testing machine and the load was applied at a rate of 140 kg/cm^2/min as specified in IS: 516 (1959). The compressive strengths of 7-, 28-, and 90-day cured cube specimens were measured for all the four concrete mixes (NAC IS, NAC PPM, RAC IS, and RAC PPM) and the results are reported in Table 5.1. The 7-day compressive strength at w/c 0.45 exhibited negligible difference (0.84% higher for NAC PPM) between NAC PPM and RAC PPM (Table 5.1 and Fig. 5.1). However, for w/c 0.40, 0.43, and 0.47, the 7-day compressive strength values of RAC PPM were 2.6–3.3% higher than that of NAC PPM. In PPM mix proportioning, apart from w/c 0.40, the 28-day compressive strength of RAC PPM was 0.2–11.3% higher than NAC PPM and the 28-day compressive strength obtained at w/c 0.40 for NAC PPM was about 3.9% higher than that of RAC PPM. This shows that the 7-day and 28-day compressive strengths of RAC PPM were slightly on the higher side than NAC PPM. However, the 90-day

Table 5.1 Compressive strength of different types of concrete

Type of concrete	Mix design method	w/c	7 days strength (MPa)	28 days strength (MPa)	90 days strength (MPa)
NAC	PPM	0.40	30.9	51.3	65.5
		0.43	28.7	44.3	58.9
		0.45	25.4	42.8	56.2
		0.47	23.7	38.2	52.5
NAC	IS: 10262 (2009)	0.45	29.7	47.7	64.9
RAC	PPM	0.40	31.7	49.3	56.3
		0.43	29.5	48.1	53.7
		0.45	25.1	42.8	47.7
		0.47	24.5	42.5	47.6
RAC	IS: 10262 (2009)	0.45	36.0	45.2	54.5

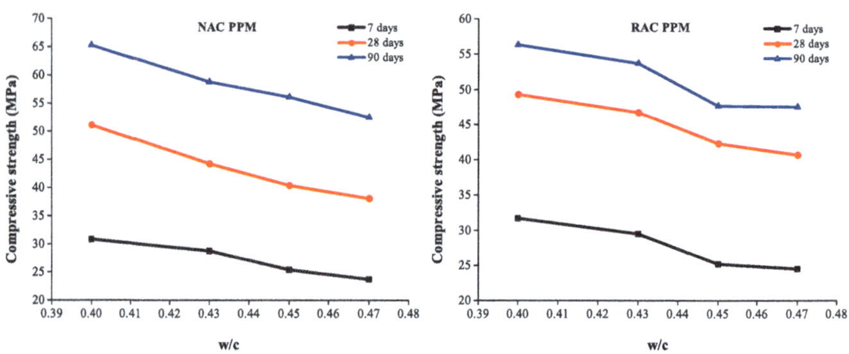

Fig. 5.1 Compressive strength of NAC PPM and RAC PPM at different w/c ratio (Authors')

curing compressive strength values of RAC PPM at different w/c were 9.6–17.7% lower than NAC PPM. There was an increase of 27–37% observed in compressive strength at different w/c of NAC PPM from 28 days to 90 days, whereas for RAC PPM only 11–14% increment was observed for the same period of curing (Table 5.1 and Fig. 5.2). The compressive strength of NAC IS and RAC IS was higher than the NAC PPM and RAC PPM, respectively, for all curing ages (Table 5.1). The 28-day compressive strengths of NAC IS and RAC IS were about 11.5% and 5.6% higher than NAC PPM and RAC PPM, respectively, at w/c 0.45. The compressive strength of RAC IS at 7 days of curing was about 21.2% higher in comparison to NAC IS (Table 5.1). However, at 28 days and 90 days of curing, the compressive strength of NAC IS was about 5.4% and 19.1% higher than RAC IS (Table 5.1).

The 7-day and 28-day curing compressive strength of RAC is very close to NAC in both the methods of design mix (Table 5.1). However, after 90 days of curing,

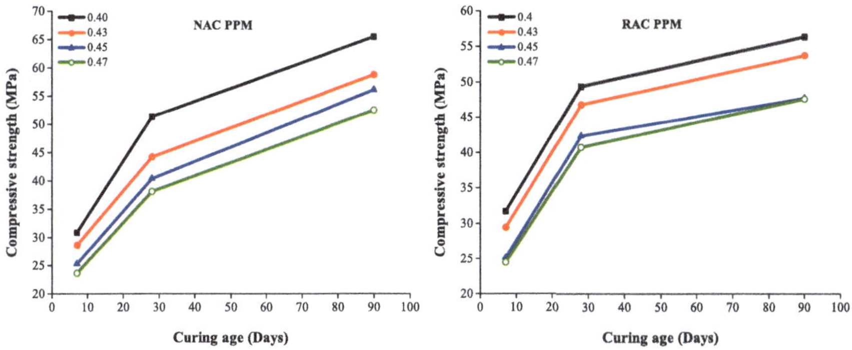

Fig. 5.2 Compressive strength of NAC PPM and RAC PPM at different curing age (Authors')

a difference of 17.7% and 19.1% was observed in NAC with respect to RAC for PPM and IS: 10262 (2009) method of mix design, respectively. From Table 5.1 and Fig. 5.2, it can be seen that, regardless of the mix design process, RAC gains less strength from 28 to 90 days after curing. This discrepancy might be caused by a variation in the amount of unhydrated cement present in the matrix of the concrete. Because RCA absorbs more water, the amount of water at the interface might not be sufficient to maintain the same rate of hydration.

Tensile Strength

The split tensile test and the flexural tensile test, which use cylindrical and cuboid specimens, respectively, are typically used to assess the tensile strength of concrete indirectly. The addition of RCA has an impact on the tensile strength of RAC, similar to the compressive strength. However, Behera et al. (2014) and Kisku et al. (2017) reported that the effect is less significant. As the replacement percentage of RCA rises, so does the tensile strength of RAC (Yang et al. 2008; Zega and Di Maio 2011; Kou et al. 2011; Elhakam et al. 2012; Kou and Poon 2013). As the replacement ratio rises, the split tensile strength of RAC falls, and the decline was up to 10% at various replacement percentages (Ajdukiewicz and Kliszczewicz 2002; Yang et al. 2008; Malešev et al. 2010). However, Chakradhara Rao et al. (2011) observed the reduction in split tensile strength was up to 24% for 100% usage of RCA. The split tensile strength of RAC increases with the increase in curing period (Sagoe-Crentsil et al. 2001; Etxeberria et al. 2007a; Kou and Poon 2008). According to Sagoe-Crentsil et al. (2001), the quality of the binding mortar has a greater impact on split tensile strength than the type of aggregate. A 14% increase in split tensile strength of RAC was seen by Mukharjee and Barai (2014b) with 100% RCA utilization and the addition of nano-silica. Matias et al. (2014) found that RAC had a higher split tensile strength than NAC and deduced that this difference was caused by the rough surface of RCA, which helped to strengthen the bond with the paste matrix.

The flexural tensile strength of RAC was reported to be relatively unaffected by the replacement ratio of RCA (Tam et al. 2005; Xiao et al. 2005; Rao et al. 2007) and the maximum reduction was observed to be up to 10% for 100% use of RCA (Topcu and Şengel 2004; Yang et al. 2008; Malešev et al. 2010). However, past investigations have determined that an increase in the percentage of RCA replacement leads to a decrease in flexural tensile strength (Katz 2003; Topcu and Şengel 2004; Padmini et al. 2009; Malešev et al. 2010; Kou et al. 2011; Chakradhara Rao et al. 2011). The flexural tensile strength of RAC is unaffected by the moisture state of RCA, i.e. dry, semi-saturated, and saturated (de Oliveira and Vazquez 1996). With increasing concrete strength comes a corresponding increase in the reduction in flexural tensile strength (Limbachiya et al. 2000; Padmini et al. 2009). Akbarnezhad et al. (2011) observed about 15% reduction in flexural tensile strength as compared to the 30% reduction in compressive strength and concluded that, the bond strength between aggregate and mortar influences the flexural tensile strength. Butler et al. (2013) reported that, the flexural tensile strength of RAC increases as the strength of parent concrete increases.

Using cylindrical specimens with a 150 mm diameter and 300 mm height, Pradhan et al. (2017) determined the split tensile strength (STS). Two 10 mm square cross-sectional rods that were positioned in diametrically opposite directions were used to support the specimen along its curved surface. The specimen was then loaded at a rate of 1.2 to 2.4 N/mm^2/min as specified in IS: 5816 (1999). Furthermore, the flexural tensile strength (FTS) was measured using cuboids of size 100 × 100 × 500 mm after 28 days of curing. The test was carried out in an universal testing machine of 100 kN capacity. The specimens were placed on two roller support spaced at a distance of 450 mm centre to centre. The load was applied at the middle third (150 mm centre to centre) with the help of two rollers at a rate of 0.7 N/mm^2/min as specified in IS: 516 (1959).

Following a 28-day curing period for the respective specimens, STS and FTS were both measured. The STS results obtained for NAC and RAC are presented in Table 5.2. The STS of RAC was found to be lower than NAC. The STS values of NAC were, respectively, 21.5% and 25.8% higher than RAC prepared by employing PPM and IS: 10262 (2009) method of mix design. Interestingly, STS values of NAC PPM and RAC PPM were, respectively, 2.6% and 6.3% higher than those of NAC IS and RAC IS. This confirms that, PPM mix design has an advantage over IS: 10262 (2009) mix design method in improving the STS of both NAC and RAC. Even though mixes employing IS: 10262 (2009) method had better NAC and RAC compressive strengths than PPM-based mixes, the STS values are higher in PPM-based concrete mixes.

The FTS values of both NAC and RAC are presented in Table 5.2. The behaviour of FTS was similar to STS for both NAC and RAC. NAC exhibited 7.2% and 23.3% higher FTS than RAC in PPM and IS: 10262 (2009) method of mix design, respectively. The FTS values of the NAC PPM and RAC PPM specimens were greater than those of the NAC IS and RAC IS, respectively, by 8.2% and 24.4%. This indicates that the FTS of RAC can be improved more effectively with the PPM mix design than for NAC. When compared to PPM mix-designed concrete, the concrete that was

Table 5.2 Tensile strength and modulus of elasticity of different types of concrete

Type of concrete	w/c	Tensile strength (MPa)		Modulus of elasticity (MPa)
		Flexural	Split	
NAC PPM	0.45	3.5	4.8	35670
NAC IS	0.45	3.4	4.4	36019
RAC PPM	0.45	2.9	4.4	35063
RAC IS	0.45	2.1	3.6	35500

proportioned using IS: 10262 (2009) mix design method had a lower FTS, even if the compressive strength was higher.

Modulus of Elasticity

The mechanical property of RAC that is most impacted is its modulus of elasticity. Because mortar is less stiff than aggregate, RAC has a lower modulus of elasticity than NAC due to a larger mortar content (new mortar in addition to the attached mortar) (Xiao et al. 2012; Ho et al. 2013). The lower modulus of elasticity is another effect of the larger porosity of RAC (Behera et al. 2014). With more RCA being substituted, the modulus of elasticity was shown to significantly decrease, and at 100% replacement level, the loss can be up to 45% (Ravindrarajah and Tam 1985; Ajdukiewicz and Kliszczewicz 2002; Xiao et al. 2005; Rahal 2007; Yang et al. 2008; Malešev et al. 2010; Chakradhara Rao et al. 2011). Ajdukiewicz and Kliszczewicz (2002) and Rao et al. (2007) noted that the elastic modulus of RAC was roughly 50—70% that of normal concrete. Furthermore, Corinaldesi (2010) observed that replacing RCA by up to 30% reduced the elastic modulus of RAC by roughly 15%. According to Casuccio et al. (2008), the lower stiffness and higher brittleness of RAC are the causes of its lower elasticity modulus.

After 28 days of moisture curing, cylindrical specimens with a 150 mm diameter and 300 mm height were used to determine the static modulus of elasticity (Pradhan et al. 2017). The test was carried out in accordance to the procedure described in IS: 516 (1959). The specimen was subjected to a compressive load at a rate of 140 kg/cm^2/min, and the longitudinal displacement was measured using an extensometer with a gauge length of 150 mm. Prior to this test, the compressive strength of cylindrical specimen of the same mix was determined, as the load was applied maximum up to $(C + 5)$ kg/cm^2 (where C is the compressive strength of the concrete cylinder). The modulus of elasticity obtained for both NAC and RAC was almost similar in both the mix design methods as reported in Table 5.2.Comparing NAC IS and RAC IS to NAC PPM and RAC PPM, only a 1—1.2% increase in modulus of elasticity value was seen. An accepted parameter to determine concrete's elastic modulus is the material's compressive strength. Therefore, the higher compressive strength of the corresponding concrete can be used to explain the higher modulus of elasticity value seen in concrete mixtures proportioned using the IS code approach.

5.3 Comparative Study

The suitable mix design method to prepare RAC out of the two discussed methods employed by Pradhan et al. (2017) was verified by comparing the obtained mechanical properties. Since compressive strength is a function of various observed mechanical parameters, such as STS, FTS, and modulus of elasticity, these properties are normalized using the power function of the corresponding compressive strength. According to the normalized values of STS, FTS, and modulus of elasticity (Table 5.3), the PPM mix design technique is superior than the IS: 10262 (2009) method of mix design for both NAC and RAC. Despite having lower compressive strengths than concrete mixes proportioned according to IS: 10262 (2009) method of mix design, both NAC and RAC showed higher experimental STS and FTS values in the PPM mix design. In the mix design approach described in IS: 10262 (2009), the modulus of elasticity values for both NAC and RAC were greater. However, the normalized values of the elastic modulus show that PPM mix design significantly improves the elastic modulus of both NAC and RAC. This improvement may be ascribed to the decrease in the quantity of less stiff mortar.

The mechanical properties of concrete, such as STS, FTS, and modulus of elasticity, are generally expressed as a function of compressive strength in different codes and literature. The available expressions as shown in Table 5.4 were considered by Pradhan et al. (2017) for comparison with experimental results. In Figs. 5.3, 5.4, and 5.5, the values obtained from established expressions are shown along with experimental results obtained by Pradhan et al. (2017). The experimental STS and FTS values are not showing a good correlation with the values calculated using expressions from various codes and literature. This implies that the STS and FTS predictions made using the current formulations for ordinary concrete may not be applicable to the aforementioned characteristics of RAC. Thus, Pradhan et al. (2017) assembled the experimental results of RAC presented in various publications for

Table 5.3 Normalized value of different mechanical properties of different types of concrete

Type of concrete	f_c (MPa)	Actual value (MPa)			Normalized value		
		f_t	f_r	E	$f_t/\sqrt{f_c}$	$f_r/\sqrt{f_c}$	$E/\sqrt{f_c}$
NAC PPM	42.8	3.5	4.8	35670	0.535	0.726	5455.55
NAC IS	47.7	3.4	4.4	36019	0.494	0.636	5216.27
RAC PPM	42.8	2.9	4.4	35063	0.440	0.677	5358.25
RAC IS	45.2	2.7	3.6	35500	0.403	0.530	5278.60

Table 5.4 Expressions to predict f_{sp}, f_t, and E from f_c

Split tensile strength (f_{sp})		Flexural tensile strength (f_r)		Modulus of elasticity (E_c)	
Codes or literature	Expression	Codes or literature	Expression	Codes or literature	Expression
CEB	$f_{sp} = 1.56\left(\frac{f_c-8}{10}\right)^{\frac{2}{3}}$	CEB	$f_r = 0.81\sqrt{f_c}$	Ravindrarajah and Tam	$E = 7770 f_c^{0.33}$
EHE	$f_{sp} = 0.21 f_c^{\frac{2}{3}}$	IS: 456 (2000)	$f_r = 0.7\sqrt{f_c}$	Dillman	$E = 634.43 f_c + 3057.6$
ACI 318 (2008)	$f_{sp} = 0.56\sqrt{f_c}$	ACI 318 (2008)	$f_r = 0.62\sqrt{f_c}$	Dhir	$E = 370 f_c + 13100$
GB 10010	$f_{sp} = 0.19 f_c^{0.75}$	DJ/TJ07	$f_r = 0.75\sqrt{f_c}$	Mellman	$E = 378 f_c + 8242$
NBR 6118	$f_{sp} = 0.3 f_c^{\frac{2}{3}}$			IS 456	$E = 5000\sqrt{f_c}$
Hueste	$f_{sp} = 0.55\sqrt{f_c}$			NBR 6118	$E = 5600\sqrt{f_c}$
Xiao	$f_{sp} = 0.24 f_c^{0.65}$			Hueste	$E = 5230\sqrt{f_c}$
Kou and Poon	$f_{sp} = 0.093 f_c^{0.8842}$				

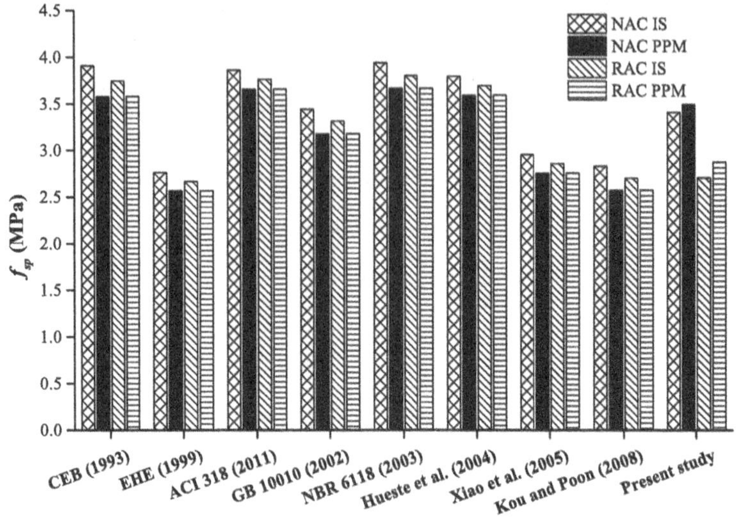

Fig. 5.3 Split tensile strength of different types of concrete (Pradhan et al. 2017)

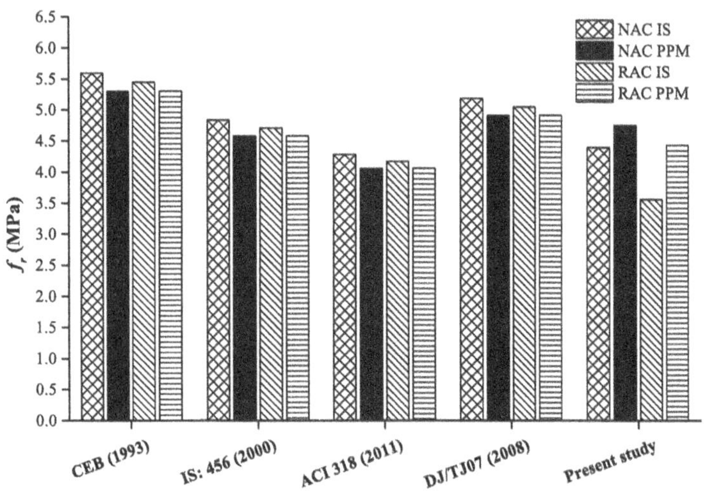

Fig. 5.4 Flexural tensile strength of different types of concrete (Pradhan et al. 2017)

additional study. The study only takes into account the test data of RAC with 100% replacement of NCA. The experimental variables, such as the mix design process, aggregate grading, specimen size and shape, number of replicas, curing condition, testing circumstances, and loading conditions, were found to vary across the literature. These variances have a significant impact on the accuracy of the test data population as a whole. The size and shape effect is a major factor among all of these experimental conditions. The influence of the size and shape of the specimen on

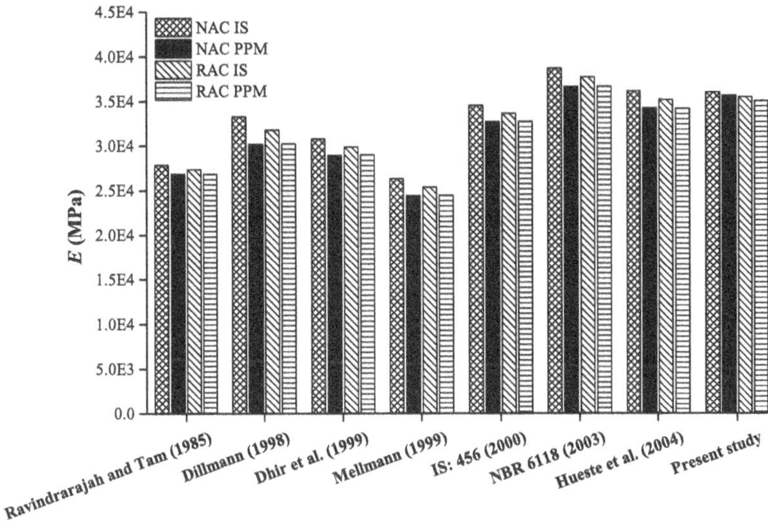

Fig. 5.5 Modulus of elasticity of different types of concrete (Pradhan et al. 2017)

compressive strength was normalized to the compressive strength of a cube of 150 mm by using Eq. 5.3.1 (Neville 2011).

$$\frac{f_c}{f_{cu,152}} = 0.56 + \frac{0.697}{\left(\frac{V}{152hd} + \frac{h}{d}\right)} \tag{5.3.1}$$

where f_c = compressive strength of the tested specimen, $f_{cu,152}$ = compressive strength of the cube of 152 mm size, V = volume of specimen, h = specimen height, and d = least lateral dimension of the specimen. Furthermore, the STS of 100 mm diameter and 200 mm high cylinder must be multiplied by a factor 0.87 to normalize its STS to that of 150 mm diameter and 300 mm high cylindrical specimens (Neville 2011). In order to produce a valid data population to develop a correlation between mechanical properties and strength of RAC, the obtained data were grouped according to compressive strength. In this method, the information gathered from the literature was arranged in groups of 5 MPa apart and classified according to compressive strength along with the matching STS, FTS, and modulus of elasticity. The mean compressive strength and related STS, FTS, and modulus of elasticity for each group were then computed. The formulas for STS, FTS, and elastic modulus of RAC as a function of its compressive strength were derived using these mean data points.

For the experimental data taken from the published literature, a methodical analysis, assessment, and synthesis was done. The relationship between the compressive strength and mechanical properties of RAC is represented in Figs. 5.6, 5.7, and 5.8. The comparative values of mechanical properties obtained from experimental investigation, derived expression (Pradhan et al. 2017), and codal expression are pre-

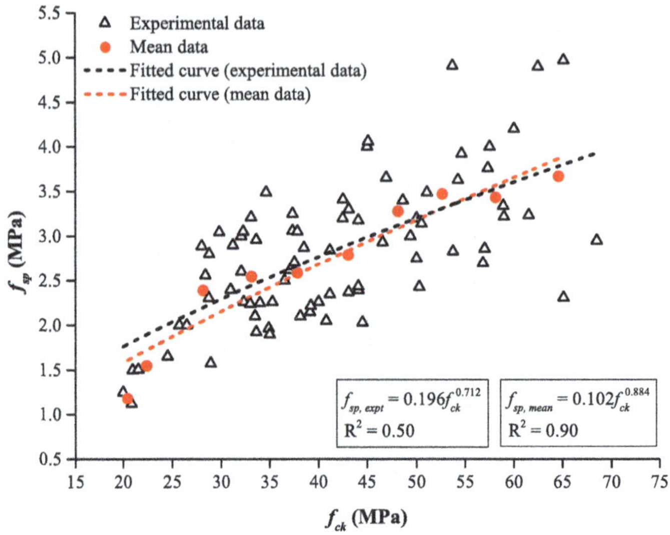

Fig. 5.6 Split tensile strength vs. compressive strength (Pradhan et al. 2017)

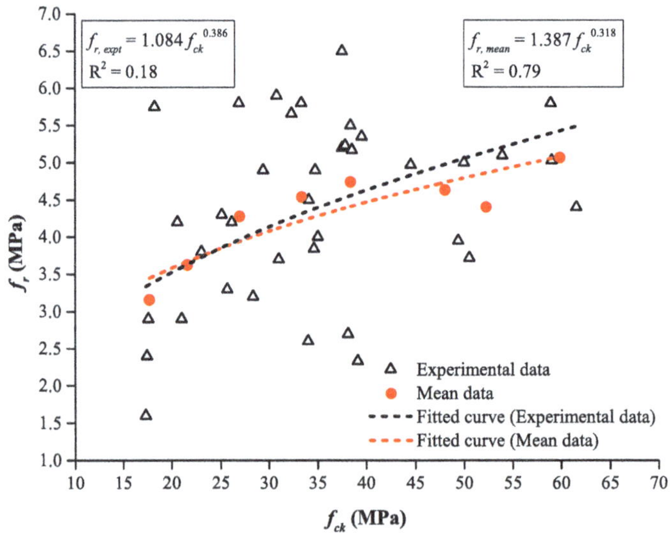

Fig. 5.7 Flexural tensile strength vs. compressive strength (Pradhan et al. 2017)

sented in Table 5.5. The relationship derived by Pradhan et al. (2017) for STS is more reliable than the expression available in ACI 318 (2008) for conventional concrete. The derived relationship and established expression in IS: 456 (2000) for FTS show no difference when compared with the experimental result. However, the equation derived for modulus of elasticity using experimental data of published literature is

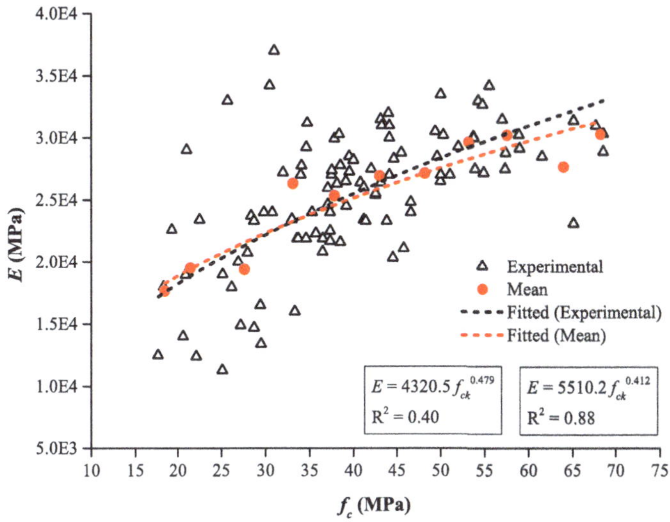

Fig. 5.8 Modulus of elasticity vs. compressive strength (Pradhan et al. 2017)

found to be more erroneous as compared to relationship available in IS: 456 (2000). This inaccuracy can be caused by the greater value of the RAC's elastic modulus that was acquired using the PPM mix design approach. This further suggests that, in comparison to other mix design techniques, the PPM methodology helps to improve the modulus of elasticity of RAC.

5.4 Statistical Analysis

To confirm that the enhanced mechanical qualities in the PPM mix design approach are not the consequence of arbitrary variations in concrete properties, a single factor ANOVA test was also carried out by Pradhan et al. (2017). The mechanical characteristics of RAC and NAC, which were proportioned using PPM and IS: 10262 (2009) methods, were compared using a single factor ANOVA to determine the significant difference.

The results of the ANOVA analysis for compressive strength, STS, FTS, and modulus of elasticity are shown in Table 5.6. Since the p-value of NAC PPM against NAC IS is smaller than the allowable error level of 0.05, the compressive strength data show that the PPM mix design strategy has a substantial impact on NAC. The mix proportioning method, however, has no appreciable impact on the compressive strength parameter for RAC, as seen by the larger p-value of 0.52 for RAC PPM compared to RAC IS. The STS and FTS p-values are 0.0004 and 0.022, respectively, indicating a considerable improvement in the tensile strength parameter of the RAC as compared to the IS: 10262 (2009) mix design procedure. For the tensile

Table 5.5 Comparison of derived equation for mechanical properties with codal expressions

Mechanical parameters	Expression	Derived expression	Experimental result	% Error in codal expression	% Error in derived expression
Split tensile strength (MPa)	$f_{sp} = 0.56\sqrt{f_{ck}} = 3.66$ (ACI 318 2008)	$f_{sp} = 0.102 f_{ck}^{0.884} = 2.83$	2.88	−21.3	1.8
Flexural tensile strength (MPa)	$f_r = 0.7\sqrt{f_{ck}} = 4.58$ (IS: 456 2000)	$f_r = 1.387 f_{ck}^{0.318} = 4.58$	4.43	−3.3	−3.3
Modulus of elasticity (MPa)	$E = 5000\sqrt{f_{ck}} = 32719$ (IS: 456 2000)	$E = 5510.2 f_{ck}^{0.412} = 25915$	35063	7.2	35.3

Table 5.6 Single factor ANOVA test result

Type of concrete	*p*-value			
	f_c	f_{sp}	f_r	E
NAC IS	0.185	0.003	0.395	0.508
NAC PPM	0.156	0.014	0.276	0.361
RAC IS	0.013	0.602	0.873	0.270
RAC PPM	0.520	0.0004	0.022	0.463

strength qualities of NAC, the PPM mix design approach has no discernible advantage over IS: 10262 (2009) method of mix proportioning because the resulting p-values (0.602 and 0.873) are substantially greater than the permitted error level of 0.05. The p-values of the modulus of elasticity range from 0.27 to 0.508, above the 0.05 tolerable error limit. Therefore, the modulus of elasticity of both NAC and RAC is unaffected by the kind of coarse aggregates (NCA or RCA) or the mix proportioning techniques (PPM or IS: 10262 (2009) method).

5.5 Constitutive Relationship of RAC

The macroscopic response of the fundamental properties of concrete under compression and tension, respectively, is inferred from the uniaxial compressive and tensile stress–strain curves. It displays the concrete's deformation properties and failure mode under various loading conditions. The features of the overall mechanical performance of the structural components are governed by the strength and deformation characteristics of concrete under uniaxial compression and tension. The concrete damaged plasticity (CDP) model was considered by Pradhan et al. (2023) to model the behaviour of concrete under compression and tension. The model assumes that the two primary damage mechanisms in concrete are tensile cracking and compressive crushing. The yield surface of the CDP model available in ABAQUS 6.14 (2014) is the modified Drucker and Prager (1952) which was initially proposed by Lubliner et al. (1989) and subsequently modified by Lee and Fenves (1998) by accounting the evolution of resistance in compression and tension. A scalar damage variable "d" is used to describe the gradual material damage or stiffness degradation. The value of d is a number between 0 and 1, where 1 indicates entirely damaged material. Consequently, the stress–strain relationship can be defined by the following expression (Eq. 5.5.1):

$$\sigma = (1 - d) D_0{}^{el} : \left(\varepsilon - \varepsilon^{pl}\right) \tag{5.5.1}$$

where $D_0{}^{el}$ = initial elasticity matrix, ε = total strain, and ε^{pl} = plastic strain. Owing to the lack of complete understanding of the failure mechanism, several empirical relationships were proposed between the stress state and damage evolution (Xiao et al. 2014; Zhang et al. 2017). For simplicity, the damage variable is determined by Pradhan et al. (2023) as a linear function of stress and represented in Eq. 5.5.2. This relationship was earlier employed by Shamass et al. (2015) to determine the damage evolution for concrete in compression as well as tension.

$$d = 1 - \frac{\sigma^*}{\sigma_0} \tag{5.5.2}$$

where σ^* = stress at the softening region and σ_0 = peak stress in compression or tension.

5.5.1 Material Model

Poisson's ratio (ν) and the modulus of elasticity (E) are the two concrete material parameters needed by the CDP model for the study. For both uncracked and cracked NAC and RAC, the model assumes a constant value of ν to be 0.2. Different compression and tension yield stresses are accounted for by the shape parameter (K_c), which modifies the yield surface in the deviatoric plane. To ensure that the yield surface is convex, the value of K_c is assumed to be 0.667 (must fulfill $0.5 \leq K_c \leq 1$). The dilation angle (ψ) and plastic potential eccentricity (ε) parameters determine the plastic flow, and ψ controls the plastic volumetric strain that results from the plastic shearing process. The value of ψ is assumed to be constant during plastic yielding and observed to be between 25° and 45° for concrete (Genikomsou and Polak 2015; Othman et al. 2017). The values of ψ considered in the present study for NAC and RAC are 36° and 32°, respectively. When confining pressure is low and the parameter ε is set to 0, it is possible to increase the parameter ψ close to the uniaxial loading condition. For both NAC and RAC, the default value of 1.16 is used to depict the state of the material under multiaxial stress conditions. This ratio (σ_{bo}/σ_{co}) compares strength in the biaxial condition to strength in the uniaxial condition. The key elements to describing the behaviour of concrete in compression and tension are its uniaxial stress–strain relationships, which are covered in detail for both NAC and RAC in the following sections.

The suitability of the stress–strain relationship of RAC proposed by Xiao et al. (2005) and Chen and Yang (2012) in compression and Chinese standard GB50010-2010 in tension is discussed comprehensively. Subsequently, appropriate modifications are suggested for complete stress–strain curves for RAC in compression and tension.

Compression

The expression suggested by Xiao et al. (2005) for the stress–strain curve of RAC is shown in Eq. 5.5.3, which accounts for the influence of RCA replacement content (r).

$$\overline{\sigma_c}^r = \begin{cases} a\overline{\varepsilon_c} + (3-2a)\overline{\varepsilon_c}^2 + (a-2)\overline{\varepsilon_c}^3 & (\overline{\varepsilon_c} \le 1) \\ \frac{\overline{\varepsilon_c}}{b(\overline{\varepsilon_c}-1)^2 + \overline{\varepsilon_c}} & (\overline{\varepsilon_c} > 1) \end{cases} \tag{5.5.3}$$

where $\overline{\sigma_c^r} = \sigma/\sigma_c$, $\overline{\varepsilon_c} = \varepsilon/\varepsilon_c^r$, σ_c = compressive strength of prism = 0.7 to 0.92 σ_{cu} ($0.7\sigma_{cu}$ considered by Pradhan et al. (2023)), σ_{cu} = compressive strength of cube, and ε_c^r = peak strain for RAC during the uniaxial compression test. As suggested by Xiao et al. (2005), $\varepsilon_c{}^r$ can be determined from Eqs. 5.5.4 and 5.5.5 and ε_0 value can be obtained from Eq. 5.5.6. Furthermore, considering the r value, the constants $a = 2.2(0.748r^2 - 1.231r + 0.975)$ and $b = 0.8(7.6483r + 1.142)$ are to be determined.

$$\varepsilon_c^r = \varepsilon_0\left(1 + \frac{r}{\beta}\right) \tag{5.5.4}$$

$$\beta = 65.715r^2 - 109.43r + 48.989 \tag{5.5.5}$$

$$\varepsilon_0 = 0.00076 + \left[(0.626f_c - 4.33) \times 10^{-7}\right]^{0.5} \tag{5.5.6}$$

The model of stress–strain relationship in compression for RAC as proposed by Chen and Yang (2012) is shown in Eq. 5.5.7, which was subsequently adopted by researchers in different studies of RAC (Yang et al. 2019; Zhao et al. 2017).

$$\overline{\sigma_c^r} = \begin{cases} \frac{n_r\overline{\varepsilon_c}}{n_r - 1 + \overline{\varepsilon_c}^{n_r}} & (\overline{\varepsilon_c} \le 1) \\ \frac{\overline{\varepsilon_c}}{\alpha_c^r(\overline{\varepsilon_c}-1)^2 + \overline{\varepsilon_c}} & (\overline{\varepsilon_c} > 1) \end{cases} \tag{5.5.7}$$

where n_r and $\alpha_c{}^r$ can be determined from Eqs. 5.5.8 and 5.5.9, respectively, for different r value.

$$n_r = \frac{E_c^r\varepsilon_c^r}{E_c^r\varepsilon_c^r - f_c} \tag{5.5.8}$$

$$\alpha_c^r = \alpha_c\left(3.062r^2 + 3.491r + 1\right) \tag{5.5.9}$$

The equation suggested by Chen and Yang (2012) is used to determine the modulus of elasticity of RAC (E_c^r), i.e. $E_c^r = E_c(0.112r^2 - 0.333r + 1)$, where $E_c = \frac{10^5}{2.2 + \left(\frac{34.7}{\sigma_{cu}}\right)}$. In Eq. 5.5.9, the parameter $\alpha_c = 0.157\sigma_c^{0.785} - 0.905$.

The expressions proposed by Xiao et al. (2005) and Chen and Yang (2012) are compared with the uniaxial compression test results reported by Xiao et al. (2005) and Yang et al. (2019) for RAC with 100% use of RCA. Figure 5.9 demonstrates the non-dimensional stress–strain curves obtained from different experimental results as well as available expressions (Xiao et al. 2005; Chen and Yang 2012; Yang et al. 2019). The expression proposed by Chen and Yang (2012) does not exhibit good correlation with the experimental results of Xiao et al. (2005), Yang et al. (2019) in the ascending as well as descending portion. The expression suggested by Xiao et al. (2005) exhibits good correlation with the experimental results of Xiao et al. (2005); Yang et al. (2019) in the ascending portion, whereas the descending portion is not comparable with the experimental result of Xiao et al. (2005). Owing to the absence of complete correlation of the available stress–strain relationship with the experimental results of Xiao et al. (2005), Yang et al. (2019) in both ascending and descending regions necessitates a more suitable expression to represent the behaviour of RAC in uniaxial compression.

The expression (Eq. 5.5.10) proposed by Pradhan et al. (2023) is similar to the expression recommended by Popovics (1973) for conventional concrete as well as the equation suggested by Chen and Yang (2012) for the ascending portion of RAC. However, the slope of the ascending and descending parts is regulated by the parameter n'. The value of n' is considered as $1.5n_r$ and $2n_r$ for ascending and descending parts, respectively, and n_r can be determined from Eq. 5.5.8 as suggested by Chen and Yang (2012). Moreover, the peak strain (ε_c^r) is determined by using the expression suggested by Xiao et al. (2005) (Eq. 5.5.4). Equation 5.5.10 exhibits good correlation with the experimental results of Xiao et al. (2005) in both ascending and descending regions (Fig. 5.9). The suggested constitutive model of RAC was then used to conduct a finite element (FE) simulation of uniaxial compression for a RAC cylinder in order to verify the validity of the proposed expression. In this instance, the C3D8R element in ABAQUS was used to model the concrete. Figure 5.9 confirms that there is evidence of a correlation between the results of the FE analysis and the experimental findings.

$$\overline{\sigma_c^r} = \frac{n'\overline{\varepsilon_c}}{n' - 1 + \overline{\varepsilon_c}^{n'}} \begin{cases} n' = 1.5n_r & (\overline{\varepsilon_c} < 1) \\ n' = 2n_r & (\overline{\varepsilon_c} \geq 1) \end{cases} \tag{5.5.10}$$

The compression damage factor for RAC (d_c^r) was obtained from (Eq. 5.5.11) by incorporating Eqs. 5.5.10 in 5.5.2.

$$d_c^r = \begin{cases} 0 & (\overline{\varepsilon_c} < 1) \\ 1 - \frac{n'\overline{\varepsilon_c}}{n'-1+\overline{\varepsilon_c}^{n'}} & (\overline{\varepsilon_c} \geq 1) \end{cases} \tag{5.5.11}$$

To demonstrate a correlation between the compressive strength of the RAC cube and cylinder, an ABAQUS analysis using a comparable size effect as in the case of ordinary concrete was carried out. However, the compressive strength of the cylinder was assumed in the FE analysis for RAC to be 0.70 times that of the cube. The compressive strengths of the concrete cube as determined by FE analysis and experimentally are 39.9 MPa and 42.8 MPa, respectively, which demonstrates a strong correlation between the results of the two methods of study. Additionally, the concrete cylinder's compressive strength, which was determined using FE analysis, is 28.0 MPa,

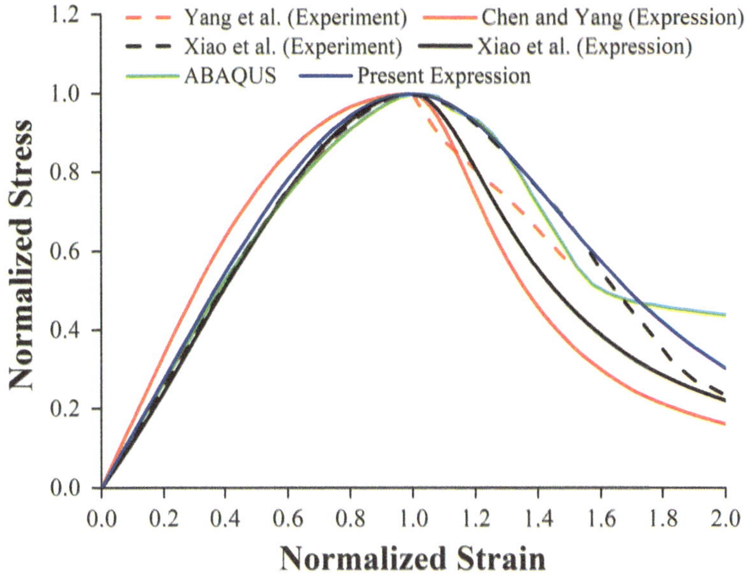

Fig. 5.9 Non-dimensional stress–strain curves in compression for RAC with 100% RCA (Pradhan et al. 2023)

or almost 0.70 times that of the concrete cube. In order to establish the stress–strain relationship for further study of RAC columns, the compressive strength of cylinder is calibrated as 0.70 times the compressive strength of RAC cube, taking into consideration the validation of the compressive strength of RAC cube derived from experimental and FE analysis.

Tension

The study on the complete stress–strain relationship of RAC in tension is very limited. Sun et al. (2018) used a stress–strain relationship (Eq. 5.5.12) for RAC in tension similar to the conventional concrete as recommended in the Chinese standard GB50010-2010.

$$\overline{\sigma_t^r} = \begin{cases} a\overline{\varepsilon_t} - (a-1)\varepsilon_t^{-6} & (\overline{\varepsilon_t} \le 1) \\ \alpha_t(\overline{\varepsilon_t} - 1)^{1.7} + \overline{\varepsilon_t} & (\overline{\varepsilon_t} > 1) \end{cases} \tag{5.5.12}$$

where $\overline{\sigma_t^r} = \sigma/\sigma_t^r$, $\overline{\varepsilon_t} = \varepsilon/\varepsilon_t^r$, σ_t^r = cracking strength of RAC = $(-0.06r + 0.24)\left(\sigma_{cu}^r\right)^{\frac{2}{3}}$ (Liu et al. 2018), ε_t^r = strain corresponds to σ_t^r and as per the Chinese standard GB50010-2010 it is estimated to be 1×10^{-4} for the calculated σ_t^r value. The values of constants a and α_t can be obtained from Sun et al. (2018) and GB50010-2010, respectively. However, a simple equation (Eq. 5.5.13) is proposed for RAC by Pradhan et al. (2023). The proposed equation assumes a linear relationship up to cracking strength, whereas for post cracking strength a power function is used similar to the conventional concrete (Hsu and Mo 2010; Belarbi and Hsu 1995). Figure 5.10 demonstrates a good correlation between the tensile stress–strain curves obtained by employing the proposed expression (Eq. 5.5.13) and the equation (Eq. 5.5.12) used by Sun et al. (2018). Consequently, the proposed tensile stress–strain relationship of RAC can be used for FE analysis of RAC structural members.

$$\overline{\sigma_t^r} = \begin{cases} E_c^r \varepsilon & (\overline{\varepsilon_t} \leq 1) \\ \sigma_t^r \left(\frac{\varepsilon_t^r}{\varepsilon}\right)^{0.6} & (\overline{\varepsilon_t} > 1) \end{cases} \tag{5.5.13}$$

The tension damage parameter of RAC (d_t^r) was determined by substituting Eqs. 5.5.13 in 5.5.2 and shown in Eq. 5.5.14.

$$d_t^r = \begin{cases} 0 & (\overline{\varepsilon_t} \leq 1) \\ 1 - \left(\frac{\varepsilon_t^r}{\varepsilon}\right)^{0.6} & (\overline{\varepsilon_t} > 1) \end{cases} \tag{5.5.14}$$

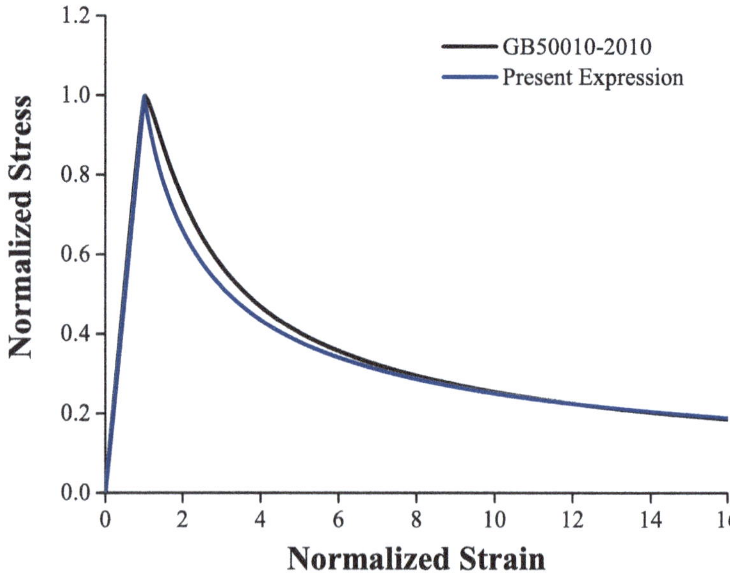

Fig. 5.10 Non-dimensional tensile stress–strain curves of RAC with 100% RCA (Pradhan et al. 2023)

5.6 Closure

The fresh and hardened properties of both NAC and RAC are discussed extensively in the present chapter. Moreover, the research conducted by Pradhan et al. (2017) using PPM and IS: 10262 (2009) mix design methods was analysed critically and following key observations on the fresh and hardened properties of RAC are outlined.

- Freshly prepared RAC showed lower workability than NAC irrespective of the mix design method. The PPM mix proportioning has no significant effect in improving the workability of both NAC and RAC.
- The compressive strength was measured for each type of concrete mix at different curing age. The 28-day curing compressive strength of RAC was comparable to that of NAC in PPM design mix at w/c 0.45. However, the gain in strength from 28 days to 90 days is less for RAC than NAC.
- The single factor ANOVA analysis indicates that PPM mix proportioning has a significant effect on the compressive strength of NAC than IS: 10262 (2009) method of design mix. The PPM mix proportioning was found out to be very effective in improving the tensile strength (STS and FTS) of both NAC and RAC and can be confirmed from the higher normalized values of STS and FTS. The single factor ANOVA analysis suggests that the PPM mix design approach is more effective in case of RAC.
- The modulus of elasticity of RAC was lower than that of NAC as expected. However, PPM mix proportioning has positive impact in enhancing the modulus of elasticity of both NAC and RAC.
- By using the data available in the published literature, expressions are proposed to predict STS, FTS, and modulus of elasticity of RAC, which are essentially a function of its compressive strength.
- The constitutive relationship is proposed to represent the strength and deformation of RAC under uniaxial compression and tension. Employing the proposed compressive and tensile stress–strain curves and CDP model in ABAQUS, the experimental results under uniaxial compression are validated.

References

ABAQUS 6.14 (2014) Analysis user's manual

ACI 318 (2008) ACI committee 318. Building code requirements for structural concrete and commentary, American Concrete Institute, Farmington Hills (MI)

Ajdukiewicz A, Kliszczewicz A (2002) Influence of recycled aggregates on mechanical properties of HS/HPC. Cement Concr Compos 24(2):269–279

Akbarnezhad A, Ong KCG, Tam CT, Zhang MH (2013) Effects of the parent concrete properties and crushing procedure on the properties of coarse recycled concrete aggregates. J Mater Civil Eng 25(12):1795–1802

Akbarnezhad A, Ong KCG, Zhang MH, Tam CT, Foo TWJ (2011) Microwave-assisted beneficiation of recycled concrete aggregates. Constr Build Mater 25(8):3469–3479

Bairagi N, Ravande K, Pareek V (1993) Behaviour of concrete with different proportions of natural and recycled aggregates. Resour Conserv Recycl 9:109–126

Bairagi NK, Vidyadhara HS, Ravande K (1990) Mix design procedure for recycled aggregate concrete. Constr Build Mater 4(4):188–193

Behera M, Bhattacharyya SK, Minocha AK, Deoliya R, Maiti S (2014) Recycled aggregate from C&D waste & its use in concrete-A breakthrough towards sustainability in construction sector: A review. Constr Build Mater 68:501–516

Belarbi A, Hsu TTC (1995) Constitutive laws of concrete in tension and reinforcing bars stiffened by concrete. ACI Struct J 91(4):465–474

Brand AS, Roesler JR, Salas A (2015) Initial moisture and mixing effects on higher quality recycled coarse aggregate concrete. Constr Build Mater 79:83–89

Butler L, West JS, Tighe SL (2013) Effect of recycled concrete coarse aggregate from multiple sources on the hardened properties of concrete with equivalent compressive strength. Constr Build Mater 47:1292–1301

Casuccio M, Torrijos MC, Giaccio G, Zerbino R (2008) Failure mechanism of recycled aggregate concrete. Constr Build Mater 22(7):1500–1506

Chakradhara Rao M (2010) Characterisation and behaviour of recycled aggregate concrete. PhD Thesis, IIT Kharagpur

Chakradhara Rao M, Bhattacharyya SK, Barai SV (2011) Influence of field recycled coarse aggregate on properties of concrete. Mater Struct 44:205–220

Chen J, Yang H (2012) Experimental investigation of concrete with saturated recycled coarse aggregate under axial compression. In: Proceedings of the 5th international symposium on innovative civil & architectural engineering, pp 147–156

Corinaldesi V (2010) Mechanical and elastic behaviour of concretes made of recycled-concrete coarse aggregates. Constr Build Mater 24(9):1616–1620

de Oliveira MB, Vazquez E (1996) The influence of retained moisture in aggregates from recycling on the properties of new hardened concrete. Waste Manag 16(1–3):113–117

Drucker DC, Prager W (1952) Soil mechanics and plastic analysis or limit design. Quart Appl Math 10(2):157–165

Elhakam AA, Mohamed AE, Awad E (2012) Influence of self-healing, mixing method and adding silica fume on mechanical properties of recycled aggregates concrete. Constr Build Mater 35:421–427

Etxeberria M, Marí AR, Vázquez E (2007) Recycled aggregate concrete as structural material. Mater Struct 40(5):529–541

Etxeberria M, Vázquez E, Marí A, Barra M (2007) Influence of amount of recycled coarse aggregates and production process on properties of recycled aggregate concrete. Cement Concr Res 37(5):735–742

Fonseca N, De Brito J, Evangelista L (2011) The influence of curing conditions on the mechanical performance of concrete made with recycled concrete waste. Cement Concr Compos 33(6):637–643

Genikomsou AS, Polak MA (2015) Finite element analysis of punching shear of concrete slabs using damaged plasticity model in ABAQUS. Eng Struct 98:38–48

Hansen TC, Narud H (1983) Strength of recycled concrete made from crushed concrete coarse aggregate. Concr Int 5:79–83

Ho NY, Lee YPK, Lim WF, Zayed T, Chew KC, Low GL, Ting SK (2013) Efficient utilization of recycled concrete aggregate in structural concrete. J Mater Civil Eng 25(3):318–327

Hsu TTC, Mo YL (2010) Unified theory of concrete structures. John Wiley & Sons

IS: 10262 (2009) Concrete mix proportioning–guidelines

IS: 456 (2000) Plain and reinforced concrete code of practice. Bureau of Indian Standards, New Delhi, India

IS: 516 (1959) Methods of tests for strength of concrete. Bureau of Indian Standards, New Delhi, India

IS: 5816 (1999) Splitting tensile strength of concrete-method of test. Bureau of Indian Standards, New Delhi, India
IS: 7320 (1974) Specification for concrete slump test apparatus. Bureau of Indian Standards, New Delhi, India
Katz A (2003) Properties of concrete made with recycled aggregate from partially hydrated old concrete. Cement Concr Res 33(5):703–711
Kisku N, Joshi H, Ansari M, Panda SK, Nayak S, Dutta SC (2017) A critical review and assessment for usage of recycled aggregate as sustainable construction material. Constr Build Mater 131:721–740
Kou S, Poon C (2008) Mechanical properties of 5-year-old concrete prepared with recycled aggregates obtained from three different sources. Mag Concr Res 60(1):57–64
Kou SC, Poon CS (2013) Long-term mechanical and durability properties of recycled aggregate concrete prepared with the incorporation of fly ash. Cement Concr Compos 37(1):12–19
Kou S-C, Poon C-S (2015) Effect of the quality of parent concrete on the properties of high performance recycled aggregate concrete. Constr Build Mater 77:501–508
Kou SC, Poon CS, Agrela F (2011) Comparisons of natural and recycled aggregate concretes prepared with the addition of different mineral admixtures. Cement Concr Compos 33(8):788–795
Kou SC, Poon CS, Chan D (2008) Influence of fly ash as a cement addition on the hardened properties of recycled aggregate concrete. Mater Struct 41(7):1191–1201
Lee J, Fenves G (1998) Plastic-damage model for cyclic loading of concrete structures. J Eng Mech 124(8):892–900
Limbachiya MC, Leelawat T, Dhir RK (2000) Use of recycled concrete aggregate in high-strength concrete. Mater Struct 33(9):574–580
Liu W, Cao W, Zong N, Wang R, Ren L (2018) Experimental study on punching performance of recycled aggregate concrete thin wallboard with single-layer reinforcement. Appl Sci 8(188):1–21
Lubliner J, Oliver J, Oller S, Oñate E (1989) A plastic-damage model for concrete. Int J Solids Struct 25(3):299–326
Malešev M, Radonjanin V, Marinković S (2010) Recycled concrete as aggregate for structural concrete production. Sustainability 2(5):1204–1225
Matias D, de Brito J, Rosa A, Pedro D (2014) Durability of concrete with recycled coarse aggregates: influence of superplasticizers. J Mater Civil Eng 26(7)
McNeil K, Kang TH-K (2013) Recycled concrete aggregates: a review. Int J Concr Struct Mater 7(1):61–69
Mukharjee BB, Barai SV (2014) Influence of incorporation of nano-silica and recycled aggregates on compressive strength and microstructure of concrete. Constr Build Mater 71:570–578
Mukharjee BB, Barai SV (2014) Influence of Nano-Silica on the properties of recycled aggregate concrete. Constr Build Mater 55:29–37
Neville AM (2011) Properties of concrete. Pearson Education Limited
Othman H, Marzouk H (2017) Finite-element analysis of reinforced concrete plates subjected to repeated impact loads. J Struct Eng 143(9):1–16
Padmini AK, Ramamurthy K, Mathews MS (2009) Influence of parent concrete on the properties of recycled aggregate concrete. Constr Build Mater 23(2):829–836
Poon CS, Kou SC, Lam L (2007) Influence of recycled aggregate on slump and bleeding of fresh concrete. Mater Struct 40(9):981–988
Poon CS, Shui ZH, Lam L (2004) Effect of microstructure of ITZ on compressive strength of concrete prepared with recycled aggregates. Constr Build Mater 18(6):461–468
Poon CS, Shui ZH, Lam L, Fok H, Kou SC (2004) Influence of moisture states of natural and recycled aggregates on the slump and compressive strength of concrete. Cement Concr Res 34(1):31–36
Popovics S (1973) A numerical approach to the complete stress-strain curve of concrete. Cement Concr Res 3(5):583–599
Pradhan S, Kumar S, Barai SV (2017) Recycled aggregate concrete: particle packing method (PPM) of mix design approach. Constr Build Mater 152:269–284

Pradhan S, Nayak TK, Kumar S, Barai SV (2023) Experimental and numerical study of recycled aggregate concrete column. Struct Concr
Rahal K (2007) Mechanical properties of concrete with recycled coarse aggregate. Build Environ 42(1):407–415
Rao A, Jha KN, Misra S (2007) Use of aggregates from recycled construction and demolition waste in concrete. Resour Conserv Recycl 50(1):71–81
Ravindrarajah SR, Tam CT (1985) Properties of concrete made with crushed concrete as coarse aggregate. Mag Concr Res 37(130)
Sagoe-Crentsil KK, Brown T, Taylor AH (2001) Performance of concrete made with commercially produced coarse recycled concrete aggregate. Cement Concr Res 31:707–712
Salem RM, Burdette EG (1998) Role of chemical and mineral admixtures on physical properties and frost resistance of recycled aggregate concrete. ACI Mater J 95(5):558–563
Shamass R, Zhou X, Alfano G (2015) Finite-element analysis of shear-off failure of keyed dry joints in precast concrete segmental bridges. J Bridge Eng 20(6):1–12
Sun C, Xiao J, Lange DA (2018) Simulation study on the shear transfer behavior of recycled aggregate concrete. Struct Concr 19(1):255–268
Tabsh SW, Abdelfatah AS (2009) Influence of recycled concrete aggregates on strength properties of concrete. Constr Build Mater 23(2):1163–1167
Tam VWY, Gao XF, Tam CM (2005) Microstructural analysis of recycled aggregate concrete produced from two-stage mixing approach. Cement Concr Res 35(6):1195–1203
Tam VWY, Tam CM, Wang Y (2007) Optimization on proportion for recycled aggregate in concrete using two-stage mixing approach. Constr Build Mater 21(10):1928–1939
Topcu IB, Şengel S (2004) Properties of concretes produced with waste concrete aggregate. Cement Concr Res 34(8):1307–1312
Xiao J, Li J, Zhang C (2005) Mechanical properties of recycled aggregate concrete under uniaxial loading. Cement Concr Res 35(6):1187–1194
Xiao J, Li W, Fan Y, Huang X (2012) An overview of study on recycled aggregate concrete in China (1996–2011). Constr Build Mater 31:364–383
Xiao JZ, Pham TL, Wang PJ, Gao G (2014) Behaviors of semi-precast beam made of recycled aggregate concrete. Struct Design Tall Special Build 23(9):692–712
Yang H, Zhao H, Liu F (2019) Compressive stress-strain relationship of concrete containing coarse recycled concrete aggregate at elevated temperatures. J Mater Civil Eng 31(9):1–8
Yang K-H, Chung H-S, Ashour AF (2008) Influence of type and replacement level of recycled aggregates on concrete properties. ACI Mater J 105(3):289–296
Zega CJ, Di Maio ÁA (2011) Use of recycled fine aggregate in concretes with durable requirements. Waste Manag 31(11):2336–2340
Zhang Z, Zhang Y, Yan C, Liu Y (2017) Influence of crushing index on properties of recycled aggregates pervious concrete. Constr Build Mater 135:112–118
Zhao H, Wang Y, Liu F (2017) Stress-strain relationship of coarse RCA concrete exposed to elevated temperatures. Mag Concr Res 69(13):649–664

Chapter 6
Macro-level Performance Assessment of Concrete: Experimental Fracture Analysis

6.1 Introduction

Natural coarse aggregate (NCA) can potentially be replaced with recycled coarse aggregate (RCA), which is derived from the leftover concrete from construction and demolition (C&D) projects. However, RCA has drawbacks of its own, notably because of the porous mortar layer and micro-cracks. This motivates academics throughout the world to reduce the impact of the RCA's inherent drawbacks in recycled aggregate concrete (RAC). Different mix design techniques, mixing procedures, the addition of fibres, and the addition of mineral admixtures are suggested in this context. The performance of the RAC is being improved in the current work by using the proven two-stage mixing approach (TSMA) and the PPM mix design approach. In chapter "Macro-level Performance Assessment of Concrete: Conventional Approach", it was determined that the proposed procedure had a favourable impact on the macro-mechanical performance of the RAC. The current chapter analyses the performance of both natural aggregate concrete (NAC) and RAC from the perspective of fracture behaviour in order to provide a more thorough understanding of the impact of the PPM mix design approach. In this context, the experimental research is used to determine the fracture energy, double-K fracture parameters, and size effect, which are then described in the next sections.

6.2 Fracture Analysis of Concrete

The analytical models, such as the fictitious crack model by Hillerborg et al. (1976), the crack band model (CBM) by Bažant and Oh (1983), two-parameter fracture model by Jenq and Shah (1985, 1986), plastic damage model by Lubliner et al. (1989), effective crack model by Karihaloo and Nallathambi (1989, 1990), size-effect model by Bažant and Kazemi (1990), and double-K fracture model by Xu and Reinhardt (1999a, b, 2000) were proposed to characterize the fracture behaviour of concrete

S. Pradhan et al., *Particle Packing Method for Recycled Aggregate Concrete*,
https://doi.org/10.1007/978-981-99-7516-7_6

and evaluate its fracture toughness parameters. The three key stages in the process of complete fracture of concrete, such as crack initiation, stable crack propagation, and unstable fracture, can be delineated using the double-K fracture model (Kumar and Barai 2008; Kumar et al. 2013; Lee and Lopez 2014). The three-point bending (TPB) test of single edge notched (SEN) specimen is one of the widely accepted practical approaches to determine the fracture energy and fracture parameters using double-K fracture model. The theoretical basis of the TPB test is the work-of-fracture method proposed by Hillerborg (1985). Subsequently, RILEM TC-50 (1985) prepared a draft recommendation on TPB test for the evaluation of fracture energy of concrete, which is also governed by the principles of work-of-fracture method. Later, JCI Standard (2003) also recommended to employ TPB test of SEN beam specimen for the estimation of the fracture energy of concrete. Moreover, in this method, the fracture energy (G_F) is obtained from the load–crack mouth opening displacement (load–CMOD) curve of the TPB test.

The fracture toughness metrics, such as fracture energy and critical stress intensity factor, are used to quantitatively reflect the concrete's fracture behaviour. These fracture parameters are influenced by the size and shape of the specimens as well as the size, shape, and type of the aggregates and the concrete's mix proportions. Bažant and Yu (2011) suggested doing a size-effect investigation on the concrete specimens because the cohesive softening rule discovered in the work-of-fracture test does not apply to them uniquely. Additionally, due to the load–CMOD curve's asymptotic nature, choosing the truncation point of the experimentally acquired curve is crucial when calculating the true fracture energy of concrete (Petersson 1981; Planas et al. 1992). The variation in the curtailment of the load–CMOD curve considerably affects the real fracture energy of concrete. However, the requirements for the load–CMOD curve truncation point identification for various-sized SEN beam specimens have not been adequately addressed.

There are experimental and analytical investigations to discern the fracture behaviour of RAC. The failure mechanism of RAC was studied by Casuccio et al. (2008) using the TPB test on SEN beam specimens. The study reported lower G_F and characteristic length (l_{ch}) of RAC as compared to the conventional concrete. Moreover, the progression of cracks exhibited less branching and tortuous path in RAC. Bordelon et al. (2009) conducted the experimental study on synthetic macro-fibre-incorporated RAC to enhance the fracture toughness parameters of resulting concrete. Arezoumandi et al. (2014) used the TPB test on SEN specimens to investigate the effects of RCA replacement levels (30%, 50%, 70%, and 100%) on G_F of RAC. According to the study, RAC's G_F decreases as RCA replacement content rises. Choubey et al. (2016) used the compressive strength of concrete to derive empirical relations for various material qualities and then used those parameters to estimate the fracture parameters of RAC with various RCA contents. In this context, the double-K fracture model and the fictional crack model were applied. The effects of temperature (room temperature, 200°C, 400°C, and 600°C) and steel fibre content (0%, 0.5%, 1%, and 1.5% by volume of concrete) on the fracture behaviour of RAC were investigated (Chen et al. 2016). The inclusion of steel fibre not only increases the initial cracking load (P_{ini}) but also restricts the opening of the cracks. This leads to increase in G_F and fracture toughness of RAC and the pattern is also reflected even after exposure to the elevated temperature. Using 30%, 65%, and 100% of RCA, Ghorbel and Wardeh (2017) investigated the effect of RCA replacement content and found that G_F declines as RCA increases.

The fracture process of RAC was numerically validated by Xiao et al. (2012). The RCA was represented for this purpose as a five-phase composite material consisting of natural aggregate, an old interfacial transition zone (ITZ) between natural aggregate and an old mortar, an old mortar, and an ITZ between an old mortar and a new mortar. Musiket et al. (2016, 2017) investigated the influence of strain rate ($10^{-5}/s$ to $10^{-2}/s$) for crack mouth opening displacement (CMOD) on the fracture properties of RAC. The test results showed that when the strain rate increases, so do G_f, critical stress intensity factor (K_{Ic}), and effective fracture process zone (c_f). To investigate the impact of strain rate as well size effect, a viscoelastic model for RAC and a four-phase composite material model for RCA (natural aggregate, old mortar, pre-treatment mortar, and new mortar) were developed. The analytical study was conducted using ABAQUS^R software (Musiket et al. 2017). Guo et al. (2017) conducted the TPB test of SEN beam specimen and used an acoustic emission technique to monitor the characteristics of the fracture process zone and micro-fracture phase (aggregate fracture, crack in mortar, and crack at ITZ) of RAC. The study found that RAC had a higher fracture propagation rate than NAC and a lower crack initiating load. The fracture energy of RAC is reduced when its RCA content rises. The acoustic emission test showed that the ITZ cracking is the main micro-fracture phenomenon in RAC and that the micro-fractures in RAC are more localized due to the weak mortar and RCA bonding. The TPB test using SEN specimens was conducted by Pradhan et al. (2018) to study the influence of aggregate type on G_F and fracture toughness parameters. Moreover, the impact of mix design method (PPM and conventional (IS: 10262 2009) approach) was also analysed by Pradhan et al. (2020). The following sections provide a detailed description of this experimental inquiry.

6.3 Experimental Investigation

The experimental program was designed to verify the effect of type of aggregate (NCA and RCA) and PPM mix design on the fracture properties of concrete. Moreover, for comparative study, the fracture behaviour of IS: 10262 (2009) mix-designed concrete was also studied. The detail procedure for concrete preparation and mix proportions is discussed in chapter "Particle Packing Method of Mix Proportioning and Modified Mixing Approach".

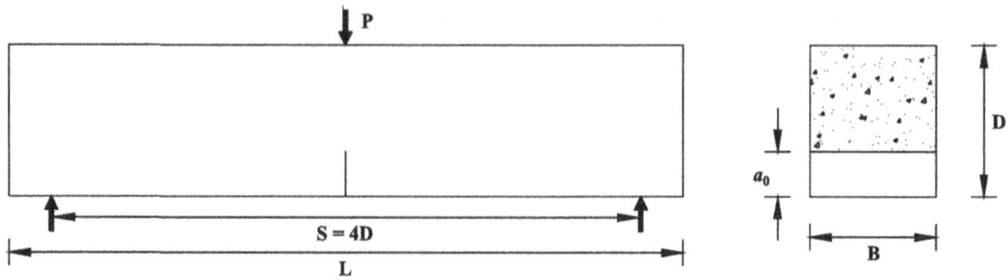

Fig. 6.1 Geometry of three-point bending specimen (Pradhan et al. 2018)

Table 6.1 Dimensions of specimens

Depth (D) (mm)	Width (B) (mm)	Length (L) (mm)	Span (S) (mm)	Notch depth (a_0) (mm)
75	150	400	300	22.5
125	150	600	500	37.5
175	150	800	700	52.5

6.3.1 Details of Test Setup and Test Procedure

Three different sizes of concrete beams were prepared. A minimum of three samples from each dimension were evaluated, for a total of 36 samples. Figure 6.1 provides an illustration of the test specimen's specifics. The beams' dimensions were chosen in accordance with the information in RILEM TC-50 (1985). The details of the tested beams are provided in Table 6.1. For each size of the specimen, a span-to-depth ratio (S/D) of 4 was maintained constant. Similar to this, 0.3 was chosen for the initial notch-to-depth ratio (a_0/D). To ensure the creation of the crack tip with a sharp edge, a metal plate with a bevelled edge was inserted at the specimen's mid-span during casting to create the notch. Less than 3 mm separated the two sides of the notch. After being prepared for 24 hours, the specimens were demoulded and put in the water tank for wet curing.

The fracture behaviour of concrete after 28 days of curing was investigated using a typical TPB test setup (Fig. 6.2). The SEN beam specimens were rested in simply supported condition. A roller with a concentrated load at its centre was used to apply the load at the specimen's mid-span, and this configuration had the benefit of avoiding torsion (Fig. 6.2). A 500 kN universal testing equipment with a closed-loop system was used to apply the load to the specimen (Fig. 6.2). The loading direction was orthogonal to the direction of casting. The loading direction was perpendicular to the casting direction. Since the testing machine's material is stiffer than the concrete specimen, a CMOD controlled displacement rate of 0.005 mm/s effectively controlled the load rate. To measure the CMOD, a clip gauge was installed below the notch.

Fig. 6.2 Three-point bending test setup (Pradhan et al. 2018)

6.4 Results

6.4.1 Load–CMOD Relationship

A data acquisition system was used to record the applied load and CMOD data. The reported results were the repercussion of the careful elimination of the unnecessary data. The aberration of the test data, especially at the beginning of the test, was attributed to the differential settling of the specimen at the two supports, stick-slip phenomena between the concrete surface and the surface of the loading roller, and initial tenacious and uneven movement of the clip gauge. Finally, the load–CMOD curves were plotted using the filtered load and associated CMOD data. In Figs. 6.3, 6.4, 6.5, and 6.6, the acquired load–CMOD curves of various-sized specimens (depth = 75 mm, 125 mm, and 175 mm) for four types of concrete (NAC IS, NAC PPM, RAC IS, and RAC PPM) are displayed. The highest load that the specimen can carry is known as the peak load. It was calculated using the load–CMOD data that was obtained while the SEN beam specimen was being tested, and it is represented in Table 6.2.

6.4.2 Fracture Energy

The energy required for the generation of a crack of unit surface area in a plane projected parallel to the direction of the notch (perpendicular to the direction of

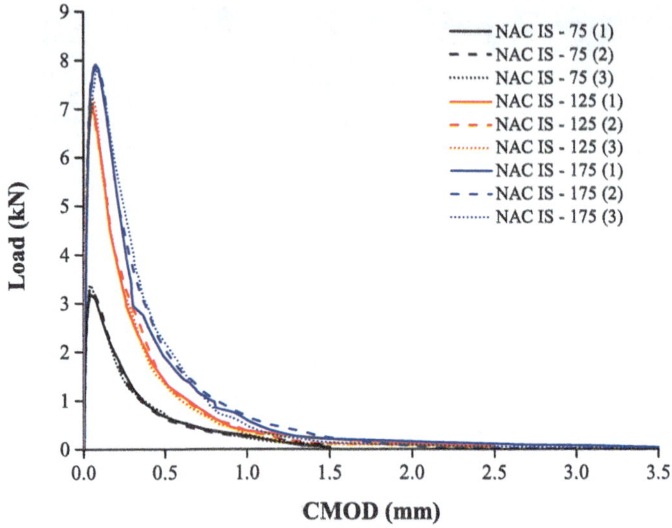

Fig. 6.3 Load–CMOD of NAC IS specimens (Pradhan et al. 2018)

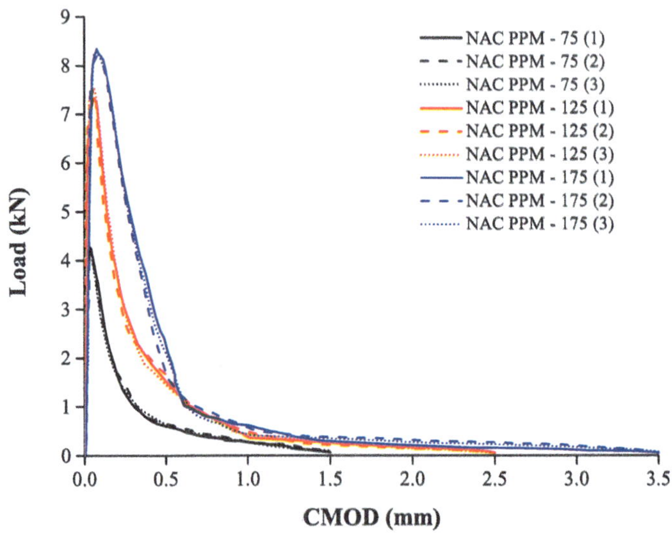

Fig. 6.4 Load–CMOD of NAC PPM specimens (Pradhan et al. 2018)

stress) is known as fracture energy. The evaluation of this fracture energy is based on the work-of-fracture method explained by Hillerborg (1985). The total energy absorbed by the SEN beam specimen before splitting into two halves during the TPB test represents the fracture energy. The amount of energy absorbed in this process is determined by calculating the area projected under the load–CMOD curve.

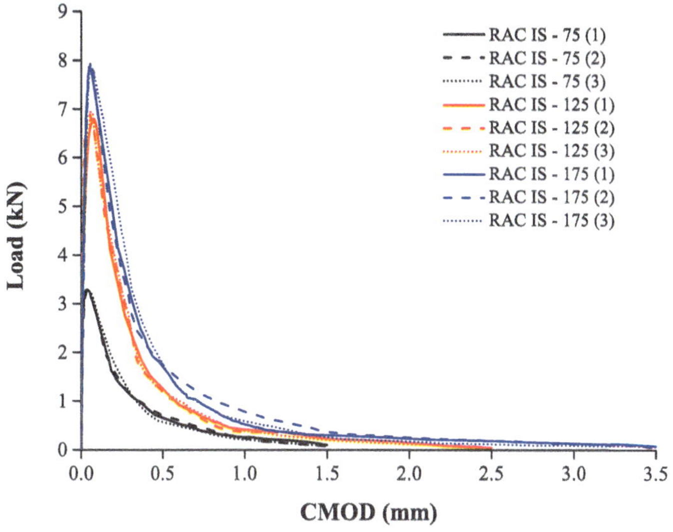

Fig. 6.5 Load–CMOD of RAC IS specimens (Pradhan et al. 2018)

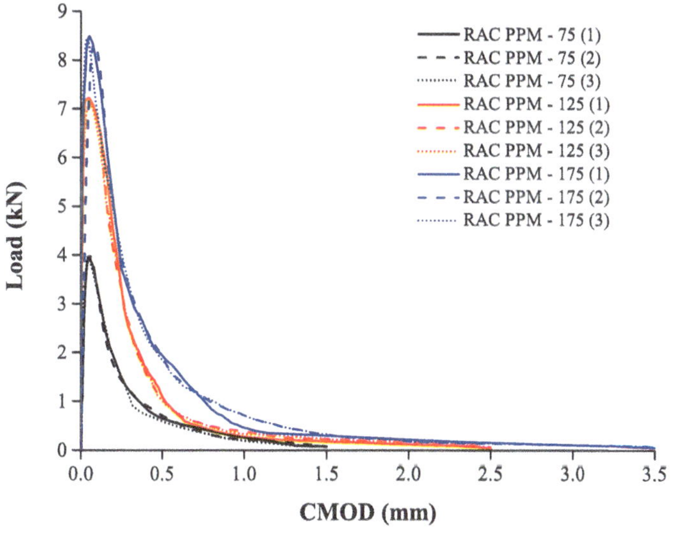

Fig. 6.6 Load–CMOD of RAC PPM specimens (Pradhan et al. 2018)

Furthermore, the fracture energy is evaluated by dividing the energy absorbed by the specimen and the fracture surface area. This approach was also adopted by RILEM TC-50 (1985).

An additional load (F_1) was added to the typical load–CMOD curve obtained from the TPB test of a SEN concrete beam, which enabled to account the bending

Table 6.2 Peak load and fracture energy of tested specimens of different types of concrete

Type of concrete	d (mm)	P_u (kN), G_F **(N/m)** of each specimen			Average P_u (kN), **Average G_F (N/m)**
		1	2	3	
NAC IS	75	3.19, **161.10**	3.32, **159.88**	3.38, **158.70**	3.30, **159.89**
	125	7.02, **236.72**	7.26, **235.09**	7.26, **228.25**	7.18, **233.35**
	175	7.94, **262.69**	7.89, **266.63**	7.84, **263.61**	7.89, **264.31**
NAC PPM	75	4.26, **164.10**	4.12, **168.60**	4.19, **165.85**	4.19, **166.18**
	125	7.35, **244.23**	7.52, **242.23**	7.58, **242.64**	7.48, **243.03**
	175	8.37, **286.01**	8.23, **289.56**	8.31, **284.36**	8.30, **286.64**
RAC IS	75	3.29, **155.58**	3.26, **155.39**	3.29, **154.21**	3.28, **155.06**
	125	6.82, **227.57**	6.93, **226.15**	6.95, **225.60**	6.90, **226.44**
	175	7.87, **254.79**	7.99, **261.75**	7.92, **257.16**	7.93, **257.90**
RAC PPM	75	3.98, **166.50**	3.99, **164.90**	4.03, **163.77**	4.00, **165.06**
	125	7.24, **228.87**	7.14, **228.62**	7.13, **231.33**	7.17, **231.54**
	175	8.51, **261.04**	8.36, **263.95**	8.48, **262.62**	8.45, **262.54**

moment induced due to the self-weight of the specimen and the weight of the testing equipment. Finally, the true fracture energy (G_F) was calculated by dividing the surface area of the ligament ($A_{lig} = (D - a_0)B$) of the SEN concrete beam specimen to the energy absorbed by the specimen (W) (Eq. 6.4.1).

$$G_F = \frac{W}{A_{lig}} = \frac{W_0 + 2F_1\delta_0}{(D - a_0)B} = \frac{W_0 + mg\left(\frac{S}{L}\right)\delta_0}{(D - a_0)B} \tag{6.4.1}$$

Sensitivity Analysis

The load–CMOD curve of SEN beam specimen approaches asymptotically to the CMOD axis. It is practically difficult to reach the zero load value. Therefore, during the bending test of SEN beam the test is terminated prior to the complete failure of the specimen. As a result of this, the selection of the end point of the load–CMOD curve is important while evaluating the fracture energy. This emphasizes the study of the curtailment of the load–CMOD curve for accurate calculation of fracture energy (G_F). The variation in G_F at different tail curtailment of the load–CMOD curve was addressed and the end point of load–CMOD curve was finally selected to achieve a consistent G_F even at higher load CMOD curve end point (Pradhan et al. 2018, 2020).

The curtailment of the tail of the load–CMOD curve was carried out to determine a critical end point, which helped in obtaining a consistent G_F independent of the long asymptotic tail of load–CMOD curve. The sensitivity analysis of the curtailment of the tail of load–CMOD curve was carried out at CMOD of 0.9 mm (suggested by Roesler et al. (2007)), 1.2% D, 1.5% D, 1.75% D, 2% D, 2.25% D, and 2.5% D and reported in Table 6.3. This exercise was executed for each size of the tested SEN beam. A significant increment in G_F is observed at end point of 1.2% D with respect to 0.9 mm. The increment in G_F is also significant up to the end point of 2% D. However, a stable increment in G_F is noticed with further increase in the end point of load–CMOD curve, i.e. at 2.25% D and 2.5% D. This increment is attributed to the asymptotic approach of the load–CMOD curve towards the CMOD axis and unavoidable while calculating G_F. Such behaviour is observed in each size of the tested specimen.

The percentage increase in G_F at 1.2% D, 1.5% D, 1.75% D, 2% D, 2.25% D, and 2.5% D tail curtailment was compared with respect to the G_F at 0.9 mm end point of load–CMOD curve. In case of 75-mm-depth specimens at 2% D tail curtailment, the increment in G_F with respect to 0.9 mm is approximately 13.5%, 11.5%, 14.4%, and 13.0% for NAC IS, NAC PPM, RAC IS, and RAC PPM, respectively. With further increase in the load–CMOD curve tail curtailment only a stable increment of about 17.0%, 15.0%, 18.2%, and 16.4% is observed at 2.5% D tail curtailment for NAC IS, NAC PPM, RAC IS, and RAC PPM, respectively. Similar behaviour was also observed for 125-mm- and 175-mm-depth specimens. For 125-mm-depth specimen, approximately 27%, 32%, 32%, and 31% increment is recorded at 2% D tail curtailment, whereas this increment goes up to about 34.0%, 39.0%, 38.4%, and 37.7% at 2.5% D tail curtailment for NAC IS, NAC PPM, RAC IS, and RAC PPM specimens, respectively. Again for 175-mm-depth NAC IS, NAC PPM, RAC IS, and RAC PPM specimens around 49.0%, 62.0%, 61.0%, and 56.6% increment in G_F is recorded at 2% D tail curtailment, whereas at 2.5% D tail curtailment 62%, 74%, 73.6%, and 69% increment is observed. It is worth mentioning that, the increment in G_F is persistent with the increase in load–CMOD curve tail curtailment owing to its asymptotic behaviour. However, for each size case the increment in G_F is noticed maximum up to 2% D tail curtailment and with the further increase in the tail length almost a stable improvement in G_F is observed (Table 6.3). Hence, this study indicates that, 2% D tail curtailment of load–CMOD curve is advisable to calculate G_F. The selection of 2% D end point is justified as it does not eliminate decisive test data and further increment in tail curtailment does not influence G_F significantly. In addition to this, the 2% D end point selection avoids the unnecessary extension of the experiment. Moreover, the selection of a constant end point for load–CMOD curve (2% D as suggested here), which is independent of the effect of the size of the specimens helps in standardizing the calculation of fracture energy and comparing the fracture energy of different types of concrete appropriately.

The load–CMOD curve was curtailed at 2% D and the estimated G_F is shown in Table 6.2. The size effect on G_F was studied and the variation in G_F with respect to the different size of the specimen is plotted in Fig. 6.7 for different types of concrete.

Table 6.3 Variation in fracture energy at different CMOD curtailment

Type of concrete	Depth of the SEN beam specimen (mm)	Fracture energy of each specimen (N/m)			Average fracture energy (N/m)
		1	2	3	
CMOD = 0.9 mm					
NAC IS	75	142.31	139.84	140.45	140.87
	125	181.67	187.04	181.67	183.67
	175	171.57	180.02	181.22	177.60
NAC PPM	75	144.98	146.52	155.51	166.18
	125	184.87	184.20	182.95	184.00
	175	187.43	180.79	182.19	183.47
RAC IS	75	131.21	137.04	136.22	134.82
	125	171.15	171.38	171.63	171.39
	175	157.70	157.54	166.02	160.42
RAC PPM	75	146.67	146.70	141.28	144.88
	125	179.30	172.10	175.28	175.56
	175	173.65	167.22	168.57	169.81
CMOD = 1.2% D					
NAC IS	75	142.31	139.84	140.45	140.87
	125	207.49	210.90	202.53	206.97
	175	220.42	230.27	225.07	225.25
NAC PPM	75	144.98	146.52	155.51	166.18
	125	212.11	211.91	211.12	211.71
	175	240.75	237.48	235.01	237.75
RAC IS	75	131.21	137.04	136.22	134.82
	125	198.21	197.32	196.84	197.46
	175	208.66	216.36	216.77	213.93
RAC PPM	75	146.67	146.70	141.28	144.88
	125	201.06	197.05	200.18	199.43
	175	223.57	220.77	222.81	222.38
CMOD = 1.5% D					
NAC IS	75	152.30	149.19	149.32	150.27
	125	219.72	220.12	212.75	217.53
	175	237.16	244.24	240.10	240.50
NAC PPM	75	153.09	156.45	154.14	154.56
	125	225.68	224.61	225.12	225.14
	175	258.69	258.73	255.01	257.48
RAC IS	75	140.03	145.70	144.52	143.42
	125	210.86	209.75	208.49	209.70
	175	227.04	235.11	233.25	231.80

(continued)

Table 6.3 (continued)

Type of concrete	Depth of the SEN beam specimen (mm)	Fracture energy of each specimen (N/m)			Average fracture energy (N/m)
		1	2	3	
RAC PPM	75	156.17	155.27	149.09	153.51
	125	212.55	210.09	213.00	211.88
	175	241.39	237.74	240.03	239.72
CMOD = 1.75% D					
NAC IS	75	157.69	155.35	154.54	155.86
	125	228.71	227.84	220.74	225.76
	175	250.20	255.79	252.15	252.71
NAC PPM	75	159.47	163.57	160.07	161.04
	125	235.61	233.98	235.40	235.00
	175	272.87	275.15	270.52	272.85
RAC IS	75	146.71	151.12	149.97	149.27
	125	219.85	218.68	217.03	218.52
	175	241.18	249.25	246.26	245.56
RAC PPM	75	161.56	161.19	153.96	158.90
	125	221.25	219.96	222.79	221.33
	175	255.05	251.26	253.52	253.28
CMOD = 2% D					
NAC IS	75	161.10	159.88	158.70	159.89
	125	236.72	235.09	228.25	233.35
	175	262.69	266.63	263.61	264.31
NAC PPM	75	164.10	168.60	165.85	166.18
	125	244.23	242.23	242.64	243.03
	175	286.01	289.56	284.36	286.64
RAC IS	75	151.96	155.39	154.21	153.85
	125	228.72	226.15	224.02	226.30
	175	254.36	260.46	258.96	257.93
RAC PPM	75	166.01	165.54	157.98	163.18
	125	229.11	228.62	231.54	229.76
	175	267.74	264.00	266.19	265.98
CMOD = 2.25% D					
NAC IS	75	163.71	162.51	161.14	162.45
	125	243.67	241.21	234.69	239.86
	175	274.22	277.79	275.04	275.68
NAC PPM	75	167.02	171.73	168.60	169.12
	125	251.05	249.00	249.10	249.72

(continued)

Table 6.3 (continued)

Type of concrete	Depth of the SEN beam specimen (mm)	Fracture energy of each specimen (N/m)			Average fracture energy (N/m)
		1	2	3	
	175	296.04	299.83	296.11	297.33
RAC IS	75	155.47	158.4	157.34	157.07
	125	233.54	232.47	231.00	232.34
	175	265.77	271.34	267.84	268.32
RAC PPM	75	168.81	168.47	161.00	166.09
	125	235.33	235.23	237.69	236.08
	175	277.54	275.12	277.59	276.75
		CMOD = 2.5% D			
NAC IS	75	166.04	164.82	163.45	164.77
	125	249.74	247.09	240.64	245.82
	175	285.3	288.79	286.06	286.72
NAC PPM	75	169.47	174.25	170.93	171.55
	125	256.96	254.89	254.85	255.57
	175	307.29	311.09	306.99	308.46
RAC IS	75	157.92	160.72	159.92	159.52
	125	236.69	238.07	236.61	237.12
	175	276.30	281.08	278.20	278.53
RAC PPM	75	171.19	170.86	163.44	168.50
	125	240.93	241.00	243.27	241.73
	175	287.94	285.62	288.09	287.22

6.4.3 *Double-K Fracture Parameter*

The double-K fracture model was used by Pradhan et al. (2018, 2020) to estimate the fracture toughness of the concrete owing to the ability of the model to delineate the entire fracture process (crack initiation, stable crack propagation, and unstable fracture) of concrete. The two important fracture controlling parameters, i.e. initial fracture toughness (K_{Ic}^{ini}) and unstable fracture toughness (K_{Ic}^{un}), were estimated from the TPB test results. In this regard, P_{ini}, a_0, P_u, and $CMOD_c$ are the initial parameters to be measured during the TPB test of SEN beam specimen. However, only a_0, P_u, and $CMOD_c$ were measured during the test. Since P_{ini} was not measured during the experiment, the K_{Ic}^{ini} parameter was evaluated using the weight-function method suggested by Kumar and Barai (2009, 2010a). The cohesive fracture toughness (K_{Ic}^{c}) was calculated analytically using the available test results and finally K_{Ic}^{ini} was estimated from Eq. 6.4.2.

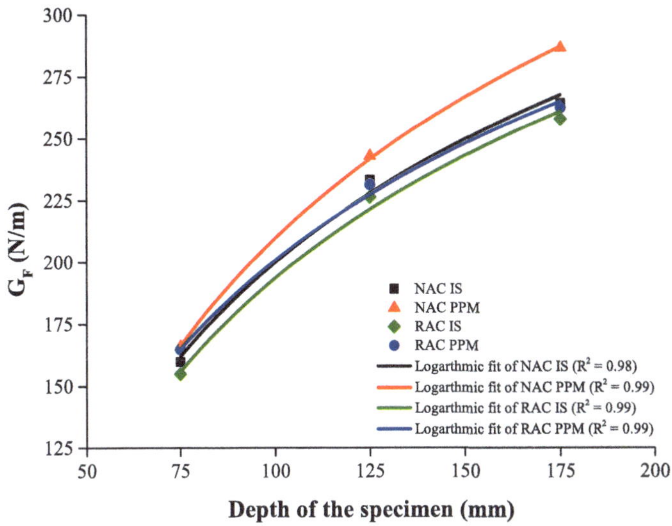

Fig. 6.7 Relationship between G_F and depth of the beam for NAC IS and NAC PPM (Pradhan et al. 2018)

$$K_{Ic}^{ini} = K_{Ic}^{un} + K_{Ic}^{c} \tag{6.4.2}$$

Determination of K_{Ic}^{un} for TPB Test

The stress intensity factor (SIF) from the TPB test of concrete SEN beam specimen can be determined using linear elastic fracture mechanics (LEFM) formulae suggested by earlier researchers (Bueckner 1970; Rice 1972; Tada et al. 1973). The SIF based on modified LEFM can be represented as follows (Eq. 6.4.3):

$$K_{Ic} = \sigma_n \sqrt{D} k(\alpha) \tag{6.4.3}$$

where k is a geometric factor and can be determined using Eq. 6.4.4 for $S/D = 4$. The σ_n in the SEN beam specimen can be determined by the following expression (Eq. 6.4.5):

$$k(\alpha) = \sqrt{\alpha}\frac{1.99 - \alpha(1-\alpha)(2.15 - 3.93\alpha + 2.7\alpha^2)}{(1+2\alpha)(1-\alpha)^{3/2}} \tag{6.4.4}$$

$$\sigma_n = \frac{3S}{4BD^2}\left[2P + w_g S\right] \tag{6.4.5}$$

where $\alpha = a/D$, a is the effective crack length, and w_g is the self-weight of the beam per unit length. Equation 6.4.3 can be used to estimate the K_{Ic}^{un} for $a = a_c$ and $P = P_u$ and a_c is the equivalent elastic critical crack length corresponding to the maximum load (P_u). The compliance (C) for the TPB test of a SEN beam specimen of standard geometry ($S/D = 4$) can be calculated from Eq. 6.4.6.

$$C = \frac{CMOD}{P} = \frac{6Sa}{BD^2E}V(\beta) \tag{6.4.6}$$

where $V(\beta)$ can be determined from the following expression (Eq. 6.4.7):

$$V(\beta) = 0.76 - 2.28\beta + 3.87\beta^2 - 2.04\beta^3 + \frac{0.66}{(1-\beta)^2} \tag{6.4.7}$$

where $\beta = \frac{a+H_0}{D+H_0}$, a = equivalent elastic crack length, and H_0 = thickness of the clip gauge holder. The modulus of elasticity (E) of concrete can be determined from the load–CMOD curve of TPB test of SEN beam specimen, the load–deflection curve of beam specimen, and compression test of cylinder specimen (Karihaloo and Nallathambi 1991). Pradhan et al. (2018, 2020) measured E value using the compression test of the cylinder (Table 5.2) which was used while estimating K_{Ic}^{un}. By solving Eq. 6.4.6 and Eq. 6.4.7 for critical condition, i.e. for $P = P_u$ and $CMOD = CMOD_c$, a_c was determined. Finally, a_c was incorporated in Eq. 6.4.4 to determine $k(\alpha)$, which was further employed in Eq. 6.4.3 to estimate the K_{Ic}^{un} of concrete.

Determination of K_{Ic}^{ini} for TPB Test

An inverse analytical approach was operated to estimate the K_{Ic}^{ini}. The difficulties associated with the determination of P_{ini} during the TPB test were the reason for adopting an indirect approach to calculate K_{Ic}^{ini}. In this approach, the cohesive fracture toughness (K_{Ic}^{c}) was first evaluated by using the weight-function method and finally K_{Ic}^{ini} was estimated by using Eq. 6.4.2. The weight-function method (Kumar and Barai 2010a) was adopted out of the four available methods (Gauss–Chebyshev integral method (Xu and Reinhardt 1999b, c), simplified equivalent cohesive force method (Xu and Reinhardt 2000), simplified Green's function method (Zhang and Xu 2011), and weight-function method (Kumar and Barai 2010a, b)) to estimate K_{Ic}^{c}. The simplified weight-function approach renders a closed-form solution for K_{Ic}^{c}, which avoids the skills required in the numerical integration technique because of the association of singularity problem at the integral boundary. In the weight-function method K_{Ic}^{c} can be estimated using Eq. 6.4.8.

$$\begin{aligned} K_{Ic}^{c} = \frac{2}{\sqrt{2\pi a}}\Bigg[& A_1 a\left(2s^{\frac{1}{2}} + M_1 s + \frac{2}{3}Ms^{\frac{3}{2}} + \frac{M_3}{2}s^2 + \frac{2}{5}M_4 s^{\frac{5}{2}}\right) \\ & + A_2 a\left(\frac{4}{3}s^{\frac{3}{2}} + \frac{M_1}{2}s^2 + \frac{4}{15}M_2 s^{\frac{5}{2}} + \frac{4}{35}M_4 s^{\frac{7}{2}} + \frac{M_3}{6}\left[1 - \left(\frac{a_0}{a}\right)^3 - 3s\left(\frac{a_0}{a}\right)\right]\right)\Bigg] \end{aligned} \tag{6.4.8}$$

where $A_1 = \sigma_s(CTOD_c)$, $A_2 = \frac{f_t - \sigma_s(CTOD_c)}{a - a_0}$, and $s = \left(1 - \frac{a_0}{a}\right)$. The crack tip opening displacement ($CTOD$) was calculated using Eq. 6.4.9 and by replacing $a = a_c$, the $CTOD$ becomes equal to critical crack opening displacement ($CTOD_c$).

$$CTOD = CMOD_c \left[\left(1 - \frac{a_0}{a}\right)^2 + \left(1.081 - 1.149\frac{a}{D}\right)\left[\frac{a_0}{a} - \left(\frac{a_0}{a}\right)^2\right]\right]^{\frac{1}{2}} \quad (6.4.9)$$

Furthermore, the cohesive stress at the crack tip of initial notch ($\sigma_s(CTOD_c)$) at $CTOD = CTOD_c$ was determined by using Eq. 6.4.10 (Reinhardt et al. 1986).

$$\frac{\sigma}{f_t} = \left[1 + \left(c_1 \frac{w}{w_0}\right)^3\right] \exp\left(-c_2 \frac{w}{w_0}\right) - \frac{w}{w_0}\left(1 + c_1^3\right)\exp(-c_2) \quad (6.4.10)$$

where c_1, c_2 and w_0 are the material constants. The parameters c_1, c_2, and w_0 were determined by satisfying the experimental w_1, f_t, and G_F for each type of concrete and each size of the specimen (Ruiz et al. 2016). In this context, different methods were suggested by researchers (Guinea et al. 1994; Bažant and Planas 1997; Elices et al. 2002) to estimate w_1. The method described in the literature (Planas et al. 1999; Elices et al. 2002) was adopted to estimate w_1. Subsequently, Eq. 6.4.10 was used to determine $\sigma_s(CTOD_c)$ for $CTOD_c > w_1$. However, for $CTOD_c \leq w_1$ the linear softening function was used to determine $\sigma_s(CTOD_c)$ by using Eq. 6.4.11 as suggested by Ruiz et al. (2016).

$$\frac{\sigma_s(CTOD_c)}{f_t} = 1 - \frac{CTOD_c}{w_1} \quad (6.4.11)$$

The five-term universal weight-function approach (Eq. 6.4.12) proposed by Kumar and Barai (2008, 2009, 2010a, b) was executed to evaluate K_{Ic}^c in order to achieve a more accurate result and it was characterized as follows:

$$m(x,a) = \frac{2}{\sqrt{2\pi(x-a)}}\left[1 + M_1\left(1 - \frac{x}{a}\right)^{\frac{1}{2}} + M_2\left(1 - \frac{x}{a}\right) + M_3\left(1 - \frac{x}{a}\right)^{\frac{3}{2}} + M_4\left(1 - \frac{x}{a}\right)^2\right] \quad (6.4.12)$$

The values of M_1, M_2, M_3, and M_4 can be obtained by using Eq. 6.4.13 and Eq. 6.4.14. For, i = 1 and 3

$$M_i = \frac{1}{\left(1 - \frac{a}{D}\right)^{\frac{3}{2}}}\left[a_i + b_i\left(\frac{a}{D}\right) + c_i\left(\frac{a}{D}\right)^2 + d_i\left(\frac{a}{D}\right)^3 + e_i\left(\frac{a}{D}\right)^4 + f_i\left(\frac{a}{D}\right)^5\right] \quad (6.4.13)$$

and for i = 2 and 4

$$M_i = \left[a_i + b_i\left(\frac{a}{D}\right)\right] \quad (6.4.14)$$

The values of the coefficients a_i, b_i, c_i, d_i, e_i, and f_i were determined from the table presented by Kumar and Barai (Kumar and Barai 2008, 2010a, b; Kumar et al. 2014). So, finally K_{Ic}^c can be estimated from Eq. 6.4.8 and using K_{Ic}^c, the value of K_{Ic}^{ini} was determined from Eq. 6.4.2. The aforementioned approach was exercised to estimate the double-K fracture parameters of the four types of concrete prepared in the present study and represented in Table 6.4.

The PPM mix-designed concrete (both NAC and RAC) mixes exhibited lower compressive strength; however, the tensile strength was higher in comparison to the IS: 10262 (2009) mix-designed concrete. Moreover, the compressive strength of NAC IS is marginally higher in comparison to NAC PPM, whereas the tensile strength results are contrary to the compressive strength. Hence, for this ambiguity the modified characteristic length ($l_{ch,mod}$) was used to examine the brittleness of the concrete, because it accounts both compressive strength and tensile strength along with the modulus of elasticity and fracture energy of concrete. The $l_{ch,mod}$ was evaluated by using Eq. 6.4.15 suggested by Rao and Prasad (2002) and incorporated in Hillerborg's expression (Hillerborg 1985) to estimate brittleness index (β_n) (Eq. 6.4.16). The calculated β_n values of tested specimens can be evidenced from Table 6.4.

$$l_{ch,mod} = \frac{EG_F}{f_c f_t} \tag{6.4.15}$$

$$\beta_n = \frac{D}{l_{ch,mod}} \tag{6.4.16}$$

The a_c and $CTOD_c$ were determined analytically and represented in Table 6.4. The P_{ini} was analytically estimated from corresponding K_{Ic}^{ini} and the variation of P_{ini} with respect to the size of the specimen for different types of concrete is represented in Fig. 6.8. The size effect on the length of the fracture process zone was studied at the critical unstable stage with the use of a non-dimensional parameter, $(a_c - a_0)/l_{ch,mod}$ and this is represented with respect to the variation of $D/l_{ch,mod}$ (Fig. 6.9).

6.5 Discussion

The failure pattern was essentially the same for the four types of concrete irrespective of the size of the SEN beam specimen. Prior to the cracking of the specimens, the CMOD was observed to be very small and the deflection was difficult to notice with naked eyes. However, after the peak load, the deflection and CMOD began to increase at a rapid rate. In each of the specimen, a single crack was developed from the tip of the notch and propagated rapidly towards the top face of the beam specimen. In the following subsections, the influence of the aggregate type, concrete mix design method, and size of the specimen on the fracture properties of concrete is discussed in detail.

Table 6.4 Double-K fracture parameters

Type of concrete	Specimen depth (mm)	K_{Ic}^{un} $(MPa\,m^{\frac{1}{2}})$			Avg. K_{Ic}^{un} $(MPa\,m^{\frac{1}{2}})$	K_{Ic}^{c} $(MPa\,m^{\frac{1}{2}})$			Avg. K_{Ic}^{c} $(MPa\,m^{\frac{1}{2}})$	K_{Ic}^{ini} $(MPa\,m^{\frac{1}{2}})$			Avg. K_{Ic}^{ini} $(MPa\,m^{\frac{1}{2}})$	β_n			Avg. β_n
		1	2	3		1	2	3		1	2	3		1	2	3	
NAC IS	75	1.205	1.215	1.257	1.226	0.365	0.386	0.368	0.373	0.840	0.829	0.889	0.853	2.10	2.12	2.13	2.12
	125	1.459	1.487	1.524	1.490	0.620	0.618	0.611	0.616	0.839	0.869	0.913	0.874	2.38	2.40	2.47	2.42
	175	1.626	1.628	1.610	1.621	0.700	0.696	0.701	0.699	0.926	0.932	0.909	0.922	3.01	2.96	3.00	2.99
NAC PPM	75	1.336	1.324	1.341	1.334	0.468	0.459	0.459	0.462	0.868	0.865	0.882	0.872	1.92	1.87	1.90	1.89
	125	1.573	1.593	1.560	1.575	0.679	0.678	0.685	0.681	0.894	0.915	0.875	0.894	2.15	2.16	2.16	2.16
	175	1.686	1.664	1.668	1.673	0.796	0.798	0.801	0.798	0.890	0.866	0.867	0.875	2.57	2.54	2.58	2.56
RAC IS	75	1.063	1.084	1.082	1.076	0.436	0.424	0.428	0.429	0.627	0.660	0.655	0.647	1.66	1.67	1.68	1.67
	125	1.327	1.308	1.312	1.316	0.508	0.514	0.514	0.512	0.819	0.794	0.798	0.804	1.90	1.91	1.91	1.91
	175	1.456	1.459	1.463	1.460	0.600	0.604	0.599	0.601	0.856	0.855	0.864	0.859	2.37	2.31	2.35	2.34
RAC PPM	75	1.210	1.197	1.218	1.208	0.364	0.375	0.364	0.368	0.846	0.822	0.854	0.840	1.58	1.60	1.61	1.60
	125	1.394	1.396	1.414	1.401	0.585	0.585	0.580	0.583	0.809	0.811	0.834	0.818	1.92	1.92	1.90	1.91

(continued)

Table 6.4 (continued)

Type of concrete	Specimen depth (mm)	K_{Ic}^{un} $(MPa\,m^{\frac{1}{2}})$			Avg. K_{Ic}^{un} $(MPa\,m^{\frac{1}{2}})$	K_{Ic}^{c} $(MPa\,m^{\frac{1}{2}})$			Avg. K_{Ic}^{c} $(MPa\,m^{\frac{1}{2}})$	K_{Ic}^{ini} $(MPa\,m^{\frac{1}{2}})$			Avg. K_{Ic}^{ini} $(MPa\,m^{\frac{1}{2}})$	β_n			Avg. β_n
		1	2	3		1	2	3		1	2	3		1	2	3	
	175	1.525	1.517	1.512	1.518	0.712	0.711	0.716	0.713	0.813	0.806	0.796	0.805	2.36	2.33	2.34	2.34
Type of concrete	Specimen depth (mm)	P_{ini} (kN)			Avg. P_{ini} (kN)	$CTOD_c$ (mm)			Avg. $CTOD_c$ (mm)	a_c (mm)			Avg. a_c (mm)	P_{ini}/P_u			Avg. P_{ini}/P_u
		1	2	3		1	2	3		1	2	3		1	2	3	
NAC IS	75	2.32	2.38	2.46	2.39	0.0234	0.0230	0.0242	0.0235	45.29	44.69	45.01	45.00	0.727	0.717	0.728	0.724
	125	5.61	5.66	5.93	5.73	0.0260	0.0261	0.0277	0.0266	63.21	62.65	64.02	63.29	0.799	0.780	0.817	0.799
	175	6.42	6.51	6.47	6.47	0.0396	0.0399	0.0393	0.0396	95.95	96.31	96.07	96.11	0.809	0.825	0.825	0.820
NAC PPM	75	3.03	3.08	3.04	3.05	0.0224	0.0227	0.0229	0.0227	41.57	42.09	42.00	41.89	0.711	0.748	0.726	0.728
	125	6.13	6.18	6.09	6.13	0.0294	0.0295	0.0279	0.0289	64.38	64.00	62.87	63.75	0.834	0.822	0.803	0.819
	175	6.39	6.49	6.55	6.48	0.0410	0.0406	0.0404	0.0407	95.21	95.38	95.01	95.20	0.763	0.789	0.788	0.780
RAC IS	75	2.41	2.52	2.57	2.50	0.0183	0.0192	0.0190	0.0188	42.15	42.77	42.54	42.49	0.733	0.773	0.781	0.762
	125	5.41	5.33	5.27	5.34	0.0221	0.0210	0.0210	0.0214	60.61	59.39	59.38	59.79	0.793	0.769	0.758	0.773
	175	6.44	6.47	6.42	6.44	0.0324	0.0320	0.0325	0.0323	90.74	90.07	90.65	90.49	0.818	0.810	0.811	0.813
RAC PPM	75	2.89	2.99	2.95	2.94	0.0200	0.0195	0.0200	0.0198	40.89	40.60	40.76	40.75	0.726	0.749	0.732	0.736
	125	5.74	5.62	5.69	5.68	0.0233	0.0238	0.0245	0.0239	60.22	60.82	61.38	60.81	0.793	0.787	0.798	0.793
	175	6.77	6.71	6.65	6.71	0.0333	0.0335	0.0328	0.0332	89.12	89.77	88.83	89.24	0.796	0.803	0.784	0.794

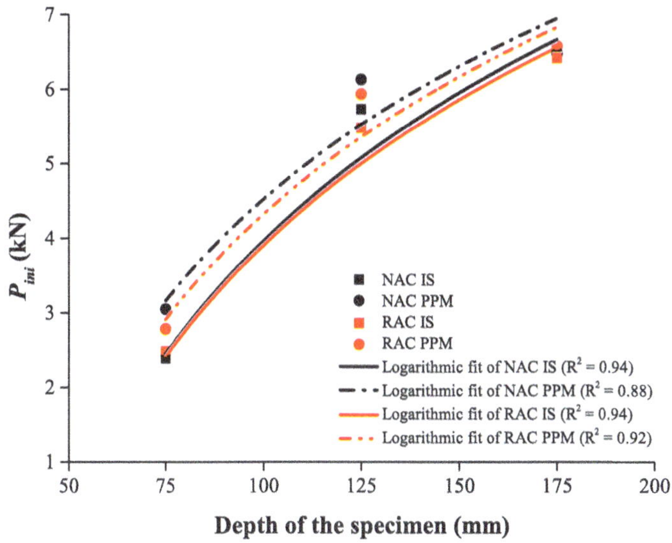

Fig. 6.8 Variation of P_{ini} with the depth of the specimens (Pradhan et al. 2018)

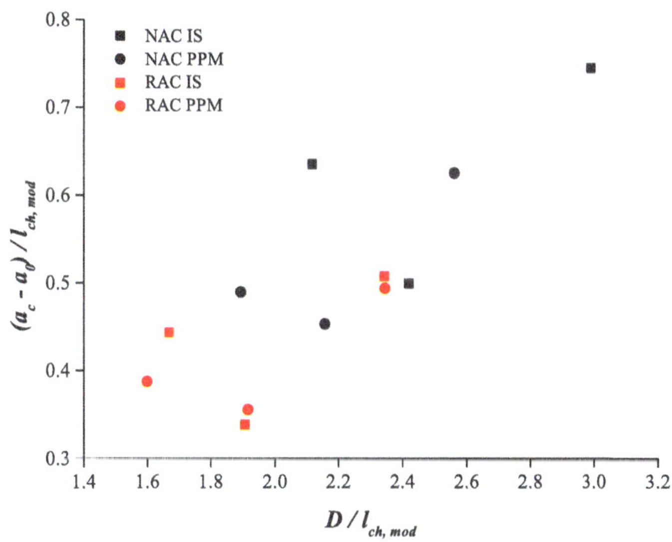

Fig. 6.9 Variation of critical fracture process zone with the depth of the specimens (Pradhan et al. 2018)

6.5.1 Influence of Aggregate Type

The inherent characteristics of RCA, such as attached mortar, micro-cracks, and old ITZ adversely, affect its quality with respect to NCA. The inferior physical and mechanical properties of RCA are also reflected in the performance of RAC (Table 5.2). Furthermore, the fracture behaviour of NAC and RAC was compared to understand the influence of RCA on it. Table 6.2 shows that, for a particular mix design method, apart from the 175-mm-depth specimens, RAC specimens exhibited lower peak load as compared to the NAC specimens. Choubey et al. (2016) also observed lower P_u for RAC with respect to NAC. The G_F of RAC specimens is lower than the NAC specimens and a maximum reduction of 10% is recorded in G_F of RAC specimen with respect to the NAC specimen (Table 6.2). Similar influence of RCA on G_F was also reported in the earlier studies (Casuccio et al. 2008; Bordelon et al. 2009; Arezoumandi et al. 2014; Chen et al. 2016; Ghorbel and Wardeh 2017; Guo et al. 2017). The estimated K_{Ic}^{un} and K_{Ic}^{ini} parameters are lower for RAC specimens with respect to the NAC specimens having same mix design method (Table 6.4). Similar observation was also reported in the earlier studies (Bordelon et al. 2009; Chen et al. 2016). A maximum reduction of 12% and 24% was observed in K_{Ic}^{un} and K_{Ic}^{ini}, respectively. This may be attributed to higher void content (Ghorbel and Wardeh 2017) and presence of two ITZ in RAC.

The β_n values of RAC specimens were observed to be lower than the NAC specimens (Table 6.4). This may be attributed to the lower tensile strength of RAC. This indicates that, RAC is relatively less brittle in comparison to NAC. For a particular type of concrete, the $CTOD_c$ increases with the increase in K_{Ic}^{un} and a similar relationship between $CTOD_c$ and K_{Ic}^{un} was also reported by Kumar et al. (2014). Because of having higher tensile strength, NAC specimens show higher P_{ini} values with respect to the same-sized RAC specimens. The ratio of P_{ini}/P_u was unaffected by the inclusion of RCA and P_{ini}/P_u values (Table 6.4) suggest that, for each type of concrete specimen the crack initiates at a load of about 70%–80% of the peak load. A similar conclusion was also reported by Zhang et al. (2007) and Zhang and Xu (2011).

6.5.2 Influence of Mix Design Method

It is also anticipated that the combined effects of PPM mix design technique (Pradhan et al. 2017) and TSMA (Tam et al. 2005) will result in better mechanical characteristics (Table 5.2) in both NAC and RAC's fracture properties. For both NAC and RAC (Pradhan et al. 2020), the combined influence of mix design and mixing strategy is compared. The peak load of the TPB test for both NAC and RAC is positively impacted by the PPM mix design technique (Table 6.2). The impact is more pronounced for specimens with a depth of 75 mm, where a peak load increase of up to 27% was noted. The peak load was 4—6.6% greater in the 125-mm- and 175-

mm-depth specimens created using the PPM mix design approach. Additionally, for both NAC and RAC, the PPM mix-designed specimens showed higher G_F in comparison to the related IS: 10262 (2009) method of mix-designed specimens, with a maximum increment of about 8%. This might be because the PPM mix-designed concrete has less void content and greater aggregate interlocking. Additionally, the PPM mix design approach's increased concentration of 20 to 10 mm coarse aggregate has a more significant impact on the mortar matrix and slows the propagation of cracks through the bigger aggregates (Bochenek and Prokopski 1989). Furthermore, there was only a small variation (maximum 0.7%) between the G_F of the RAC PPM specimens and the NAC IS specimens. This implies that the drawbacks of the weaker RCA characteristics are lessened by combining the PPM mix design approach and TSMA.

The K_{Ic}^{un} and K_{Ic}^{ini} of PPM mix-designed specimens were observed to be higher in comparison to the specimens prepared using IS: 10262 (2009) mix design approach (Table 6.4). A maximum increment of 9% and 2% was observed in K_{Ic}^{un} and K_{Ic}^{ini}, respectively, for NAC PPM specimens, whereas RAC PPM specimens exhibited 12% and 30% higher K_{Ic}^{un} and K_{Ic}^{ini} values, respectively. This tendency may be attributed to the PPM mix-designed concrete's lower void content and greater aggregate interlocking. In contrast to IS: 10262 (2009) approach, more coarse aggregate between the sizes of 20 mm and 10 mm is used in PPM mix design, while less coarse aggregate between the sizes of less than 10 mm and the fine aggregate is utilized overall. The specific surface area of the aggregate mixture increases due to the large proportion of smaller sized coarse aggregate ($<$ 10 mm) and fine aggregate, which lowers the bond stress at the cement paste and aggregate (Rao and Prasad 2002; Bochenek and Prokopski 1989). As a result, possibilities of failures between the aggregate and matrix are less likely to occur when concrete is prepared following the IS: 10262 (2009) method. Therefore, the crack propagation path in concrete with a PPM mix design took a longer route and absorbed more energy. Similar to NAC IS, RAC-PPM specimens displayed higher K_{Ic}^{un} and K_{Ic}^{ini} values due to a larger concentration of 20—10 mm coarse aggregate.

The NAC PPM specimens' lower β_n values as compared to the NAC IS specimens of the same size (Table 6.4) show that the PPM mix design enhances the ductile behaviour of concrete. The β_n values of the RAC PPM specimens, however, were the same as those of the RAC IS specimens aside from the 75 mm specimens. The ductile behaviour of NAC is more significantly impacted by the PPM mix design than that of RAC. Evidently, a tortuous fracture path and a superior aggregate interlocking mechanism cause PPM specimens to break in a less brittle way.

Regardless of the type of aggregate, 2.5% to 11.5% greater $CTOD_c$ was detected for PPM mix-designed specimens compared to specimens of 75 mm depth from NAC PPM. In addition, the a_c of the PPM mix-created specimens was lower than the specimens made using the IS: 10262 (2009) method, except the 125-mm-depth SEN beams.

Due to their higher tensile strength, the PPM mix-created specimens for a specific size of SEN beam had higher P_{ini} values than the IS: 10262 (2009) mix-designed specimens. According to the P_{ini}/P_u ratio shown in Table 6.4, the fracture starts to form for each type of concrete specimen at a load that is between 70% and 80% of the peak load. Both Zhang et al. (2007) and Zhang and Xu (2011) reported a related finding. Except for the 175-mm-depth NAC PPM and 75-mm- and 175-mm-depth RAC PPM specimens, a significantly higher P_{ini}/P_u value was seen for PPM mix-created concrete specimens compared to the specimens prepared using IS: 10262 (2009) method of mix design. This suggests that the crack initiation process is slightly delayed in specimens made from PPM mix, and this could be explained by the same factors that account for the higher K_{Ic}^{ini} values shown for NAC PPM.

6.5.3 *Influence of the Size of the Specimen*

The primary cause of the size effect in concrete is the existence of a fracture process zone (FPZ) near the crack tip of a quasi-brittle material. Regardless of the kind of aggregate and mix design method, the peak load of SEN beam specimens increases with the increase in beam depth (Table 6.4). In prior investigations (Kumar and Barai 2010b; Zhang and Xu 2011; Musiket et al. 2016, 2017), a similar effect of specimen size on the peak load of SEN specimens in TPB test was also noted. This is because as the depth of the beam increases, the second moment of area also grows, boosting the beam's ability to carry more moments. For each type of concrete, the true fracture energy (G_F) increases as the specimen size does (Table 6.4). For 125-mm- and 175-mm-depth specimens, the increase was found to be 41—47% and 60—73%, respectively, in comparison to 75-mm-depth specimens. This is consistent with the findings of the preceding investigations (Kumar et al. 2014; Issa et al. 2000a, b). Regardless of the kind of the aggregate and mix design approach, the logarithmic variation of G_F with regard to the size of the specimens shows a good agreement (R^2 value is 0.98–0.99) (Fig. 6.7).

The cohesive fracture toughness (K_{Ic}^{c}) was observed to be increased with the size of the specimen for each type of concrete mixes, which was similar to the observations of earlier studies (Kumar et al. 2014; Issa et al. 2000a, b). This indicates that, the larger sized concrete specimen is more brittle in comparison to the smaller sized specimen, which also satisfies the size effect on the brittleness behaviour of concrete. The $CTOD_c$ increased with the increase in size of the specimen for each type of concrete (Table 6.4). Such variation of $CTOD_c$ parameter with the size of the specimen was also reported in the earlier studies (Xu and Zhang 2008; Zhang and Xu 2011; Kumar et al. 2014). The a_c also increased linearly with the increase in the size of the beam for each type of concrete specimens (Table 6.4), which was similar to the observations of Zhang and Xu (2011). Furthermore, P_{ini} increased with the increase in the size of the specimen irrespective of the type of concrete, which showed a good agreement with the observations of earlier research (Xu and Zhang 2008; Zhang and Xu 2011). A logarithmic relationship between P_{ini} and the depth of the specimen exhibited a good correlation (R^2 value is 0.88–0.94) as shown in Fig. 6.8. The earlier studies (Zhang and Xu 2011; Kumar and Barai 2012) reported that, $(a_c - a_0)/l_{ch,mod}$ parameter increases with the increase in $D/l_{ch,mod}$. However, for each type of concrete mix, the trend was not observed (Fig. 6.9). The critical FPZ of 125-mm-depth specimens was observed to be lower than that of 75-mm-depth specimens (Fig. 6.9). This may be due to the higher critical crack length (a_c) observed in 75-mm-depth specimens.

6.6 Comparative Study on G_F

A comparative study was carried out to investigate the appropriateness of the existing expressions to predict the G_F of prepared concrete mixes. Basically, the available expressions (Table 6.5) to predict G_F are function of the compressive strength of concrete and apart from this, the aggregate size, aggregate shape, and w/c ratio are also considered in some of the expressions. The expression proposed by Oh et al. (1999) considers the aggregate size and tensile strength of concrete. However, the discussed expressions (Table 6.5) do not consider the size effect on G_F. Hence, the experimental results of only 75-mm-depth specimen were compared with the predicted results. Figure 6.10 confirms that the expression suggested in fib MC (2010) exhibits a better correlation with the experimental results as compared to the other expressions (CEB FIP Model Code 1993; Xu 1999; Oh et al. 1999; Bažant and Becq-Giraudon 2002; JSCE 2007). However, the dependency of fib MC (2010) expression on compressive strength of concrete resulted in an underestimation of G_F for the PPM mix-designed specimens as compared to that of IS: 10262 (2009) mix-designed specimens. The similar trend was also observed for other expressions mentioned in Table 6.5 apart from the equation suggested by Oh et al. (1999). Due to its reliance on the tensile strength of concrete as well as the aggregate size and modulus of elasticity, the equation presented by Oh et al. (1999) provides higher G_F of PPM mix-designed specimens. Therefore, it is advised to take into account the concrete's tensile strength in addition to the aggregate size, aggregate shape, w/c, compressive strength, and modulus of elasticity for proper prediction of G_F. In addition, the size effect needs to be taken into account for accurate G_F prediction of concrete specimens of various sizes.

6.7 Closure

The study by Pradhan et al. (2018, 2020) compared the fracture behaviour of NAC and RAC, which were prepared using IS: 10262 (2009) method and PPM mix design approach. Moreover, SEN beam specimens of three different sizes were tested to study the size effect.

- The SEN beam specimens prepared using PPM mix design approach exhibited higher peak load and G_F in comparison to the specimens of IS: 10262 (2009) method.

Table 6.5 Expressions to predict G_F of concrete

Codes or literature	Expressions	
CEB FIP Model Code (1993)	$G_F = \left(0.0469d_a^2 - 0.5d_a + 26\right)\left(\frac{f_c}{10}\right)^{0.7}$	N/m
Xu (1999)	$G_F = \left(0.0204 + 0.0053\frac{d_a^{0.95}}{8}\right)\left(\frac{f_c}{f_{c0}}\right)^{0.7}$	N/mm
Oh et al. (1999)	$G_F = 56.24 f_t \frac{d_a}{E}$	N/mm
	$f_t = 0.324 f_c^{\frac{2}{3}}$ or $f_t = 0.74 f_r$	
Bažant and Becq-Giraudon (2002)	$G_F =$ $2.5\alpha_0\left(\frac{f_c}{0.051}\right)^{0.46}\left(1+\frac{d_a}{11.27}\right)^{0.22}\left(\frac{w}{c}\right)^{-0.3}$	N/m
JSCE (2007)	$G_F = 10 d_a^{0.33} f_c^{0.33}$	N/m
fib MC (2010)	$G_F = 73 f_c^{0.18}$	N/m

where d_a = maximum size of the aggregate (mm), f_c = compressive strength of concrete (MPa), f_{c0} = 10 MPa, E = modulus of elasticity of concrete (MPa), f_t = tensile strength of concrete (MPa), f_r = modulus of rupture (MPa), w/c = water-to-cement ratio, and $\alpha_0 = 1$ for round-shaped aggregate and 1.44 for crushed or angular-shaped aggregate

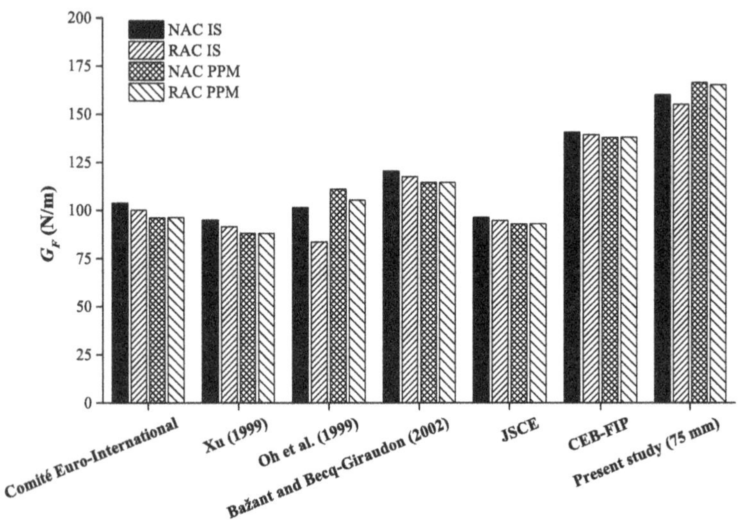

Fig. 6.10 Variation of critical fracture process zone with the depth of the specimens (Pradhan et al. 2018)

- The RAC specimens showed lower G_F in comparison to the corresponding NAC specimens.
- The double-K fracture parameters of the material constants were obtained using nonlinear softening function for $CTOD_c > w_1$, whereas for $CTOD_c < w_1$ linear softening function was used. The c_1, c_2, and w_0 parameters were determined for each type of concrete and each size of the specimen by satisfying the experimental w_1, G_F, and f_t.

- The fracture toughness parameters were observed to be inferior for RAC. However, the improvement was witnessed for PPM mix-designed concrete.
- The size-effect analysis showed that the K_{Ic}^{ini} parameter was not influenced significantly with the increase in the depth of the specimens, whereas increment was observed in K_{Ic}^{un} with the increase in the size of the specimens.
- The a_c, $CTOD_c$, critical fracture process zone $(a_c - a_0)$, β_n, and $l_{ch,mod}$ parameters were influenced by the size of the specimen.
- The available expressions in different codes and literature were considered to predict G_F of different types of concrete. The prediction of G_F of concrete using fib MC (2010) code proved to be better than other available expressions.

References

Arezoumandi M, Drury J, Volz JS (2014) Effect of recycled concrete aggregate replacement level on the fracture behavior of concrete. 3:1–8

Bažant ZP, Becq-Giraudon E (2002) Statistical prediction of fracture parameters of concrete and implications for choice of testing standard. Cement Concr Res 32(4):529–556

Bažant ZP, Kazemi MT (1990) Determination of fracture energy, process zone length and brittleness number from size effect, with application to rock and concrete. Int J Fract 44:111–131

Bažant ZP, Oh B (1983) Crack band theory of concrete. Mater Struct 16(93):155–177

Bažant ZP, Planas J (1997) Fracture and size effect in concrete and other quasibrittle materials. CRC Press

Bažant ZP, Yu Q (2011) Size-effect testing of cohesive fracture parameters and nonuniqueness of work-of-fracture method. J Eng Mech 137(8):580–588

Bochenek A, Prokopski G (1989) The investigation of aggregate grain size effect on fracture toughness of ordinary concrete structures. Int J Fract 41(3):197–205

Bordelon A, Cervantes V, Roesler J (2009) Fracture properties of concrete containing recycled concrete aggregates. Mag Concr Res 61(9):665–670

Bueckner HF (1970) Novel principle for the computation of stress intensity factors. Zeitschrift fuer Angewandte Mathematik & Mechanik 50(9):529–546

Casuccio M, Torrijos MC, Giaccio G, Zerbino R (2008) Failure mechanism of recycled aggregate concrete. Constr Build Mater 22(7):1500–1506

CEB FIP Model Code (1993). Comité Euro-International du Béton. Bulletin D'Information

Chen GM, Yang H, Lin CJ, Chen JF, He YH, Zhang HZ (2016) Fracture behaviour of steel fibre reinforced recycled aggregate concrete after exposure to elevated temperatures. Constr Build Mater 128:272–286

Choubey RK, Kumar S, Chakradhara Rao M (2016) Modeling of fracture parameters for crack propagation in recycled aggregate concrete. Constr Build Mater 106:168–178

Elices M, Guinea GV, Gómez J, Planas J (2002) The cohesive zone model: advantages, limitations and challenges. Eng Fract Mech 69(2):137–163

Fib MC (2010) Fib model code for concrete structures. Wilhelm Ernst & Sohn, Berlin

Ghorbel E, Wardeh G (2017) Influence of recycled coarse aggregates incorporation on the fracture properties of concrete. Constr Build Mater 154:51–60

Guinea GV, Planas J, Elices M (1994) A general bilinear fit for the softening curve of concrete. Mater Struct 27(2):99–105

Guo M, Alam SY, Bendimerad AZ, Grondin F, Rozière E, Loukili A (2017) Fracture process zone characteristics and identification of the micro-fracture phases in recycled concrete. Eng Fract Mech 181:101–115

Hillerborg A (1985) The theoretical basis of a method to determine the fracture energy GF of concrete. Mater Struct 18(4):291–296
Hillerborg A, Modéer M, Petersson PE (1976) Analysis of crack formation and crack growth in concrete by means of fracture mechanics and finite elements. Cement Concr Res 6(6):773–781
IS: 10262 (2009). Concrete mix proportioning–guidelines
Issa MA, Issa MA, Islam MS, Chudnovsky A (2000) Size effects in concrete fracture-part II: Analysis of test results. Int J Fract 102(1):25–42
Issa MA, Issa MA, Islam MS, Chudnovsky A (2000) Size effects in concrete fracture: Part I, experimental setup and observations. Int J Fract 102(1):1–24
JCI Standard (2003) Method of test for fracture energy of concrete by use of notched beam. JCI-S-001-2003
Jenq BY, Shah SP (1986) Two parameter fracture model for concrete. J Eng Mech 111(10):1227–1241
Jenq YS, Shah SP (1985) A Fracture toughness criterion for concrete. Eng Fract Mech 21(5):1055–1069
JSCE (2007) Standard specifications for concrete structures–2007 design, 15th edn. Japan Society of Civil Engineers
Karihaloo B, Nallathambi P (1989) An improved effective crack model for the determination of fracture toughness of concrete. Cement Concr Res 19(4):603–610
Karihaloo B, Nallathambi P (1990) Effective crack model for the determination of fracture toughness (KIce) of concrete. Eng Fract Mech 35(4/5):637–645
Karihaloo BL, Nallathambi P (1991) Test methods for determining mode I fracture toughness of concrete. Springer, Netherlands, Dordrecht, pp 91–124
Kumar S, Barai SV (2008) Influence of specimen geometry on determination of double-K fracture parameters of concrete: A comparative study. Int J Fract 149(1):47–66
Kumar S, Barai SV (2009) Determining double-K fracture parameters of concrete for compact tension and wedge splitting tests using weight function. Eng Fract Mech 76(7):935–948
Kumar S, Barai SV (2010) Determining the double-K fracture parameters for three-point bending notched concrete beams using weight function. Fatigue Fract Eng Mater Struct 33(10):645–660
Kumar S, Barai SV (2010) Size-effect prediction from the double-K fracture model for notched concrete beam. Int J Damage Mech 19(4):473–497
Kumar S, Barai SV (2012) Size-effect of fracture parameters for crack propagation in concrete: A comparative study. Comput Concr 9(1):1–9
Kumar S, Pandey SR, Srivastava AKL (2013) Analytical methods for determination of double- K fracture parameters of concrete. Adv Concr Constr 1(4):319–340
Kumar S, Pandey SR, Srivastava AKL (2014) Determination of double-K fracture parameters of concrete using peak load method. Eng Fract Mech 131:471–484
Lee J, Lopez MM (2014) An experimental study on fracture energy of plain concrete. Int J Concr Struct Mater 8(2):129–139
Lubliner J, Oliver J, Oller S, Oñate E (1989) A plastic-damage model for concrete. Int J Solids Struct 25(3):299–326
Musiket K, Rosendahl M, Xi Y (2016) Fracture of recycled aggregate concrete under high loading rates. J Mater Civil Eng 28(6):04016018
Musiket K, Vernerey F, Xi Y (2017) Numeral modeling of fracture failure of recycled aggregate concrete beams under high loading rates. Int J Fract 203(1–2):263–276
Oh B-H, Jang S-Y, Byun H-K (1999) Prediction of fracture energy of concrete. KCI Concr J 11(3):211–221
Petersson P (1981) Crack growth and formation of fracture zones in plain concrete and similar materials. Report TrBm21006
Planas J, Elices M, Guinea GV (1992) Measurement of the fracture energy using three-point bend tests: Part 2-Influence of bulk energy dissipation. Mater Struct 25(5):305–312
Planas J, Guinea GV, Elices M (1999) Size effect and inverse analysis in concrete fracture. Int J Fract 95(1–4):367–378

Pradhan S, Kumar S, Barai SV (2017) Recycled aggregate concrete: particle packing method (PPM) of mix design approach. Constr Build Mater 152:269–284

Pradhan S, Kumar S, Barai SV (2018) Impact of particle packing mix design method on fracture properties of natural and recycled aggregate concrete. Fatigue Fract Eng Mater Struct 42(4):943–958

Pradhan S, Kumar S, Barai SV (2020) Impact of particle packing method of design mix on fracture behavior of concrete: critical analysis. J Mater Civil Eng 32(4):1–13

Rao GA, Prasad BKR (2002) Fracture energy and softening behavior of high-strength concrete. Cement Concr Res 32(2):247–252

Reinhardt HW, Cornelissen HAW, Hordijk DA (1986) Tensile tests and failure analysis of concrete. J Struct Eng 112(11):2462–2477

Rice JR (1972) Some remarks on elastic crack-tip stress fields. Int J Solids Struct 8(6):751–758

RILEM TC-50 (1985) Determination of the fracture energy of mortar and concrete by means of three-point bend tests on notched beams. Mater Struct 18(106):285–290

Roesler JR, Paulino GH, Park K, Gaedicke C (2007) Concrete fracture prediction using bilinear softening. Cement Concr Compos 29(4):300–312

Ruiz G, Ortega JJ, Yu RC, Xu S, Wu Y (2016) Effect of size and cohesive assumptions on the double-K fracture parameters of concrete. Eng Fract Mech 166:198–217

Tada H, Paris PC, Irwin GR (1973) The stress analysis of cracks handbook, vol 2. Del Research Corporation, Hellertown, PA

Tam VWY, Gao XF, Tam CM (2005) Microstructural analysis of recycled aggregate concrete produced from two-stage mixing approach. Cement Concr Res 35(6):1195–1203

Xiao J, Liu Q, Wu YC (2012) Numerical and experimental studies on fracture process of recycled concrete. Fatigue Fract Eng Mater Struct 35(8):801–808

Xu S (1999) Determination of parameters in the bilinear, Reinhardt's non-linear and exponentially non-linear softeming curves and their physical meanings. Werkstoffe und Werkstoffprüfung im Bauwesen, Hamburg 15:410–424

Xu S, Reinhardt HW (1999) Determination of double-K criterion for crack propagation in quasi-brittle fracture Part I: experimental investigation of crack propagation. Int J Fract 98(2):111–149

Xu S, Reinhardt HW (1999) Determination of double-K criterion for crack propagation in quasi-brittle fracture, Part II: Analytical evaluating and practical measuring methods for three-point bending notched beams. Int J Fract 1984:151–177

Xu S, Reinhardt HW (1999) Determination of double-K criterion for crack propagation in quasi-brittle fracture, Part III: Compact tension specimens and wedge splitting specimens. Int J Fract 98(2):179–193

Xu S, Reinhardt HW (2000) A simplified method for determining double-K fracture parameters for three-point bending tests. Int J Fract 104:181–209

Xu S, Zhang X (2008) Determination of fracture parameters for crack propagation in concrete using an energy approach. Eng Fract Mech 75(15):4292–4308

Zhang X, Xu S (2011) A comparative study on five approaches to evaluate double-K fracture toughness parameters of concrete and size effect analysis. Eng Fract Mech 78(10):2115–2138

Zhang X, Xu S, Zheng S (2007) Experimental measurement of double-k fracture parameters of concrete with small-size aggregates. Front Architect Civil Eng China 1(4):448–457

Chapter 7
Performance Assessment of Concrete: Meso-, Micro-, Nano-level, and Physio-chemical Analysis

7.1 Introduction

The effect of microstructural characteristics on the macro-mechanical properties of concrete has been the subject of research over the last few decades. In chapters "Macro-level Performance Assessment of Concrete: Conventional Approach" and "Macro-level Performance Assessment of Concrete: Experimental Fracture Analysis", it was examined how the type of aggregates, mix design techniques, and mixing methodologies affected macro-mechanical properties, which are thought to also affect meso-, micro-, and nano-level characteristics. Therefore, the current chapter focuses on how the aforementioned parameters affect the features and degree of hydration at the microstructure and mesostructure level. Additionally, both the ITZ and the entire specimen's pore content and distribution are examined. The impact of each of these micro- and meso-level variables on the macro-mechanical properties is then examined at various curing times.

7.2 Multi-scale Characterization of Concrete

Concrete is made up of a heterogeneous mixture of aggregates, including coarse and fine aggregate, as well as products that result from the hydration of cement or other cementitious materials. Ettringite, calcium silicate hydrate, and calcium hydroxide are manifested as a heterogeneous and anisotropic combination together with scattered holes of erratic size and shape throughout the hydration process. In addition to the kind and quality of the aggregate, concrete's macro-mechanical properties are also influenced by the quantity and distribution of the hydration process's byproducts (hydration products and pores). The application of nanoindentation, scanning electron microscope (SEM) image analysis of back-scattered electrons (BSE) images, thermogravimetric analysis (TGA), and image analysis of X-ray μ-CT images techniques used by Pradhan et al. (2020a) are discussed in this context.

S. Pradhan et al., *Particle Packing Method for Recycled Aggregate Concrete*,
https://doi.org/10.1007/978-981-99-7516-7_7

7.2.1 Thermogravimetric Analysis (TGA)

The four major compounds present in ordinary Portland cement (OPC) are tricalcium silicate (C_3S), dicalcium silicate (C_2S), tricalcium aluminate (C_3A), and tetracalcium aluminoferrite (C_4AF). The hydration products of these compounds are the bridging units between the aggregates, as well as the strength contributors. Thermogravimetric analysis (TGA) can be used to analyse the chemical compounds produced as a result of hydration reactions in concrete. The TGA method calculates the amount of mass lost as a result of the breakdown of hydration products like ettringite, calcium silicate hydrate (CSH), calcium hydroxide (CH), and carbonated calcium hydroxide, or calcium carbonate ($CaCO_3$). The detected mass loss is the result of the water that was chemically bonded and present in the hydration products decomposing. As a result, the mass loss observed during the TGA test can be used to infer indirectly how much hydration has occurred. TGA is one of the most important analytical techniques for determining the degree of hydration due to its accuracy and speed in detecting small chemical changes.

Tam et al. (2009) used the differential scanning calorimetry (DSC) technique to investigate the advantages of TSMA with respect to the NMA. In this context, the hydrated cement paste in the vicinity of aggregate was analysed by heating the collected samples from room temperature to 620 °C at a rate of 20 °C/min. The summation of heat flow required for the degeneration of different hydration products was observed to be higher for the concrete prepared using TSMA as compared to NMA. Furthermore, the summation of the reported hydration products was higher in NAC with respect to RAC. However, TGA is the most preferred technique to estimate the degree of hydration of cement and cementitious material in concrete or mortar. In this context, different methods were proposed by Bhatty (1986), Monteagudo et al. (2014), and Deboucha et al. (2017) to estimate the degree of hydration from the measured chemically bound water in TGA test of the hydrated cement paste.

7.2.2 Nanoindentation Technique

The investigation of the potential impact of ITZ on the macro-mechanical characteristics of concrete uses nanoindentation as one of its methodologies. ITZ thickness and its micro-mechanical properties are influenced by the kind of aggregate, mixing techniques, hydration time, and integration of pozzolanic elements, as documented in the literature. Concrete specimens made utilizing the two-stage mixing approach (TSMA) and the normal mixing approach (NMA) were compared by Li et al. (2012). When compared to the NMA, the volume fraction of porosity and CH in ITZ is effectively decreased as a result of the reduction in wall effect and micro-bleeding in TSMA. Additionally, more CSH was reported in TSMA than NMA. As a result, the fluctuation in the ITZ indentation modulus in TSMA decreases. Xiao et al. (2013) looked into how the quality of ITZ was affected by the aggregate type, mix proportion, and hydration time.

The old ITZ and new ITZ of the RAC were found to be 40 to 50 μm and 55 to 65 μm, respectively. In addition, the indentation moduli of the old and new ITZ were calculated to be 70–80% and 80–90% of old and new paste matrix, respectively. The thickness of ITZ was found to be unaffected by the curing time. Because hydrates produce more with age, the ITZ does, however, become denser as the curing time lengthens. The influence of w/c and aggregate type was studied by Sidorova et al. (2014). ITZ thickness and porosity rise with an increase in w/c, whereas the indentation modulus of ITZ and paste matrix decreases as w/c increases. Sidorova et al. (2014) reported that it was difficult to determine the ITZ thickness accurately using nanoindentation technology.

The effect of RCA replacement ratio on the properties of ITZ characteristics at different age was studied by Zhang and Zhao (2015) and concrete mixes prepared with RCA replacement ratio of 0, 50, and 100% were investigated. It was found that as the RCA replacement ratio rises, the thickness of the ITZs (the ITZs between the NCA and the paste matrix and between the RCA and the new paste matrix) increases while the micro-mechanical characteristics of these ITZs diminish. With increasing curing age, an improvement in the micro-mechanical characteristics of ITZs was seen. Zhang et al. (2015) examined how nano-silica slurry affected the characteristics of RAC. Because of the enhancement of the RCA's surface by the nano-silica slurry, the new ITZ displayed superior micro-mechanical characteristics. Due to the weak old ITZ in the RCA, the macro-mechanical characteristics, however, showed less substantial improvement. According to the research done by del Bosque et al. (2017), the constituent materials (such as brick, clay, asphalt, wood, and glass) have a deleterious impact on the RAC's macro-mechanical qualities via influencing its micro-mechanical properties. This is as a result of the constituent materials' thicker ITZ and lower elastic modulus compared to RCA.

7.2.3 Image Analysis of Back-Scattered Electrons (BSE) Images

To assess the elements of microstructure (voids, CH, CSH, and unhydrated cement), scanning electron microscope (SEM) images of back-scattered electrons (BSE) are analysed. BSE images can be used to investigate the effects of the kind of aggregate, w/c, mix design process, mixing techniques, integration of mineral admixtures, and curing period on ITZ features. Tam et al. (2005) investigated the impact of the mixing techniques (NMA and TSMA) on the ITZ properties of RAC. For RAC constructed using TSMA, fewer voids and a denser ITZ were noted. The addition of stone covered in pozzolanic powder (SEPP) led to denser ITZ for RAC, as observed by Li et al. (2009). By decreasing the micro-bleeding phenomena, the suggested mixing method with the application of SEPP improved the ITZ properties. The proposed triple mixing method (TM) was used to coat a layer of pozzolanic material (Kong et al. 2010). The SEM image analysis demonstrated the improvement in ITZ brought about by TM by using hydration agents to seal the micro-cracks. Additionally, the pozzolanic coating ate up the CH in the RCA and enhanced the density of the ITZ with secondary hydration byproducts.

Chakradhara Rao (2010) and Mukharjee (2014) estimated the components of microstructure by image analysis technique of the BSE images. The components of microstructure were measured for both NAC and RAC after the BSE images were split into a series of sequential strips of 10 μm. It was observed that, owing to the wall effect and micro-bleeding (Ollivier et al. 1995; Scrivener et al. 2004; Xiao et al. 2013; Gao et al. 2018), the voids and CH content decrease and CSH and unhydrated cement content increase as the distance increases in the lateral direction from the aggregate boundary. Chakradhara Rao (2010) and Mukharjee (2014) reported higher voids content at same distance from aggregate boundary in RAC with respect to NAC. Mukharjee (2014) and Mukharjee and Barai (2014) observed that the incorporation of nano-silica lowered the voids content in the ITZ of both NAC and RAC and the improved ITZ characteristics was also reflected in macro-mechanical properties of concrete. This improvement in ITZ properties may be the result of nano-silica filling voids along the aggregate boundary, which reduces the wall effect and micro-bleeding phenomena and densifies the ITZ via secondary hydration products.

7.2.4 Image Analysis of X-Ray Microtomographic Images

The methods to study the surface properties at the micro-level include nanoindentation and image analysis of BSE images. These methods are mostly employed to examine the ITZ features, the weakest area of concrete. The macro-mechanical performance of concrete is influenced by a number of other factors, not just the weak ITZ. Therefore, it is crucial to look into how meso-level variables affect concrete's behaviour at a macro-level. In this context, X-ray microtomography (XRT) is a non-destructive imaging technique in which the three-dimensional (3D) structure of the specimen is reconstructed from very high-resolution two-dimensional (2D) images of cross sections.

Leite and Monteiro (2016) used the XRT image processing technology to examine how RCA and its water absorption affected the microstructure of RAC. In this context, Leite and Monteiro (2016) prepared RAC using RCA under dry and saturated surface dry (SSD) conditions. The network of macropore was evidenced in RAC prepared using dry RCA, which was not observed by using SSD RCA. The formation of macropores close to the RCA was decreased by the application of TSMA (Tam et al. 2005). To control the production of macropores, however, the TSMA's mixing duration is insufficient. In this situation, longer TSMA mixing times were recommended, and it was found that this procedure worked better for RCA made from low-strength concrete (Leite and Monteiro 2016). Zuo et al. (2018) studied the chloride ion penetration in RAC using X-ray computed tomography images. It was found that RAC with lower w/c and curing in a CO_2 environment reduced chloride ion penetration, however fly ash incorporation had little effect on resisting chloride ion penetration. The XRT imaging technique was used by Thomas et al. (2018) to estimate the volume of adhered mortar layer to recycled aggregate. Analyses were done on the micro-level effects that successive recycling had on the qualities of recycled aggregate. In contrast to the loss of natural aggregate and the amount of adhering mortar, it was shown that the compaction capacity of recycled aggregate diminishes as recycling processes are repeated.

7.3 Experimental Investigation

The concrete mixes were proportioned using conventional mix design IS: 10262 (2009) and Particle Packing Method (PPM) of mix design approaches by Pradhan et al. (2017). For the conventional mix and PPM mix, the standard mixing method and TSMA were utilized, respectively. Comparing the macro-mechanical properties of hardened concrete to the conventional method, it is thought that the PPM mix design methodology has better packing characteristics, and TSMA has a favourable influence on ITZ quality. However, more research was done to determine the function of aggregate type, mix design methodology, and mixing technique. In this context, TGA, nanoindentation, image analysis of BSE images, and X-ray microtomographic images were conducted.

7.3.1 Sample Preparation for Thermogravimetric Analysis

After a curing time of 7, 28, and 90 days, the samples for the TGA test were collected. A minimum of three samples were examined for each curing age. To gather the material for the TGA test, cubes and cylinders prepared for the compressive strength and tensile strength tests were employed. In this context, the core portion of the tested cubes and cylinders was preferred. The hydrated cement paste was collected from the vicinity of aggregate as suggested by Tam et al. (2009). The collected powder sample was then sieved through 75 μm sieve (Vedalakshmi et al. 2003). Isopropanol solvent was used to seize the hydration of the collected material. Zhang and Scherer (2011) evaluated the various physical and chemical methods for stopping the hydration of cement paste, the author ultimately suggested using isopropanol solvent since it causes less disruption to the cement paste's microstructure. To prevent air moisture from having an impact on the hydration process, the sample was then collected in a closed micro-centrifuge tube and put in a vacuum desiccator.

7.3.2 Sample Preparation for Nanoindentation and SEM

To keep the properties of concrete unharmed and to assure reproducibility, particular care must be taken when preparing the sample for nanoindentation and taking SEM images. Due to the complicated heterogeneity of concrete, it is crucial to identify concrete that is acceptable for microscopic inspection. To get the most information, concrete surfaces with significant paste coverage should be preferred with those with substantial aggregate coverage. Approximately 5-mm-thick slices were cut from 100 × 100 × 500 mm prisms using a precision diamond saw and kerosene as lubricant. From each slice again approximately 20 × 20 mm rectangular sections were cut. The dried samples were vacuum coated with a low-viscosity epoxy called Epoxil-43 and a hardener called Epoxil-MH43 in a 3:1 ratio, and they were then left to harden at room temperature for a period of 1–6 h. The specimen is shielded by the epoxy resin, which stops damage to it during grinding and polishing. The pores, hydration products, and unhydrated cement are better contrasted because of the epoxy resin's ability to fill in the gaps.

After that, the samples were meticulously ground and polished. In order to remove epoxy from the specimens' surfaces, coarse polishing was performed using #500 and #1200 grit paper. Then, using a succession of successively finer grade diamond pastes, namely, 9, 3, 1, and 0.25 μm, the surface was subjected to fine polishing using an automated polishing equipment. Petroleum is used as a lubricant for polishing. Each specimen was typically polished for 3 to 4 h to produce a nice surface. After being cleaned in an ultrasonic bath, the polished specimens were dried in a vacuum to eliminate any last traces of lubricant from the surface. The specimens had a thin gold coating applied before BSE imaging in order to keep them from charging.

7.3.3 Sample Preparation for X-Ray Microtomography

The concrete was poured in a cylindrical specimen of diameter 65 mm and height 100 mm. After 24 h the specimen was demoulded and placed in water tank. The scanning operation was conducted after 7, 28, and 90 days of curing.

The test was conducted in "GE Phoenix v|tome|x s", an industrial high-resolution computed tomography and X-ray system, which has a humidity- and temperature-controlled chamber. It is equipped with a flat X-ray detector, also known as a digital detector array (DDA), and a monochromatic X-ray source. On a rotating stage, a cylindrical concrete specimen was put. The tube current and voltage of the X-ray source were tuned to 150 kV and 100 μA, respectively. The scanned specimens were turned into two-dimensional (2D) 8-bit grayscale pictures for image analysis. The scanned specimen's grayscale images were produced in three separate planes (bottom to top, left to right, and front to back), with 1000 images produced in each plane. For the analysis, images of the bottom-to-top plane were taken into account.

7.4 Thermogravimetric Analysis

The stoichiometry of cement hydration is highly complex due to the heterogeneity of cement and mineral additives. To further understand how OPC and mixed cement evolved during the hydration process, researchers adopted the TGA technique (Bhatty 1986; Pane and Hansen 2005; Monteagudo et al. 2014; Deboucha et al. 2017; Zeng et al. 2012; Ye et al. 2007; Vedalakshmi et al. 2003; Hemalatha et al. 2016). Furthermore, the degree of hydration of the cementitious material was correlated with the macro-mechanical properties of the concrete and observed a direct relationship with the compressive strength parameter (Monteagudo et al. 2014; Deboucha et al. 2017; Hemalatha et al. 2016). Pradhan et al. (2020b) estimated the degree of hydration quantitatively for different types of concrete and correlated it with the macro-mechanical properties of concrete. In this context, the TGA test was carried out for the hydrated cement paste collected around the aggregate surface. The TGA test was performed using Perkin Elmer Pyris Diamond TG-DTA with a balance accuracy 0.1 μg. About 10–15 mg of sample was placed in an alumina crucible. This is heated in the dynamic heating ramp between 30 and 1000 °C at a heating rate of 10 °C/min. The test was conducted in Ar atmosphere.

It is possible to monitor the temperature difference between a sample material and a reference material as a function of time or temperature using the differential thermal analysis (DTA) method. Through the use of an isothermal mode or a temperature range, the TGA technique measures the change in mass of the sample material over time. In this test, both the sample and the reference materials are exposed in the same environment while being heated or cooled at a preferred rate. The hydrated cement paste when subjected to TGA test the decomposition of cement hydrates is observed which can be witnessed from the descending TGA curve (Fig. 7.1). Endothermic reactions are responsible for the physio-chemical alteration of cement hydrates, and endothermic peaks may be seen in both the DTA curve and the derivative of the TGA curve.

The mass loss of the hydrated cement paste subjected to 105–1000 °C in the TGA test was used to assess the degree of hydration (Bhatty 1986; Deboucha et al. 2017; Scrivener et al. 2015). In accordance with the way cement hydrates decompose, the endothermic effects can be divided into three main phases. The dehydration (Ldh) of ettringite and CSH, the dehydroxylation (Ldx) of CH, and the decarbonation (Ldc) of $CaCO_3$ define the first, second, and third phases, respectively (Bhatty 1986, 1991; Monteagudo et al. 2014; Deboucha et al. 2017). The dehydration, dehydroxylation, and decarbonation phases generally occur in the temperature range of 25–400 °C, 400–600 °C, and 600–800 °C, respectively (Deboucha et al. 2017). The temperature ranges considered for different phases (Ldh, Ldx and Ldc) by different authors are represented in Table 7.1. Apart from Pane and Hansen (2005), the initial temperature for the dehydration phase was considered as 105 °C (Bhatty 1986; Monteagudo et al. 2014; Deboucha et al. 2017; Scrivener et al. 2015). Monteagudo et al. (2014) observed that the mass loss associated between temperature 105 and 140 °C due to the physically bound water contributes in improving the degree of saturation of dissolution

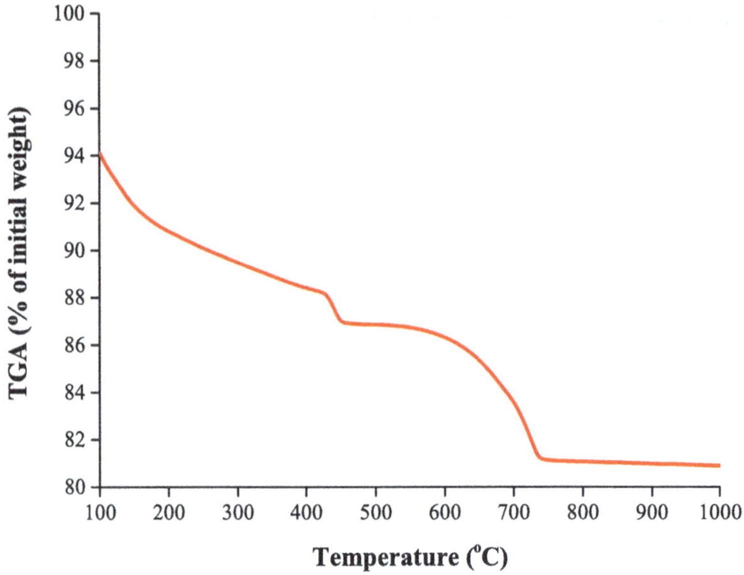

Fig. 7.1 A typical TGA curve of cement paste (Pradhan et al. 2020b)

Table 7.1 Temperature ranges considered by different authors

Region	Temperature range (°C) by different authors				
	Bhatty (1986)	Pane and Hansen (2005)	Monteagudo et al. (2014)	Deboucha et al. (2017)	Present study
Ldh	105–440	140–440	105–430	105–400	105–420
Ldx	440–580	440–520	430–530	400–600	420–500
Ldc	580–1000	520–1100	530–1100	600–1000	500–1000

and precipitation of the hydration products. Monteagudo et al. (2014) determined the temperature range for the dehydroxylation from the second derivative of the DTA curve for each sample.

The initial temperature for dehydration phase was considered as 105 °C by Pradhan et al. (2020b). From the first derivative of the DTA curve (dDTA), the temperature range of the dehydroxylation phase was identified for each individual sample. A typical dDTA curve is shown in Fig. 7.2, where between the temperature 400 and 600 °C, the temperature corresponds to the minimum and maximum values of dDTA provided the lower limit and upper limit of the dehydroxylation phase. From the dDTA curves, the temperature range for the dehydroxylation phase was observed to be between 420 and 500 °C for all the sample. Hence, the temperature range for Ldh, Ldx, and Ldc can be broadly specified between 105 and 420, 420 and 500, and 500 and 1000 °C. The next subsections address the methods that can be used to calculate

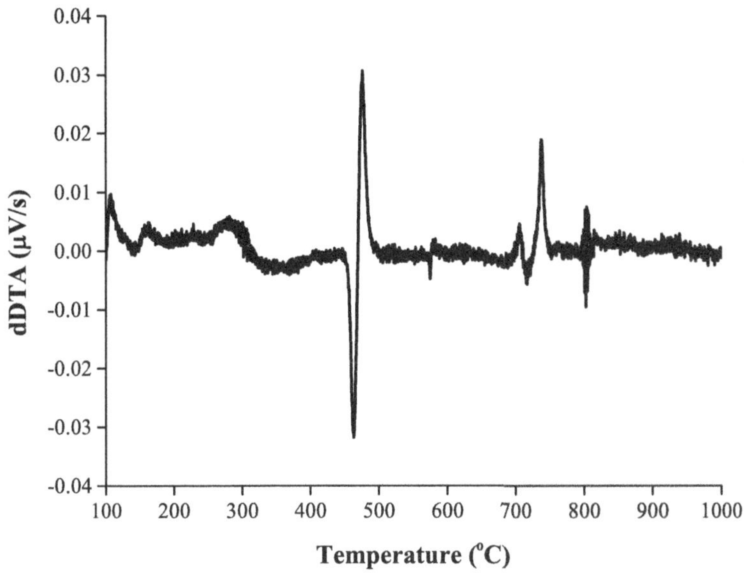

Fig. 7.2 A typical dDTA curve of cement paste (Pradhan et al. 2020b)

the degree of hydration of the hydrated cement paste based on the mass loss that is seen throughout three separate phases of the TGA test.

7.4.1 Estimation of Degree of Hydration

Bhatty (1986)

Bhatty (1986) presented a method to gauge the degree of hydration from the mass loss discovered during the hydrated cement paste's TGA test. In this context, the mass loss resulting from the dehydration of ettringite and CSH, the dehydroxylation of CH, and the decarbonation of carbonated CH between 105 and 1100 °C is used to compute the chemically bound water (W_B). However, this method does not account for the carbonated compound of anhydrous material. A conversion factor of value 0.41 (ratio of the molecular mass of H_2O and CO_2) is multiplied with Ldc to satisfy the assumption that the derived chemically bound water owing to the decarbonation is from carbonated CH. Equation 7.4.1 represents the expression to calculate W_B, and subsequently the degree of hydration (α) can be estimated from Eq. 7.4.2 as suggested by Bhatty (1986). At infinite time, the maximum chemically bound water ($W_{B\infty}$) estimated according to the stoichiometry of cement is 0.23 to 0.25 gram per gram of cement paste (Bhatty 1986; Copeland et al. 1960; Young and Hansen 1986). Hence, Bhatty (1986) assumed the value 0.24 to estimate α of the hydrated cement paste from Eq. 7.4.2.

$$W_B = Ldh + Ldx + 0.41Ldc \quad (7.4.1)$$

$$\alpha = \frac{W_B}{0.24} \times 100 \quad (7.4.2)$$

Pane and Hansen (2005)

A method to estimate α was later put forth by Pane and Hansen (2005), and it took into account mass loss during TGA tests conducted between 140 and 1100 °C. The mass loss resulting from the decarbonation of carbonated anhydrous material was taken into account in this procedure. In order to calculate W_B, the adjustment for the carbonated anhydrous material (Ldc_a) is therefore deducted from the total decarbonation loss and shown in Eq. 7.4.3. However, the conversion factor of 0.41 is not considered by Pane and Hansen (2005) to calculate the chemically bound water of the carbonated CH. In contrast to Bhatty (1986), a fixed value is not assumed for $W_{B\infty}$. In this context, a three-parameter equation (Eq. 7.4.4) was suggested, which fits the experimental data of W_B obtained at different curing age (t). The intercept and curvature of the graph, expressed in a logarithmic scale, are controlled by the parameters τ and a of the three-parameter equation shown in Eq. 7.4.4. Finally, α can be estimated from Eq. 7.4.5 by using W_B and $W_{B\infty}$.

$$W_B = Ldh + Ldx + (Ldc - Ldc_a) \quad (7.4.3)$$

$$W_B = W_{B\infty} \times \exp\left[-\left(\frac{\tau}{t}\right)^a\right] \quad (7.4.4)$$

$$\alpha = \frac{W_B}{W_{B\infty}} \times 100 \quad (7.4.5)$$

Monteagudo et al. (2014)

The method suggested by Monteagudo et al. (2014) to calculate W_B is based on the contributions of Bhatty (1986) and Pane and Hansen (2005). Furthermore, similar to Pane and Hansen (2005) there is no fixed value set for $W_{B\infty}$ from the stoichiometry of cement. For this, an equation similar to Michaelis–Menten equation was suggested. The experimental data of W_B at different curing age (t) is fitted using Eq. 7.4.7, and subsequently $W_{B\infty}$ is obtained. Consequently, using the calculated values of W_B and $W_{B\infty}$, the α is estimated in accordance to Eq. 7.4.8.

$$W_B = Ldh + Ldx + 0.41(Ldc - Ldc_a) \quad (7.4.6)$$

$$W_B = \frac{W_{B\infty} \times t}{t + K} \quad (7.4.7)$$

$$\alpha = \frac{W_B}{W_{B\infty}} \times 100 \quad (7.4.8)$$

Pradhan et al. (2020b)

The methods listed above to determine the level of hydration have benefits and drawbacks. The approach recommended by Bhatty (1986) is straightforward and simple to use. However, the carbonated anhydrous material content is not taken into account while calculating W_B. Additionally, $W_{B\infty}$ is assumed to be 0.24 based on findings from cement's stoichiometric study. The value 0.24 for $W_{B\infty}$ is practically difficult to attain. Pane and Hansen (2005) and Monteagudo et al. (2014) also observed that, $W_{B\infty}$ value is less than 0.24 for OPC cement and for blended cement the $W_{B\infty}$ value is even lower than that of OPC cement. Therefore, a fixed value of 0.24 cannot be taken into account for $W_{B\infty}$. In this case, Pane and Hansen (2005) took into consideration the carbonated anhydrous material content and employed the three-parameter equation to determine $W_{B\infty}$ from the available W_B data at various curing periods. However, while calculating W_B the conversion factor 0.41 is not applied to consider only the chemically bound water correspond to the CH equivalent in carbonated portlandite. The method suggested by Monteagudo et al. (2014) included the conversion factor (0.41) while calculating W_B. A Michaelis–Menten-type equation is also suggested in order to compute the value of $W_{B\infty}$ from the calculated W_B values at various curing periods. Although there is only one solution of $W_{B\infty}$ for the Michaelis–Menten-type equation, this value is far lower than what is actually seen. This results in a scientifically erroneous estimation of cement paste hydration that is even higher than 100% at advanced curing age.

The method used by Pradhan et al. (2020b) to calculate W_B and $W_{B\infty}$ is the synergy of the methods suggested by Pane and Hansen (2005) and Monteagudo et al. (2014). In this context, the expression proposed by Monteagudo et al. (2014) to calculate W_B is adopted because it only takes into consideration the chemically bound water and incorporates the conversion factor of 0.41 to convert it to the CH equivalent in carbonated portlandite. The three-parameter equation proposed by Pane and Hansen (2005) was used to forecast $W_{B\infty}$ value rather than the Michaelis–Menten-type equation, preventing the underestimating of $W_{B\infty}$. Hence, α is estimated from Eq. 7.4.11, where W_B and $W_{B\infty}$ are calculated by conjugating the contributions of Bhatty (1986), Pane and Hansen (2005), and Monteagudo et al. (2014).

$$W_B = Ldh + Ldx + 0.41(Ldc - Ldc_a) \quad (7.4.9)$$

$$W_B = W_{B\infty} \times \exp\left[-\left(\frac{\tau}{t}\right)^{a}\right] \quad (7.4.10)$$

$$\alpha = \frac{W_B}{W_{B\infty}} \times 100 \quad (7.4.11)$$

7.4.2 Calculation of Mass Loss at Different Phases and W_B

The mass of the sample at different stages of the decomposition of hydrated cement paste during the TGA test is specified in Table 7.2. In addition to the mass of the sample at 105 °C, initial and final mass of the dehydroxylation phase (between 420 and 500 °C) and 1000 °C, the mass of the sample corresponds to 140 °C is also reported in Table 7.2 to estimate the degree of hydration using Pane and Hansen (2005) method.

The recorded mass at a specified temperature was used to calculate the mass loss in dehydration, dehydroxylation, and decarbonation phases and reported in Table 7.3. The correction in the decarbonation phase due to the carbonated anhydrous cement (Ldc_a) was used to calculate W_B in Pane and Hansen (2005) and Monteagudo et al. (2014) methods. The Ldc_a value was recorded as 1.055% for the anhydrous cement. Subsequently, W_B was calculated for each sample using the expressions suggested by Bhatty (1986), Pane and Hansen (2005), and Monteagudo et al. (2014) and represented in Table 7.3. The W_B content calculated by Pane and Hansen (2005) method exhibited larger variation in comparison to Bhatty (1986) and Monteagudo et al. (2014) for different samples of a particular type of concrete after a certain curing period. This variance in the W_B content of the several replications using Pane and Hansen (2005) approach was brought on by the absence of conversion factor 0.41 for the decarbonation component. The W_B content increases as the curing age of the sample increases. After 7 days of curing, it was observed that RAC mixes had greater W_B content than NAC mixes for a certain design mix method, while after 28 days of curing, NAC mixtures had slightly higher W_B content than RAC mixes. In contrast to RAC samples, increased W_B concentration was found in NAC samples after 90 days of curing. The higher W_B content at the early stage (7 days) signifies the formation of a higher amount of hydration products. During the mixing of RAC, the attached mortar present in RCA gets disintegrated. The unhydrated cementitious material increases the overall cementitious material content, which in turn raises the early hydration product content of RAC. Contrarily, because of the increased surface area of RAC mixes brought about by the disintegrated unhydrated cementitious material and fine aggregate, the effective w/c of RAC mixes drops below 0.45. The heat of hydration of cement paste increases as the w/c increases, according to a comparison study for w/c of 0.35 to 0.45 (Pane and Hansen 2005; Lura et al. 2017). The increase of hydration products at lower w/c is slowed down by the availability of less capillary space, according to Pane and Hansen (2005). This could be the reason why hydration products don't grow as well in RAC, which lowers the W_B content at a later stage (28 and 90 days of curing).

7.4.3 Estimation of $W_{B\infty}$

The calculated W_B content from Pane and Hansen (2005) and Monteagudo et al. (2014) methods was employed to estimate the $W_{B\infty}$ for each type of concrete. In

Table 7.2 Summary of TGA test results

Types of concrete	Curing age (Days)	Sample number	$W_{105\,°C}$ (μg)	$W_{140\,°C}$ (μg)	$W_{400\,°C}$ (μg)	$W_{520\,°C}$ (μg)	$W_{1000\,°C}$ (μg)
NAC IS	7	1	10076.059	9942.293	9493.637	9385.004	8804.532
		2	11503.441	11339.184	10796.957	10660.621	10019.078
		3	13758.598	13587.957	12937.777	12821.051	11912.078
	28	1	12262.965	12087.848	11366.371	11280.184	10382.674
		2	12538.078	12337.563	11745.563	11612.852	10549.457
		3	12605.269	12423.676	11677.379	11576.285	10611.614
	90	1	10867.001	10693.586	9934.063	9864.827	8638.685
		2	9737.731	9576.866	8850.263	8776.079	7743.006
		3	10270.589	10146.887	9510.621	9473.348	8008.677
NAC PPM	7	1	12431.477	12254.188	11757.438	11663.617	10805.313
		2	10803.102	10656.445	10164.289	10095.938	9467.547
		3	10746.844	10603.844	10136.438	10050.391	9372.477
	28	1	10519.359	10375.453	9771.766	9719.078	9030.273
		2	12264.438	12064.258	11498.508	11345.305	10535.557
		3	10714.672	10518.578	9932.867	9798.570	9311.711
	90	1	11962.695	11810.000	11122.227	11052.734	9574.172
		2	11549.516	11371.211	10693.508	10587.578	9397.569
		3	11831.585	11680.000	10991.527	10917.574	9456.328
RAC IS	7	1	11966.555	11787.125	11283.023	11165.484	10207.692
		2	10793.648	10672.758	10176.179	10083.625	9225.564
		3	12100.563	11893.750	11332.781	11219.789	10435.212
	28	1	10763.695	10639.414	10115.281	10008.664	9106.087
		2	12198.320	12155.664	11335.594	11240.383	10481.962
		3	12223.719	12076.109	11335.313	11213.289	10528.777
	90	1	15309.566	15133.777	14262.496	14142.089	12477.591
		2	10606.589	10400.805	9677.627	9601.809	8826.999
		3	10283.094	10182.461	9660.352	9600.4375	8175.665
RAC PPM	7	1	11304.043	11115.254	10606.113	10524.230	9915.390
		2	11775.406	11593.063	11089.68	11019.555	10224.652
		3	13902.035	13728.019	13079.269	13007.777	12115.794
	28	1	11370.043	11186.184	10673.145	10529.863	9772.804
		2	10040.429	9938.500	9446.367	9416.016	8450.339
		3	11143.086	10956.617	10430.313	10286.453	9621.531
	90	1	10828.328	10689.156	9968.898	9864.891	9133.349
		2	12195.641	11998.359	11179.734	11047.063	10365.252
		3	14056.754	13882.559	13117.113	12983.566	11591.556

Table 7.3 Weight losses (%) and chemically bound water at different phases of TGA test

Types of concrete	Curing age (Days)	Sample number	Ldh (%)	Ldx (%)	Ldc (%)	$W_{B,Bhatty}$	$W_{B,Pane}$	$W_{B,Monteagudo}$
NAC IS	7	1	5.780	1.078	5.761	9.220	10.297	8.788
		2	6.142	1.185	5.571	9.611	10.483	9.178
		3	5.966	0.848	6.607	9.523	11.185	9.090
	28	1	7.311	0.703	7.319	11.015	12.935	10.582
		2	6.321	1.059	8.481	10.857	13.283	10.424
		3	7.361	0.802	7.653	11.301	13.407	10.868
	90	1	8.585	0.637	11.283	13.848	17.968	13.416
		2	9.114	0.762	10.609	14.225	17.903	13.793
		3	7.399	0.363	14.261	13.609	19.839	13.177
NAC PPM	7	1	5.422	0.755	6.904	9.008	11.222	8.575
		2	5.913	0.633	5.817	8.931	10.013	8.498
		3	5.680	0.801	6.308	9.067	10.462	8.634
	28	1	7.107	0.501	6.548	10.292	11.812	9.859
		2	6.245	1.249	6.602	10.201	11.486	9.769
		3	7.297	1.253	4.544	10.413	10.311	9.980
	90	1	7.026	0.581	12.359	12.674	17.709	12.242
		2	7.412	0.917	10.304	12.553	16.126	12.121
		3	7.100	0.625	12.350	12.789	17.815	12.356
RAC IS	7	1	5.712	0.982	8.004	9.976	12.208	9.543
		2	5.721	0.858	7.949	9.838	12.405	9.405
		3	6.345	0.934	6.484	9.937	11.079	9.505
	28	1	6.024	0.991	8.385	10.453	13.247	10.020
		2	7.073	0.781	6.217	10.402	12.689	9.969
		3	7.268	0.998	5.599	10.562	11.678	10.129
	90	1	6.839	0.787	10.872	12.083	16.361	11.651
		2	8.758	0.715	7.305	12.468	13.918	12.036
		3	6.056	0.583	13.856	12.319	18.511	11.887
RAC PPM	7	1	6.174	0.724	5.386	9.107	9.636	8.674
		2	5.823	0.596	6.751	9.187	10.633	8.754
		3	5.918	0.514	6.416	9.063	10.601	8.631
	28	1	6.129	1.260	6.658	10.119	11.449	9.687
		2	5.917	0.302	9.618	10.162	13.817	9.729
		3	6.397	1.291	5.967	10.134	11.007	9.702
	90	1	7.937	0.961	6.756	11.667	13.484	11.235
		2	8.330	1.088	5.591	11.710	12.362	11.278
		3	6.685	0.950	9.903	11.695	15.312	11.262

this case, the W_B values of each sample at various curing times were operated for a specific type of concrete mix. Plotting the computed W_B values for a specific type of concrete mix and fitting the data points using the expressions from both Pane and

Table 7.4 Calculated parameters of different methods

Types of concrete	Pane and Hansen (2005)			Monteagudo et al. (2014)		Pradhan et al. (2020b)		
	$W_{B\infty}$	τ	a	$W_{B\infty}$	K	$W_{B\infty}$	τ	a
NAC IS	0.2372	93.93	0.3721	0.1325	89.10	0.2384	163.50	0.1975
NAC PPM	0.2341	109.80	0.3393	0.1200	75.45	0.2337	196.90	0.1677
RAC IS	0.2282	34.86	0.2627	0.1147	39.55	0.2187	43.67	0.1155
RAC PPM	0.2244	42.93	0.1792	0.1107	50.12	0.2162	90.33	0.1283

Hansen (2005) (three-parameter equation) and Monteagudo et al. (2014) (Michaelis–Menten-type equation) were conducted. Consequently, $W_{B\infty}$ values were determined for Pane and Hansen (2005) method, Monteagudo et al. (2014) method, and Pradhan et al. (2020b) method and represented in Table 7.4. The $W_{B\infty}$ values obtained in Monteagudo et al. (2014) method are significantly low with respect to that of Pane and Hansen (2005) method and Pradhan et al. (2020b) method. A similar observation was also reported by Monteagudo et al. (2014).

The estimated $W_{B\infty}$ values of RAC mixes are lower than that of NAC mixes in each of the discussed method (Table 7.4). The earlier studies (Pane and Hansen 2005; Monteagudo et al. 2014; Deboucha et al. 2017; Zeng et al. 2012) observed lower $W_{B\infty}$ values for the blended cement. The parent concrete of the RCA was prepared by mixing pozzolanic material with the OPC, as indicated by this. During the mixing, the attached mortar of RCA containing unhydrated cement and pozzolanic material disintegrated. The RCA's unhydrated cementitious material participates in the hydration process, and the potential inclusion of pozzolanic material resulted in lower $W_{B\infty}$ values for RAC mixtures. Furthermore, the $W_{B\infty}$ of PPM mixes was lower than conventional mixes for both NAC and RAC. Furthermore, for both NAC and RAC, the $W_{B\infty}$ of PPM mixes was lower than that of traditional mixes. The lower $W_{B\infty}$ values of PPM mixes may be attributed to their higher fine aggregate content. The larger fine aggregate concentration in PPM mixes increases surface area, which lowers the interface w/c. Due to the constrained amount of capillary space, Pane and Hansen (2005) found that w/c 0.35 had a lower $W_{B\infty}$ than w/c 0.45. As a result, the increase in surface area in PPM mixes results in a reduction in the effective w/c at the interface and lower $W_{B\infty}$ yields.

7.4.4 *Degree of Hydration (α)*

The computed $W_{B\infty}$ values were utilized to determine the degree of hydration of various mixes at various curing periods using various approaches and are displayed

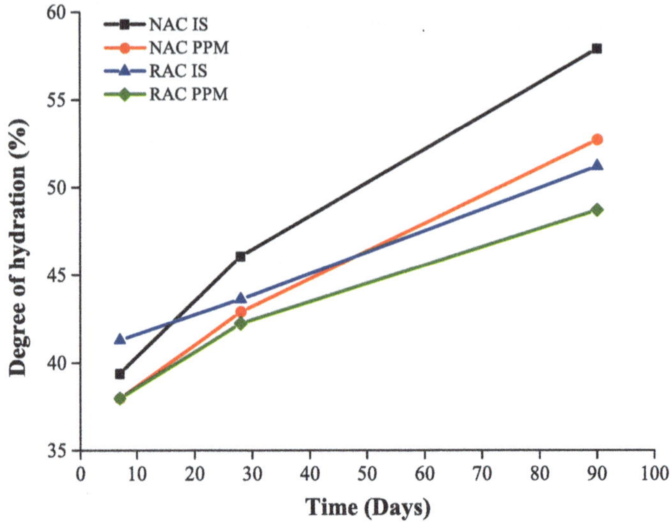

Fig. 7.3 Calculated degree of hydration using Bhatty's method (Pradhan et al. 2020b)

in Figs. 7.3, 7.4, 7.5, and 7.6. Due to the assumption of a larger $W_{B\infty}$ value (0.24), the α values produced using Bhatty (1986) technique were lower than those acquired using the other stated methods. Additionally, the lower $W_{B\infty}$ values achieved using Monteagudo et al. (2014) technique resulted in samples that were more than 100% hydrated after 90 days of curing. It's interesting to note that using Monteagudo et al. (2014) method, the α values of RAC mixes were found to be greater than those of NAC mixes. For both NAC and RAC, the α values of PPM mix-designed concrete were lower than those of conventional mix-designed concrete. This could be caused by the same factor that led to the lower $W_{B\infty}$ values seen in concrete made with a PPM mix.

7.5 Investigation on the Presence of Pozzolanic Material in RCA

7.5.1 CH Bound Water and Free CH Content

In this section, it is explored whether pozzolanic material is present in RCA and how it affects the hydration of RAC. For both NAC and RAC mixes, only OPC was used by Pradhan et al. (2017). The existence of pozzolanic material in RCA, which controls the secondary hydration process in RAC, is shown by the fact that RAC mixtures have a reduced $W_{B\infty}$ content. The established expressions were used to determine the CH bound water and free CH content in this context. The CH bound

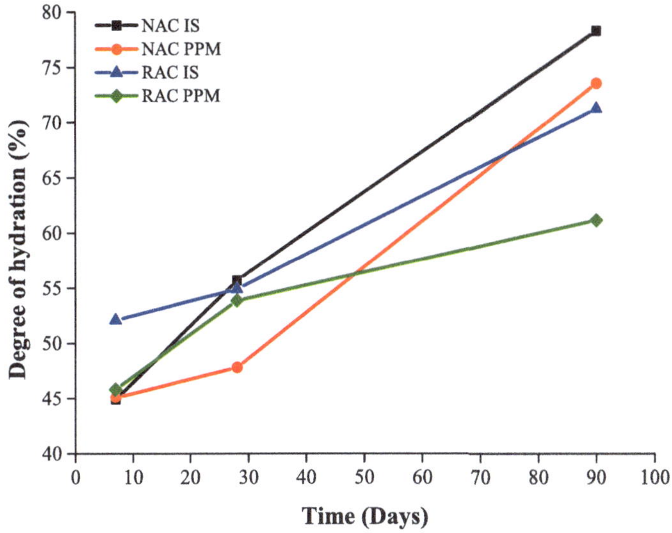

Fig. 7.4 Calculated degree of hydration using Pane and Hansen method (Pradhan et al. 2020b)

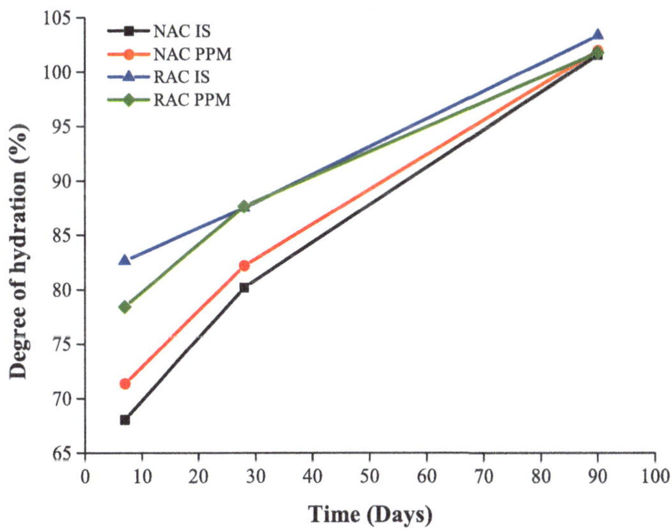

Fig. 7.5 Calculated degree of hydration using Monteagudo et al. method (Pradhan et al. 2020b)

water was calculated by using Eq. 7.5.1 and represented in Table 7.5. For each type of concrete, the CH bound water content rises with the length of the curing process. Apart from the silica fume blended OPC mix, a rising trend of CH bound water for both blended cement (OPC + ground granulated blast furnace slag and OPC + fly ash) as well as OPC alone was seen by Monteagudo et al. (2014). This shows

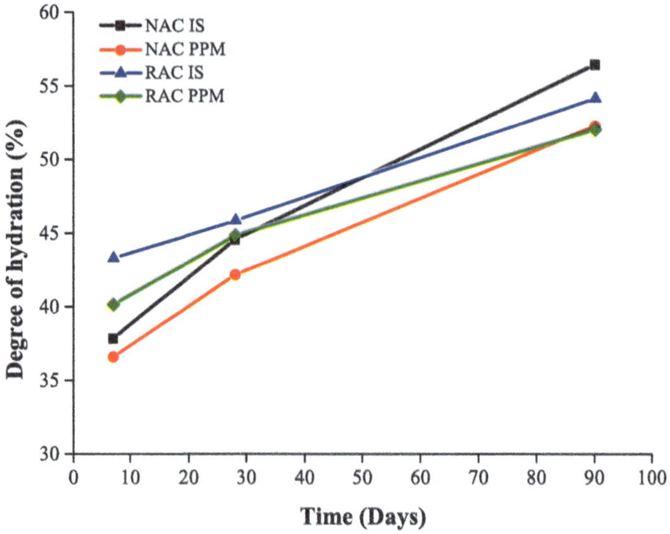

Fig. 7.6 Calculated degree of hydration using present method (Pradhan et al. 2020b)

that RCA does not contain silica fume. However, based on the estimated CH bound water content, the existence of other pozzolanic elements such as fly ash and ground granulated blast furnace slag (GGBS) cannot be established. For this, the free CH content was estimated by using Eq. 7.5.2 (El-Jazairi and Illston 1980) as well as the modified expression suggested by Monteagudo et al. (2014) (Eq. 7.5.3).

$$\text{CH bound water} = Ldx + 0.41(Ldc - Ldc_a) \tag{7.5.1}$$

$$\text{Free CH} = 4.11Ldx + 1.68Ldc \tag{7.5.2}$$

$$\text{CH}_{\text{Monteagudo}} = 4.11Ldx + 1.68(Ldc - Ldc_a) \tag{7.5.3}$$

The free CH content estimated from Eqs. 7.5.2 and 7.5.3 is shown in Table 7.5. The predicted free CH content of NAC mixes is found to be larger than RAC mixes for a particular type of mix design method (IS code method or PPM approach). Monteagudo et al. (2014) reported a lower free CH content for GGBS and fly ash blended cement with respect to OPC. As a result, the similar association between the free CH content of RAC mixes and the corresponding NAC mixes supports the hypothesis that RCA contains pozzolanic material (GGBS, fly ash, or both). Additionally, the RAC mixes had a lower age-related rise in free CH concentration. This suggests the existence of very little pozzolanic material, which is from the RCA mortar that has decomposed.

Table 7.5 Total equivalent CH bound water and free CH bound water

Types of concrete	Curing age (Days)	CH bound water	Free CH	Free $CH_{Monteagudo}$
NAC IS	7	3.056	14.308	12.536
	28	3.627	16.645	14.873
	90	5.096	22.659	20.887
NAC PPM	7	2.897	13.654	11.882
	28	2.987	14.023	12.251
	90	5.060	22.516	20.744
RAC IS	7	3.558	16.365	14.592
	28	3.252	15.107	13.335
	90	4.640	20.793	19.021
RAC PPM	7	2.714	12.902	11.130
	28	3.559	16.366	14.593
	90	3.608	16.567	14.795

7.5.2 *Fourier Transform Infrared Spectroscopy Analysis*

Fourier transform infrared spectroscopy (FT-IR) was used to assess the presence of unhydrated cement. For the investigation, hydrated cement paste from conventional concrete made with OPC and cement paste from the connected mortar of RCA were both employed. For the purpose of creating pellets, the collected samples were combined with KBr (which serves as a beam splitter) at a weight ratio of $1/100$. The prepared pellets were scanned using the spectrometer (Nicolet 6700 FT-IR spectrometer) in the mid-infrared spectra (wavelength ranges from 400 to 4000 cm^{-1}).

The FT-IR spectra for the aforementioned samples are represented in Fig. 7.7 and following major peaks are discussed. The peak corresponds to the wavelength 3640 cm^{-1} is related to the stretching of $O-H$ bond of $CaOH_2$ (Mendes et al. 2011; Pan et al. 2015; Bhat and Debnath 2011; Peyvandi et al. 2014). For the sample collected from RCA the peak at 3640 cm^{-1} was absent (Fig. 7.7). This is possibly due to the transformation of $CaOH_2$ to $Ca(CO)_3$ because of the exposure of RCA to atmospheric CO_2. The bands correspond to the peaks approximately at 2983, 2875, 2510, and 1795 cm^{-1} are due to the stretching of $C-O$ bond of CO_3^{-2} (Mendes et al. 2011; Ylmén et al. 2009; Delgado et al. 1996; Nasrazadani et al. 2016). The peaks observed at wavelength 1425, 874, and 712 cm^{-1} are, respectively, due to the asymmetric stretching, out-of-plane bending, and doubly degenerate planar bending of CO_3^{-2} (Pan et al. 2015; Peyvandi et al. 2014; Nasrazadani et al. 2016). The broad peak between 2800 and 3700 cm^{-1} is due to the stretching vibration of $O-H$ bond of H_2O, which is strongly associated with cementitious material by hydrogen bond (Pan et al. 2015; Bhat and Debnath 2011; Choudhary et al. 2015). Moreover, the peak corresponds to 1640 cm^{-1} of conventional concrete sample is due to the bending vibration of $H-O-H$ bond of H_2O molecule (Mendes et al. 2011; Pan et al. 2015;

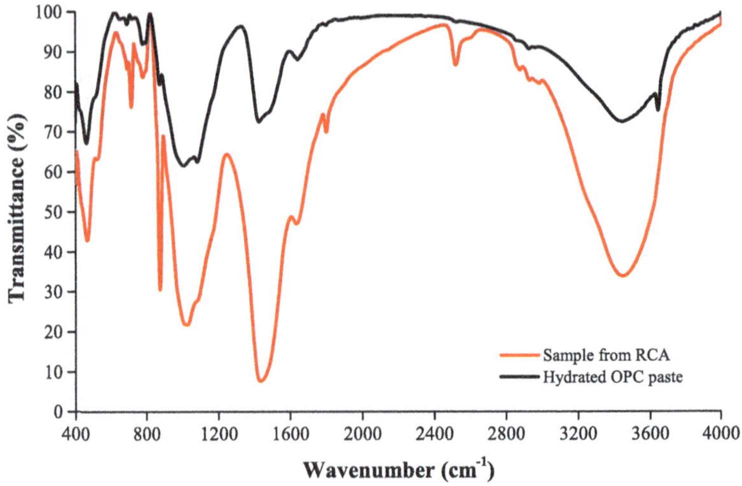

Fig. 7.7 FT-IR spectra of samples collected from RCA and NAC (Pradhan et al. 2020b)

Nasrazadani et al. 2016). In addition to the band associated with water between 1500 and 1700 cm^{-1}, the presence of a hump between 970 and 1100 cm^{-1} is the indicative of the presence of CSH (Peyvandi et al. 2014; Ylmén et al. 2009).

The components of hydrated cement paste are linked to the bands or peaks of FT-IR spectra outlined above. However, in both the samples, the peak observed at wavelength 525 cm^{-1} is associated with the presence of unhydrated cement. The peak corresponds to 525 cm^{-1} is resulted due to the out-of-plane bending vibration of Si – O bond of silicate (Bhat and Debnath 2011; Govin et al. 2006). This confirms the presence of unhydrated cement in RCA. Interestingly, a significant peak was observed at wavelength 1630 cm^{-1} along with the peaks at 1035 and 3450 cm^{-1} for the sample collected from RCA. It is worth mentioning that, in FT-IR spectra, the peaks correspond to 1035, 1630, and 3450 cm^{-1} are observed in fly ash (Zhang et al. 2012; Chindaprasirt et al. 2009; Rożek et al. 2018). The peak at 1035 cm^{-1} is due to the asymmetric stretching vibration of Al – O or Si – O bonds, whereas the peaks at 1630 cm^{-1} 3450 cm^{-1} are due to the stretching and deformation vibration of OH and H – O – H groups. As a result, the existence of fly ash in the adhering mortar of RCA is supported, which is attributable to the potential usage of fly ash in the preparation of parent concrete.

7.6 Relationship Between Degree of Hydration and Compressive Strength

The compressive strength of different types of concrete recorded by Pradhan et al. (2017) is represented in Fig. 7.8. Irrespective of the method, the estimated α value increases with the increase in curing age. This is similar to the usual trend of com-

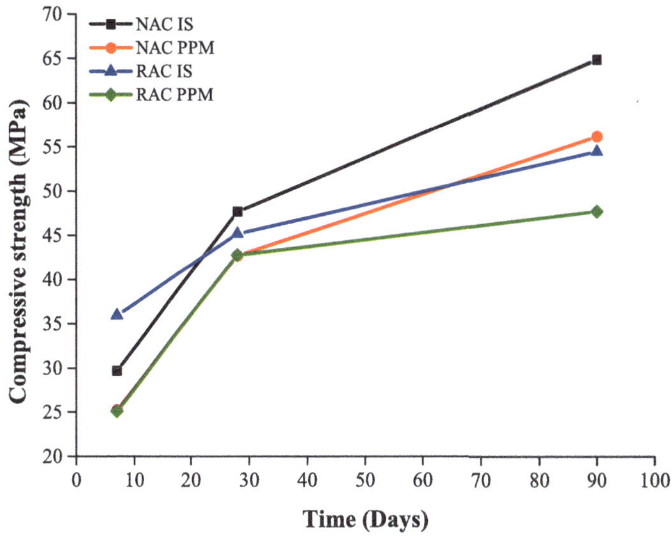

Fig. 7.8 Compressive strength at different curing period (Pradhan et al. 2020b)

pressive strength of concrete with age. The relationship between α and compressive strength was further analysed in detail. The trend of compressive strength of concrete (Fig. 7.8) is very similar to the relationship of α with curing age apart from the relationship between α and curing age obtained from Monteagudo et al. (2014) method. The degree of hydration estimated by Monteagudo et al. (2014) method was observed to be higher for RAC mixes with respect to the NAC mixes (Fig. 7.5), which is contrary to the recorded compressive strength results. The α of RAC IS mix was observed to be significantly higher than the other concrete mixes after 7 days of curing, which is also reflected in the compressive strength behaviour. For 28-day cured mixes, the conventional mix-designed concrete exhibited higher compressive strength than PPM mix-designed concrete (Fig. 7.8). The α values obtained in Bhatty (1986) method (Fig. 7.3) and Pane and Hansen (2005) method (Fig. 7.4) and Pradhan et al. (2020b) (Fig. 7.6) also exhibited similar relationship for 28-day cured mixes. Furthermore, higher compressive strength values were witnessed for NAC mixes in comparison to the RAC mixes after 90 days of curing. This observation is correlated with the α values estimated by using the Bhatty (1986) method (Fig. 7.3) and Pane and Hansen (2005) method (Fig. 7.4). In Pradhan et al. (2020b) method, the α value of 90-day cured NAC IS mix was observed to be significantly higher, whereas for RAC IS mix the α value was marginally lower than the NAC PPM mix, which is contrary to compressive strength results after 90 days of curing.

The relationship between the normalized degree of hydration and normalized compressive strength (both are normalized with respect to the values of 90-day cured sample) is shown in Fig. 7.9. Similar to the earlier studies Pane and Hansen (2005), Monteagudo et al. (2014), Deboucha et al. (2017), the study by Pradhan et al. (2020b)

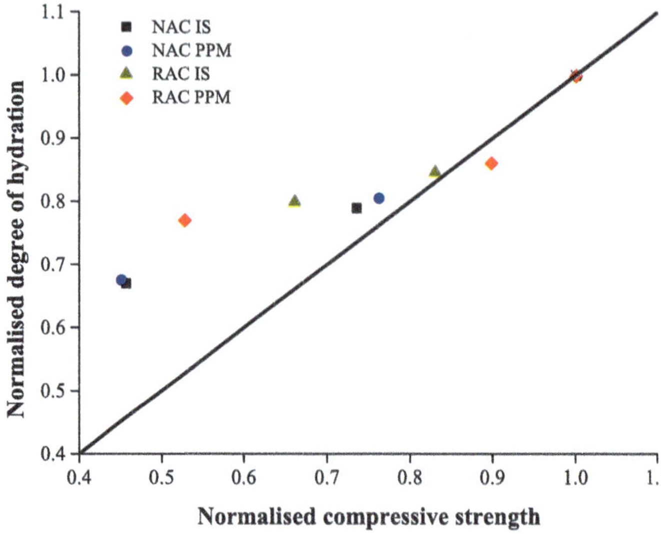

Fig. 7.9 Relationship between degree of hydration and compressive strength (Pradhan et al. 2020b)

also observed a direct relationship between α and compressive strength at different curing age. However, this was reflected only for the same concrete mix. A consistent correlation between compressive strength and α of various concrete mixtures was not found. Therefore, the degree of cement hydration cannot be the only factor affecting the mechanical properties of concrete. The ITZ characteristics, interface voids and their distribution, aggregate quality, and degree of hydration are further factors that affect the mechanical properties of concrete. After 7 days and 28 days of curing, it was found that the normalized degree of hydration of RAC mixes was higher than the NAC mixes. This suggests that, up to 28 days of curing, RAC mixes showed higher degrees of hydration than the NAC mixes, which may be attributable to the involvement of RCA's pozzolanic substance in the hydration process.

7.7 Nanoindentation

An interface between the mortar matrix and aggregate surface is known as the interfacial transition zone (ITZ). The characteristics and thickness of ITZ are dependent on factors such as the kind of aggregate, aggregate size, cement particle grain size distribution, w/c, mixing method, and curing time. With the aid of nanoindentation, the study by Pradhan et al. (2020a) examined the effects of aggregate type (NCA and RCA), mix design method (IS: 10262 (2009) method and PPM) and mixing approach (normal mixing and TSMA) on the properties (elastic modulus and hardness), thickness, and ITZ properties (Fig. 7.10).

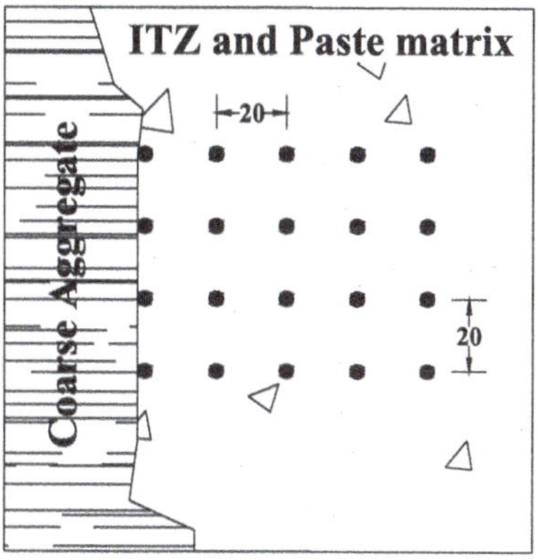

Fig. 7.10 Indentation matrix across ITZ and paste matrix (Pradhan et al. 2020a)

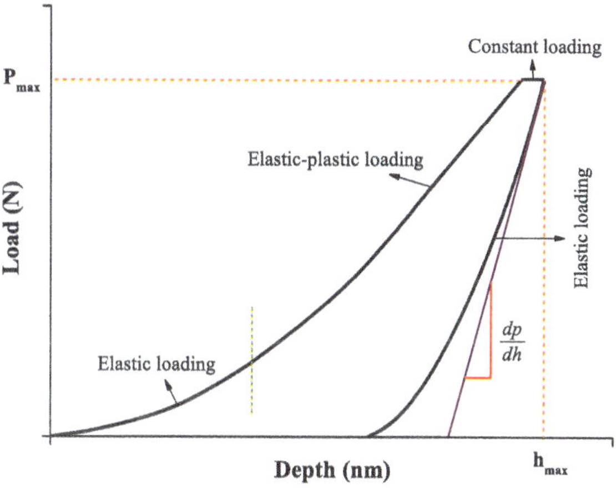

Fig. 7.11 Typical load–depth curve of nanoindentation test (Pradhan et al. 2020a)

The Hysitron TI 950 triboindenter was used to conduct the nanoindentation test on samples that had been prepared after 28 and 90 days of cure. The indenter has a Berkovitch tip with a radius of 100 nm and an angle of 142.3°. Each test was programmed for a trapezoidal load function in a single cycle loading at a constant rate of 120 μN/s up to a maximum load of 1200 μN. When the load was at its maximum, the indenter had a dwell time of 2 seconds before being withdrawn at a constant rate of 120 μN/s until the load decreased to zero. The indent area for each sample was 80 μm × 60 μm across the ITZ. Grids of 5 × 4 with 20 μm spacing in both the X- and Y-directions were used to assure the statistical significance of the indents. Two specimens were evaluated for each type of concrete and each curing interval.

A typical indentation load and indentation depth curve generated from the trapezoidal loading function is shown in Fig. 7.11. The parameters recorded in the test were peak indentation load (P_{max}), indentation depth corresponds to peak load (h_{max}), and final indentation depth after unloading (h_f). By applying a continuum scale model, the initial unloading stiffness (S) can be represented by Eq. 7.7.1 (Oliver and Pharr 1992).

$$S = \frac{dp}{dh} = \frac{2}{\sqrt{\pi}} E_r \sqrt{A} \tag{7.7.1}$$

where A is the projected contact area at the peak load and E_r is the reduced elastic modulus, which can be determined from Eq. 7.7.2.

$$\frac{1}{E_r} = \frac{(1 - \nu^2)}{E} + \frac{(1 - \nu_i^2)}{E_i} \tag{7.7.2}$$

where E and ν represent the modulus of elasticity and Poisson's ratio of the sample, respectively, and E_i and ν_i are the modulus of elasticity and Poisson's ratio of the indenter. $E_i = 1140$ GPa and $\nu_i = 0.07$ for the indenter material used by Pradhan et al. (2020a). From Eq. 7.7.3, the E of the material can be determined. In this context, the ν of calcium silicate hydrate (CSH), calcium hydroxide (CH), and unhydrated cement particle were considered to be 0.24 (Constantinides and Ulm 2004; Huang et al. 2013; Mondal et al. 2007; Bernard et al. 2003), 0.31 (Constantinides and Ulm 2004; Huang et al. 2013; Bernard et al. 2003), and 0.3 (Huang et al. 2013; Bernard et al. 2003; Sanahuja et al. 2007; Stefan et al. 2010), respectively. The hardness (H) of the material was measured by using Eq. 7.7.4.

$$E = (1 - \nu^2) \times \left[\frac{1}{E_r} - \frac{(1 - \nu_i^2)}{E_i} \right]^{-1} \tag{7.7.3}$$

$$H = \frac{P_{max}}{A} \tag{7.7.4}$$

7.7.1 Results and Discussion

The nanoindentation was conducted using the prepared samples, which were sourced from the 28- and 90-day cured specimens. The area in the vicinity of the coarse aggregate was chosen for the indentation. The contour map of indentation modulus and hardness of the 28- and 90-day cured samples are shown in Figs. 7.12, 7.13, 7.14, and 7.15.

The regions of the contour maps with lower and higher indentation modulus and hardness are depicted, respectively, by the dark blue and yellow hues. CH is harder and has a greater indentation modulus than CSH (Bernard et al. 2003; Constantinides and Ulm 2004; Mondal et al. 2008; Xiao et al. 2013). Moreover, unhydrated cement is having higher indentation modulus and hardness than CH (Bernard et al. 2003; Constantinides and Ulm 2004). Figures 7.12 and 7.13, respectively, depict the indentation modulus and hardness of the 28-day cured sample of the four types of

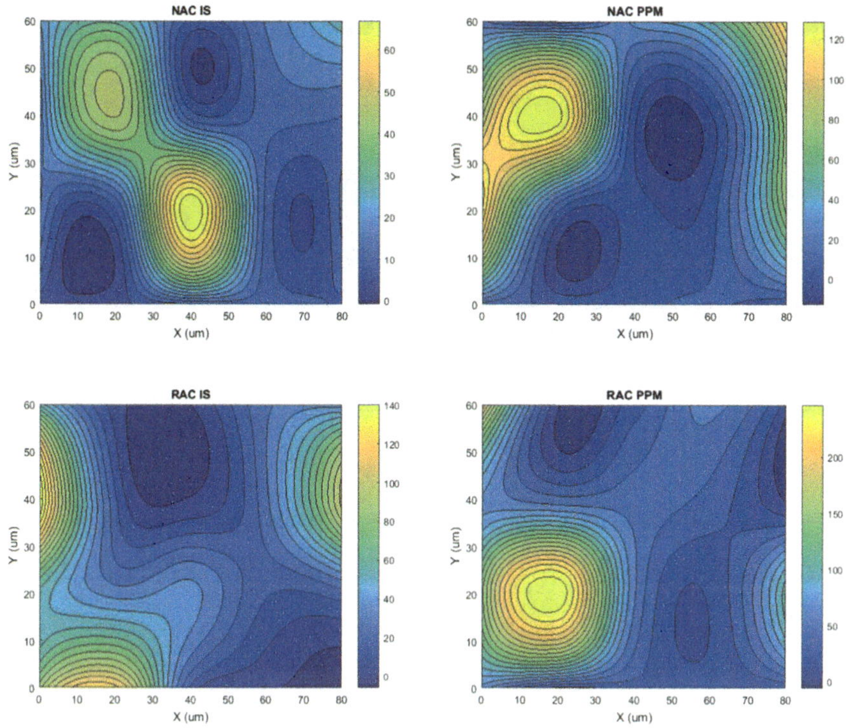

Fig. 7.12 Contour map of indentation modulus (GPa) of 28-day cured samples (Pradhan et al. 2020a)

concrete prepared by PPM and IS methods. In the contour map, voids are represented by the dark blue colour, while CH and unhydrated cement particles are represented by the yellow tint. Both Figs. 7.12 and 7.13 manifest the substantial presence of dark blue and yellow colour regions within 50 to 70 μm from the aggregate surface. This implies that, because to the wall effect and micro-bleeding, voids and CH are localized more towards the aggregate surface (Ollivier et al. 1995; Scrivener et al. 2004; Xiao et al. 2013; Gao et al. 2018). Similarly, for the samples after 90 days of curing period, the distribution of voids and CH is also concentrated within 50–70 μm from the aggregate surface (Figs. 7.14 and 7.15).

The wall effect and micro-bleeding, which result in a higher variance in indentation modulus and hardness, cause the distribution of voids and CH to be more concentrated in the vicinity of the aggregate surface (Ollivier et al. 1995; Scrivener et al. 2004; Xiao et al. 2013; Gao et al. 2018). Due to the decrease in voids and CH content relative to the area close to the aggregate surface, relatively little modification in indentation modulus and hardness was seen as the ITZ approached the paste matrix (Ollivier et al. 1995; Scrivener et al. 2004; Xiao et al. 2013; Gao et al. 2018). It is possible to calculate the average thickness of ITZ using this phenomena (Xiao et al.

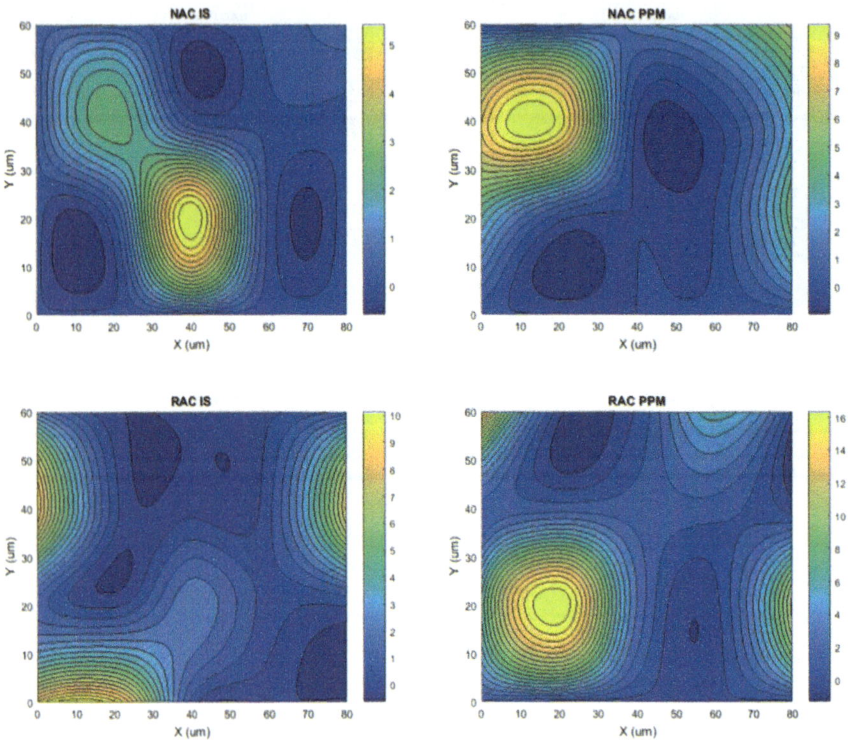

Fig. 7.13 Contour map of hardness (GPa) of 28-day cured samples (Pradhan et al. 2020a)

2013). In this context, average and the variation in indentation modulus and hardness at different distance from the aggregate surface of the tested samples are shown in Figs. 7.16, 7.17, 7.18, and 7.19.

In earlier research, the typical indentation modulus of paste matrix was found to range between 15 and 30 GPa (Xiao et al. 2013; Li et al. 2012). For NAC specimens (NAC IS and NAC PPM), the indentation modulus after 40 μm distance from the aggregate face was observed to be 15–25 GPa (Figs. 7.16 and 7.18). Moreover, the variation in indentation modulus and hardness of NAC specimens in the lateral direction generally reduces after 40 μm distance from the aggregate boundary (Figs. 7.16, 7.17, 7.18, and 7.19). This indicates that the ITZ thickness for NAC specimens is between 40 and 60 μm, which is comparable to the reported ITZ thickness of 20–50 μm in earlier investigations (Scrivener et al. 1988, 2004; Elsharief et al. 2003; Xiao et al. 2013). Furthermore, for RAC specimens also the variation in indentation modulus and hardness values was less significant after 40 μm distance from the aggregate boundary. However, the indentation modulus about 15 to 30 GPa was observed close to 60 μm (between 60 and 80 μm) distance from the aggregate face (Figs. 7.16 and 7.18). Xiao et al. (2013) also observed higher ITZ thickness (55–65 μm) for RAC specimens than the NAC specimens (40–50 μm). According to Pradhan et al. (2020a)

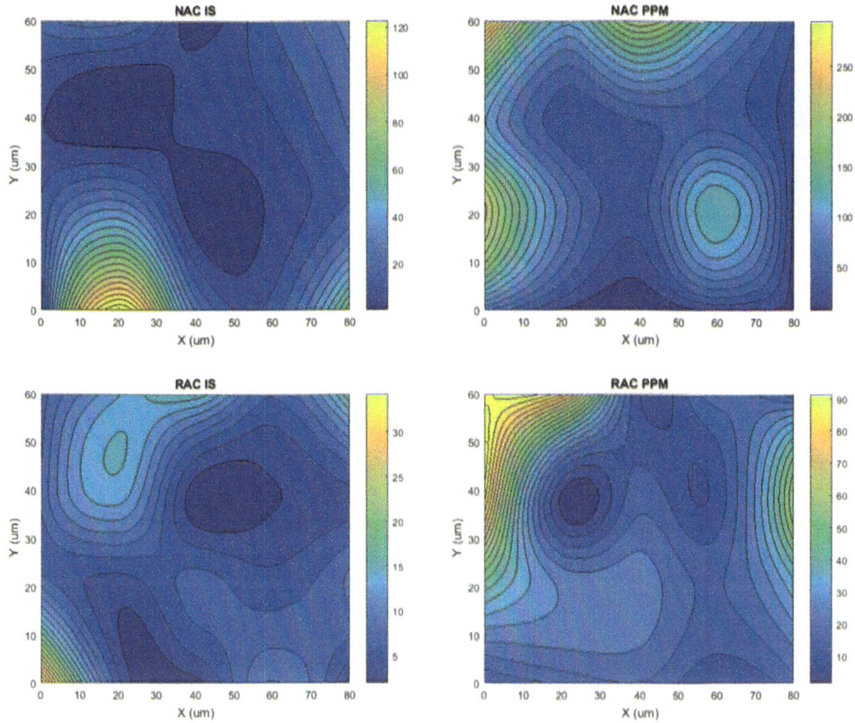

Fig. 7.14 Contour map of indentation modulus (GPa) of 90-day cured samples (Pradhan et al. 2020a)

research (contour maps and distribution of indentation modulus and hardness in the lateral direction of aggregate boundaries), the thickness of RAC specimens is roughly 60–70 μm, which is in line with Xiao et al. (2013)'s findings.

The mix design method's effect on ITZ thickness was marginal and could not be distinguished (Pradhan et al. 2020a). Indentation modulus and hardness of the PPM mix-designed specimens, however, were higher than those of the mix-designed specimens made using IS: 10262 (2009) method. When compared to specimens that had only been cured for 28 days, the paste matrix's indentation modulus for specimens that had been cured for 90 days was marginally higher. Xiao et al. (2013) also showed a similar effect of curing age, which is because hydrate yield increases with age. The ITZ of RAC IS for both 28- and 90-day cured specimens showed the lowest average indentation modulus and hardness. However, the ITZ and paste matrix hardness of the RAC PPM specimens were almost similar to those of the NAC IS and NAC PPM specimens. In addition, NAC PPM specimens had a higher indentation modulus and harder surface than NAC IS specimens for both the curing ages. This implies that the TSMA and PPM mix design method help to improve the ITZ and paste matrix's micro-mechanical properties in both RAC and NAC.

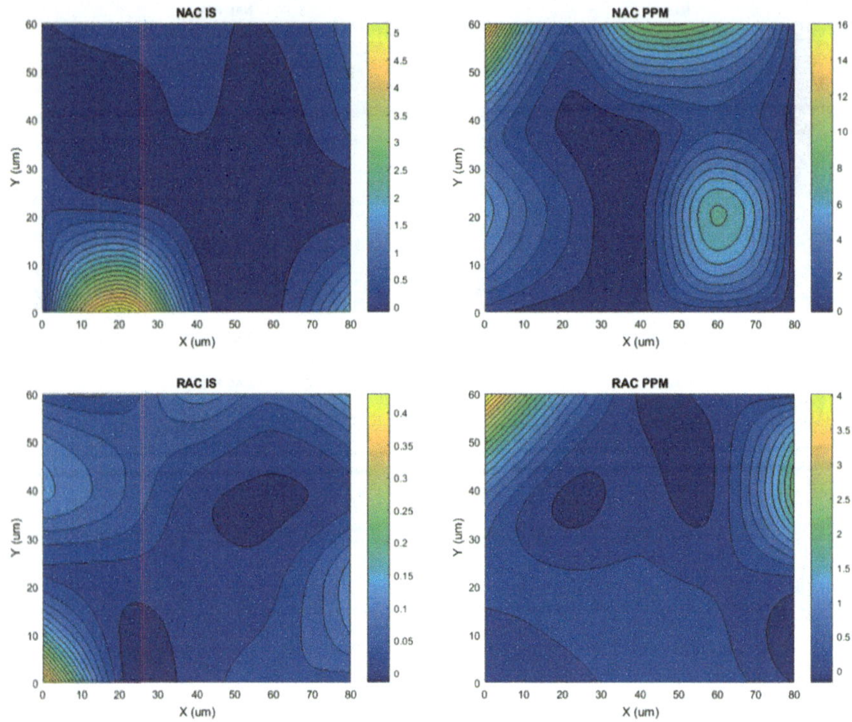

Fig. 7.15 Contour map of hardness (GPa) of 90-day cured samples (Pradhan et al. 2020a)

7.8 Image Analysis of BSE Images

The concrete's ITZ is its most vulnerable area. However, the type of aggregate, w/c, cement content, cementitious material content, mixing method, mix design process, and curing period all affect its thickness as well as other characteristic features (voids content and quantity of different hydration products). The aforementioned parameters control the wall effect and micro-bleeding phenomena noticed near aggregates, which results in the development of ITZ. Therefore, a quantitative investigation of ITZ's constituent parts is necessary to comprehend how it affects the macro-mechanical characteristics of concrete. The BSE image of ITZ can be analysed in this context utilizing a scanning electron microscope.

Pradhan et al. (2020a) used the samples made after 28 and 90 days of cure to take BSE images of the polished surface. With the focus on the area around the coarse aggregate, all of the pictures were retrieved using a MERLIN (ZEISS) field emission scanning electron microscope (FE-SEM). Each image occupied part of coarse aggregate, ITZ, and paste matrix. A typical image is presented in Fig. 7.20. The images captured at an accelerating voltage of 15 kV with 1000× magnification and each pixel is approximately 0.27 μm in both the directions.

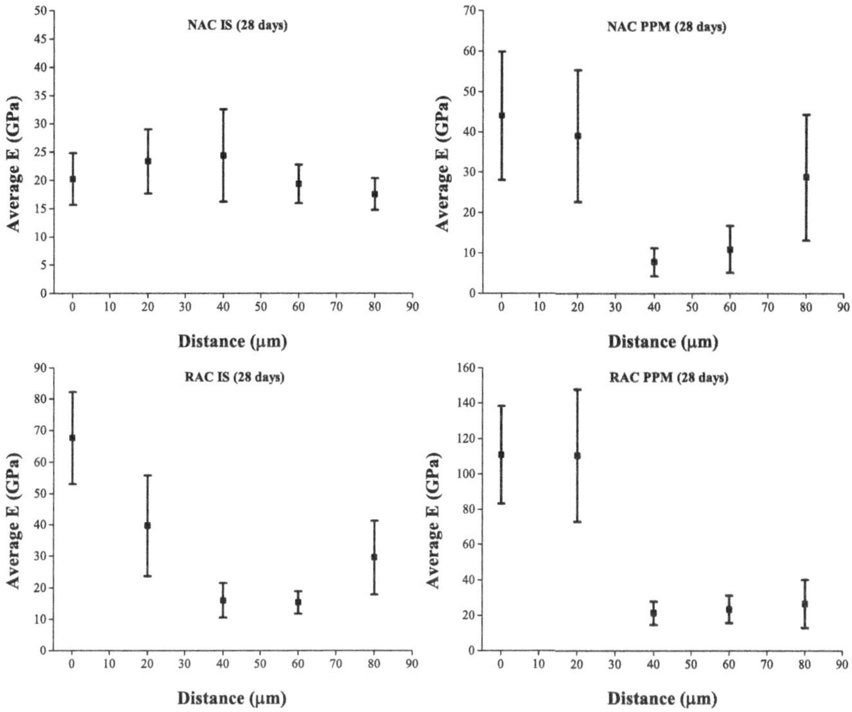

Fig. 7.16 Distribution of indentation modulus across ITZ after 28 days of curing (Pradhan et al. 2020a)

By using an image processing technique, it was possible to quantify the variation in the ITZ components' content (voids and unhydrated cement) in relation to the aggregate's boundary. In this instance, ten strips of 10 μm width beginning at the aggregate boundary were extracted using segmentation from the BSE image and are displayed in Fig. 7.21. The change in voids and the amount of unhydrated cement up to 100 μm from the aggregate boundary were therefore examined. The 8-bit grayscale photos were analysed using a Python script. The grey value range for an 8-bit picture is 0–255 (2^8 tonnes), which depicts various shades of grey from black to white. For example, a grey value of 0 indicates the colour of pure black (tone of grey), while a grey value of 255 represents the colour of pure white. The BSE pictures were used to identify the lowest and highest grey values for each sample to represent the voids and unhydrated cement, respectively. The lowest and highest grey values for voids and unhydrated cement were then obtained, and they were applied to a series of grayscale strips to estimate their quantity in each strip.

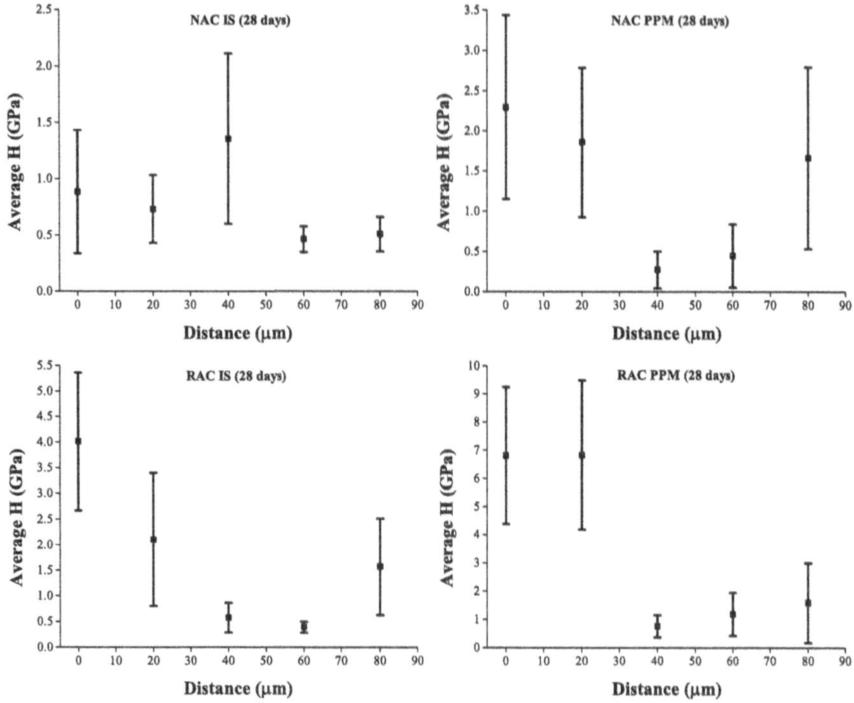

Fig. 7.17 Distribution of hardness across ITZ after 28 days of curing (Pradhan et al. 2020a)

7.8.1 Results and Discussion

The voids and unhydrated cement contents (in terms of % area) in the vicinity of coarse aggregate were estimated for the four types of concrete cured for 28 and 90 days and represented in Figs. 7.22 and 7.23, respectively. As a result, it is possible to determine how aggregate type, mixing technique, mix design technique, and curing age affect ITZ properties. The trend of the voids content (Fig. 7.22) and unhydrated cement content (Fig. 7.23) in the lateral direction from the aggregate boundary is in decreasing and increasing order, respectively, which is similar to the earlier studies (Diamond 2001; Scrivener 2004; Chakradhara Rao 2010; Mukharjee 2014). Additionally, it was shown that each type of concrete has less voids and unhydrated cement as the curing period lengthens (Figs. 7.22 and 7.23), which was consistent with previously published results (Scrivener et al. 1988, 2004; Chakradhara Rao 2010).

The voids content of NAC IS specimens was estimated to be lowest for the considered distance and curing period. This is followed by the voids content of NAC PPM, RAC IS, and RAC PPM specimens. Although RAC PPM specimens were found to have the highest void content, the difference between RAC IS and RAC PPM specimens was less significant. The void content of the four varieties of concrete showed

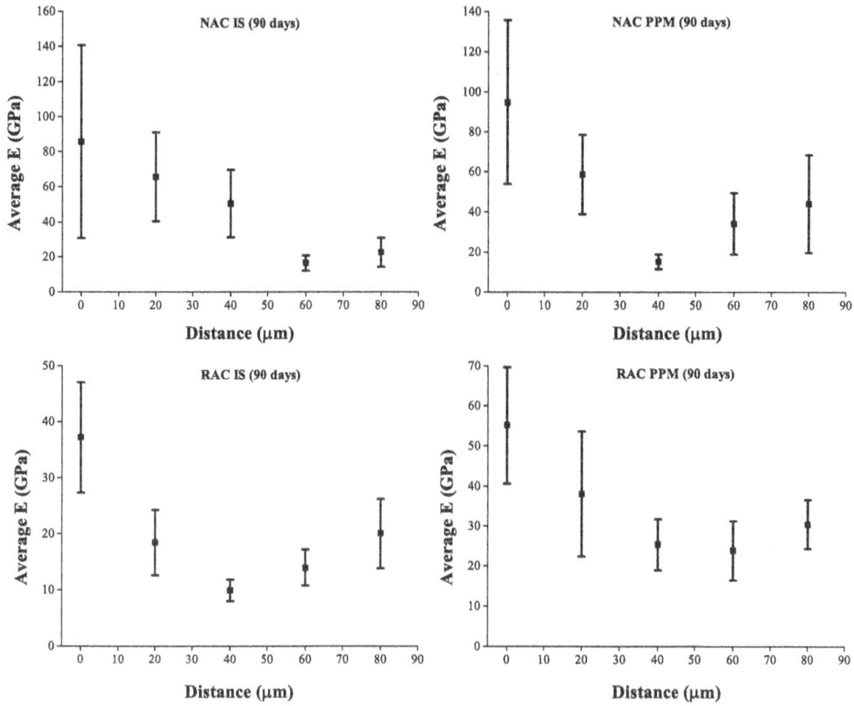

Fig. 7.18 Distribution of indentation modulus across ITZ after 90 days of curing (Pradhan et al. 2020a)

a slight variance across a distance of around 70 μm. NAC and RAC have ITZ thicknesses of roughly 20–50 and 55–65 μm, respectively, as was previously mentioned. This shows that, compared to NAC specimens, RAC specimens had marginally higher void content in the paste matrix.

The gradient of voids content in the lateral direction can be used to estimate the approximate thickness of ITZ (Fig. 7.22). The reduction in voids content in the NAC specimen strips (NAC IS and NAC PPM) was more noticeable up to 40–50 μm, beyond which just a slight drop was seen. This suggests that the ITZ thickness for the NAC specimens is around between 40 and 50 μm, which is almost identical to the outcomes obtained using the nanoindentation approach. In a similar manner, the drop in void content for RAC specimens up to 60–70 μm from the aggregate boundary was quite significant in comparison to beyond this lateral distance from the aggregate margin. Therefore, it can be determined that the ITZ thickness for RAC specimens is somewhere between 60 and 70 μm, which is also consistent with the outcomes of nanoindentation.

As can be seen from Fig. 7.23, the wall effect and micro-bleeding phenomena lead to decreased unhydrated cement concentration close to the aggregate boundary as compared to the paste matrix. The unhydrated cement content near the aggregate border in the 90-day cured specimen of NAC PPM and RAC PPM specimens was

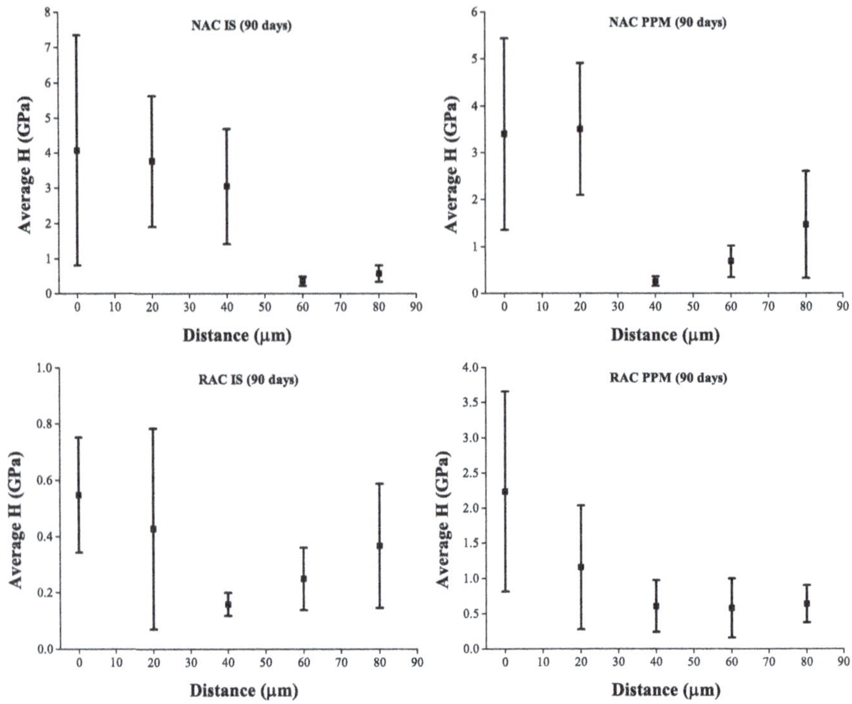

Fig. 7.19 Distribution of hardness across ITZ after 90 days of curing (Pradhan et al. 2020a)

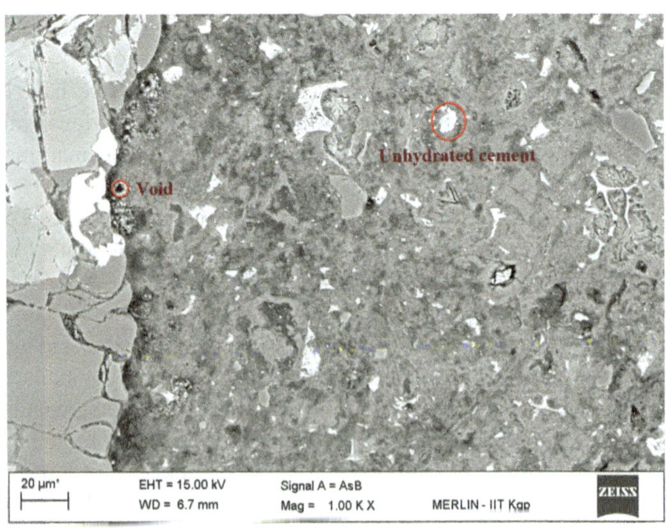

Fig. 7.20 A representative BSE image (Pradhan et al. 2020a)

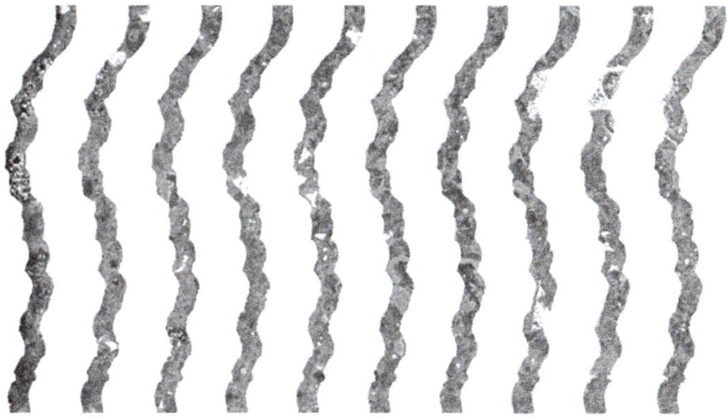

Fig. 7.21 Segmented strips of width 10 µm (Pradhan et al. 2020a)

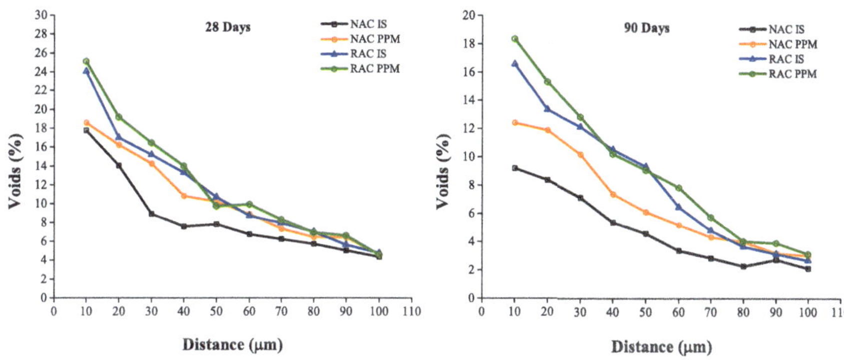

Fig. 7.22 Distribution of voids (%) from the aggregate boundary (Pradhan et al. 2020a)

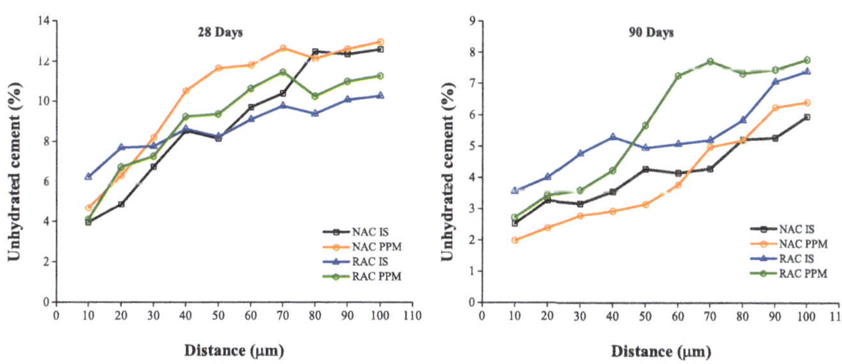

Fig. 7.23 Distribution of unhydrated cement (%) from the aggregate boundary (Pradhan et al. 2020a)

less than that of the corresponding mix-designed concrete according to IS: 10262 (2009), as shown in Fig. 7.23. This shows that, when compared to concrete made using the IS: 10262 (2009) method, the packing of cement particles is less disrupted in the PPM mix due to the wall effect. However, the TSMA may be responsible for this characteristic of concrete made using PPM mix.

7.9 X-Ray Microtomography

The approaches to examine concrete's surface features at the micro-level, particularly of the ITZ, include nanoindentation and image analysis of BSE images. For a thorough understanding of the behaviour of concrete, the influence of meso-level and volumetric properties can be investigated. X-ray microtomography (XRT), a sophisticated non-destructive imaging technology, is used in this situation to generate extremely high-resolution cross-sectional images and integrate those images to rebuild the specimen's three-dimensional (3D) structure. The impact of interface voids on the macro-mechanical characteristics of concrete was studied by Pradhan et al. (2020a). The effect of aggregate type, mix design approach, and curing time on the amount of voids in the paste matrix of concrete was investigated. The scanned specimens' two-dimensional (2D) 8-bit grayscale images were examined using image processing technique; the specific process is covered in the following subsections.

7.9.1 Methodology of Image Analysis of XRT Images

Pre-processing of the raw images is the initial step before beginning any analysis on a digital image. Very little pre-processing is necessary in order to preserve the greatest amount of information in the images. Because of its ability to maintain edges, the median filtering was chosen by Pradhan et al. (2020a). The median of the adjacent grey values was used to replace each grey value using this technique. In order to do this, a square kernel of size 21×21 was used to apply the median filtering approach to the original image Fig. 7.24a. Figure 7.24b represents the image after removal of noise using median filtering. Even after filtering, there were a lot of non-uniformity at the edges. The outer edge of the image must be uniform in order to estimate voids accurately. To do this, a circle with a radius that was slightly less than the specimen radius was cropped of the filtered image, and the remaining pixels were substituted with dummy grey value (255 in the current study) so that they could still be distinguished from the actual specimen image. Figure 7.24c represents the image with white background (grey value of 255) and uniform circular edge.

The grey values in a pixel region of a grayscale image are a representation of the X-ray absorption intensity of the sample material at that specific zone, which is a function of the region's material density. Therefore, the material density at a

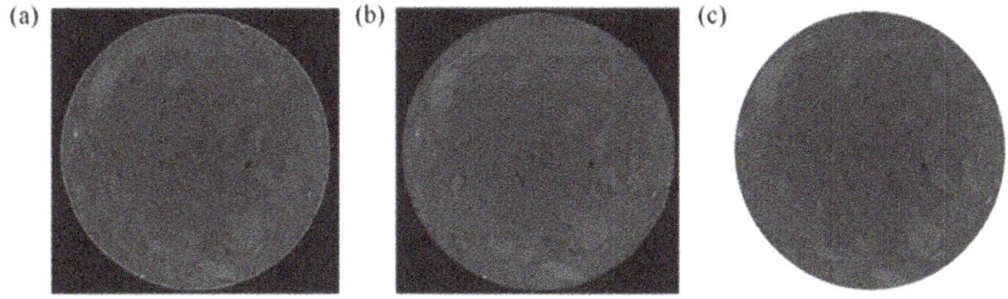

Fig. 7.24 **a** Original image; **b** Noise removal using median filtering; **c** Edge non-uniformity and background removal (Pradhan et al. 2020a)

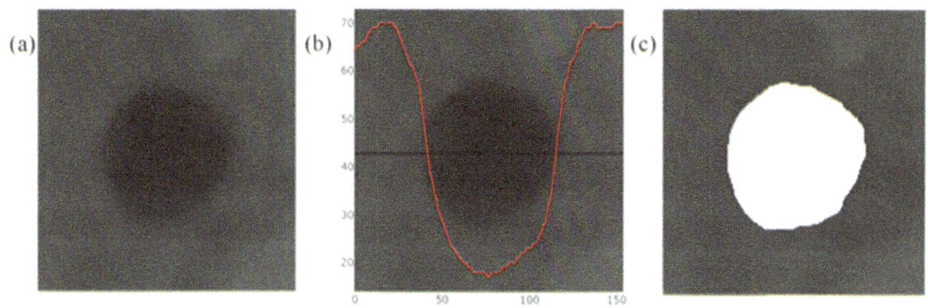

Fig. 7.25 **a** A representative void; **b** Grey value profile; **c** Detected void (Pradhan et al. 2020a)

given place in the sample increases as a pixel's grey value increases. Due to the stark discrepancy in material density between the air voids and solid phases (aggregates and mortar) of concrete, there is a notable distinction in their X-ray absorption levels and, consequently, the grey value distributions. Therefore, a straightforward thresholding technique can be used to find air spaces in the scanned image of concrete.

In this method, at first, a representative void (Fig. 7.25a) was cropped from the pre-processed image (Fig. 7.24c) and then the grey value profile of the pixels located along a horizontal line (black coloured line in Fig. 7.25b) passing through the air void was plotted. From this profile (shown in red in Fig. 7.25b), it can be observed that the grey value steadily decreases as we get closer to a void's edge from the outside, and that it always remains smaller inside the void than it is at the edges. This change in the void's grey value throughout its diameter is caused by the fact that its depth is shallower at its edge than at its centre, which in turn causes a little reduction in contrast at the void's perimeter due to various X-ray absorption levels. In order to segment the air void, the grey value at the void's edge can be chosen as the threshold grey value (Fig. 7.25c). Prior to applying it to the entire image, the accuracy at the void boundary is checked. Depending on the contrast of the image and the kind of specimen, the threshold grey value may change. Figure 7.26 shows the flowchart for the entire process of locating and measuring air voids.

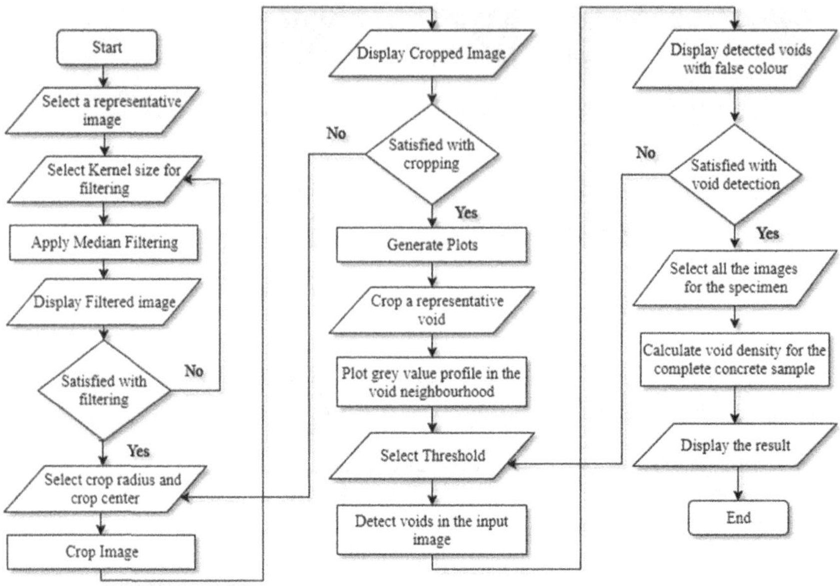

Fig. 7.26 Flowchart of the procedure involved in void detection and estimation (Pradhan et al. 2020a)

7.9.2 *Results and Discussion*

The threshold grey value of the air void was calculated using the earlier discussed method. Furthermore, the determined threshold grey value was verified on one image representing a section of concrete specimen (Fig. 7.27a) by representing the voids with false colour (Fig. 7.27b). Figure 7.27b shows that the voids in the concrete specimen can be accurately identified by the present methodology. Following the precise identification of air voids in one section, the established threshold grey value was applied to all of the XRT images that were available for the specimen to estimate the void content of the entire concrete specimen.

The estimated voids content and mean radius for the whole volume of the specimen are represented in Table 7.6. It was found that when the curing period lengthens, the volume of voids and mean radius for each type of concrete decrease. In addition, the PPM mix-designed concrete showed less void volume and mean radius than the concrete that was proportioned using IS: 10262 (2009).

Figure 7.28 shows the distribution of voids of various volumes and confirms the decrease in frequency of smaller sized voids as the cure period lengthens. For specimens that had been cured for 7 and 28 days, the smallest void size was 0.02 mm^3; however, for specimens that had been cured for 90 days, it was 0.01 mm^3.The PPM mix-designed concrete displayed more smaller voids of size 0.01–0.04 mm^3 in comparison to IS: 10262 (2009) method, but the frequency of voids of size 0.04–0.20 mm^3 was less frequent (Fig. 7.28). This phenomenon is brought on by the greater pack-

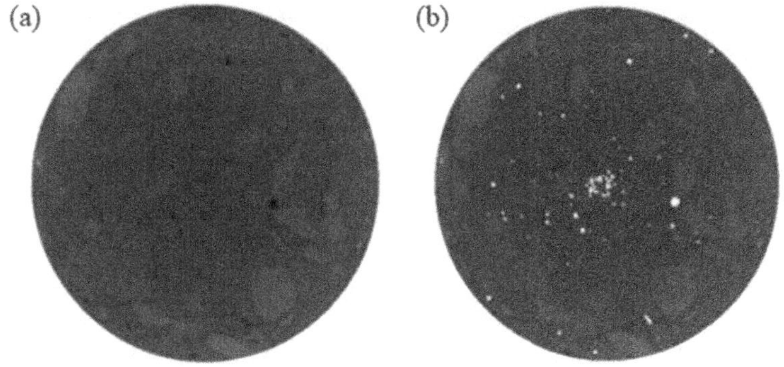

Fig. 7.27 **a** Specimen section; **b** Detected voids represented with false colour (Pradhan et al. 2020a)

Table 7.6 Volume and mean radius of voids for different types of concrete

Type of concrete	Volume of voids (%), **(Mean radius (mm))**		
	7 days	28 days	90 days
NAC IS	0.77, **(2.71)**	0.65, **(2.68)**	0.59, **(2.65)**
NAC PPM	0.76, **(2.62)**	0.62, **(2.60)**	0.52, **(2.58)**
RAC IS	0.99, **(2.93)**	0.87, **(2.89)**	0.77, **(2.79)**
RAC PPM	0.95, **(2.90)**	0.84, **(2.85)**	0.72, **(2.73)**

ing density observed in concrete made using PPM mix. The voids between 0.01 and 0.04 mm in diameter were also less frequent in the RAC specimens than in the NAC specimens, while the voids between 0.04 and 0.20 mm in diameter were more frequent in the RAC specimens (Fig. 7.28).

7.10 Relationship Between Micro-structural and Macro-mechanical Properties

Concrete is made up of a dispersed pore system, which is a heterogeneous and anisotropic mixture of particles (aggregates, cement hydration products, and unhydrated cement). Concrete pores are a result of chemical shrinkage, water evaporation, and insufficient compaction. The variables that control the pore system and its quantity in concrete include the w/c, water absorption of aggregate, and compaction effort. Given that w/c primarily controls the capillary and ITZ porosity, a number of relationships between strength and w/c of concrete are evidenced (Kumar and Bhattacharjee 2003). Along with the pore system, the degree of cement hydration also affects strength metrics and is influenced by curing age and environmental exposure circumstances. Additionally, the aggregate characteristics have a signifi-

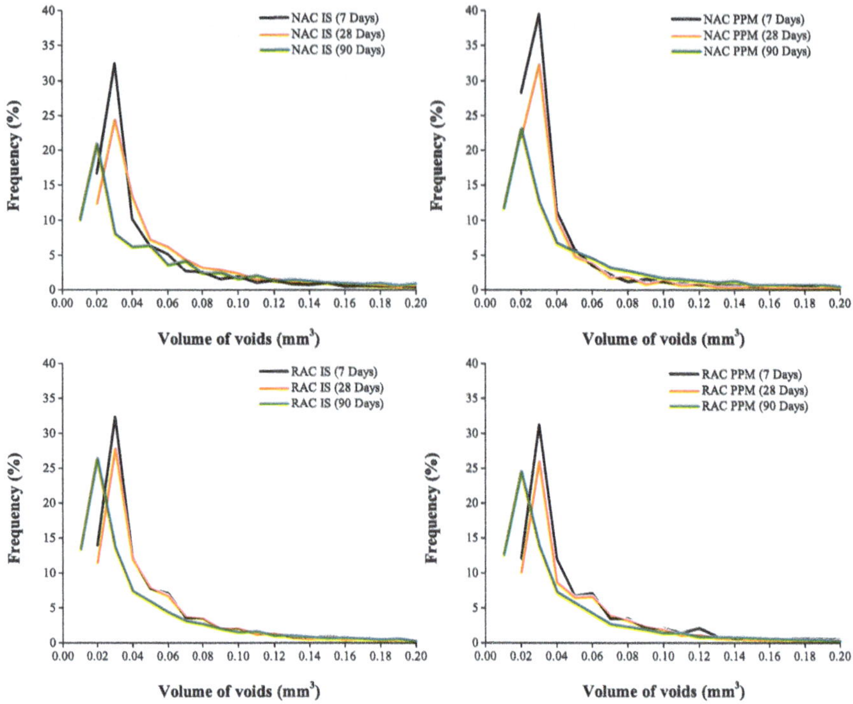

Fig. 7.28 Distribution of voids of different volume (Pradhan et al. 2020a)

cant role in affecting the macro-mechanical characteristics since they have an impact on the micro-mechanical properties (particularly those of ITZ). In this context, the relationship between the crushing value of the aggregate and the macro-mechanical parameters (compressive strength, tensile strength, and static modulus of elasticity) of concrete was explored by Zhang et al. (2017). Additionally, past studies (Mukharjee 2014; Mukharjee and Barai 2014) examined the impact of the void content in the ITZ on macro-mechanical characteristics. The formulas those are currently used to forecast the compressive strength of concrete based on its microstructural features are represented in Table 7.7.

The ability of the formulas in Table 7.7 to forecast compressive strength was examined, and the generated values were compared with the actual data obtained at various curing times by Pradhan et al. (2020a) and shown in Fig. 7.29. It can be seen that the formulas proposed by Kumar and Bhattacharjee (2003) and Kondraivendhan and Bhattacharjee (2010) considerably underestimate the compressive strength of RAC (Fig. 7.29). The aforementioned formulas, however, more accurately forecast the compressive strength of NAC. Furthermore, the expression provided by Mukharjee (2014) overestimates the compressive strength of both NAC and RAC for both a 28-day and a 90-day curing period. In this situation, a sophisticated expression is

Table 7.7 Expressions to predict compressive strength from porosity

Authors	Expressions
Balshin (1949)	$\sigma = \sigma_0 (1 - p)^b$
Ryshkewitch (1953)	$\sigma = \sigma_0 e^{-kp}$
Hasselman (1969)	$\sigma = \sigma_0 - cp$
Schiller (1971)	$\sigma = q \ln\left(\frac{p_0}{p}\right)$
Atzeni et al. (1987)	$\sigma = K \frac{\sigma_0(1-p)}{\sqrt{r_m}}$
Kumar and Bhattacharjee (2003)	$\sigma = KC\frac{(1-p)}{\sqrt{r_m}}$
Kondraivendhan and Bhattacharjee (2010)	$\sigma = KC\alpha\frac{(1-p)}{\sqrt{r_m}}$
Chen et al. (2013)	$\sigma = \sigma_0\left[\left(\frac{p_c-p}{p_c}\right)^{1.85}\left(1-p^{2/3}\right)\right]^{1/2}$
Mukharjee (2014)	$\sigma = 0.2181p_i^2 - 11.091p_i + 176.04$

where σ = compressive strength of concrete; σ_0 = compressive strength at zero porosity; p = porosity of the specimen (%); b, k, c, q, and K = empirical constants; C = fraction of cement content in the concrete mixture; α = degree of hydration; p_c = critical porosity corresponds to the percolation limit of the solid phase; and p_i = porosity of the ITZ (%)

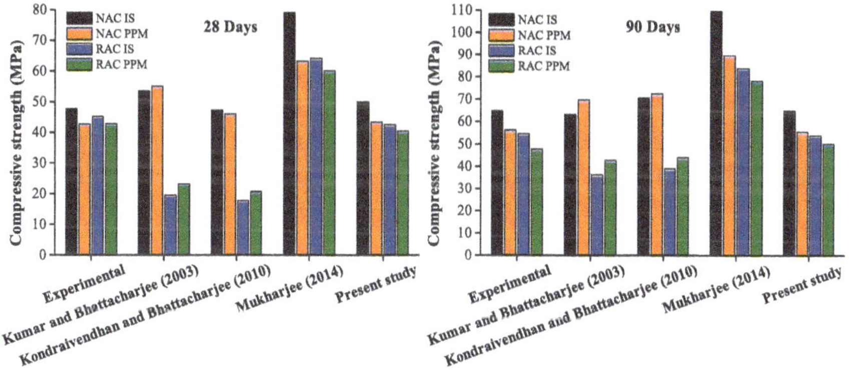

Fig. 7.29 Comparison of experimental and predicted compressive strength (Pradhan et al. 2020a)

required, one that takes into account the aggregate's microstructural qualities as well as its volume fraction and mechanical properties.

The expressions in Table 7.7 take into account the specimen's porosity, mean radius of the pores, fraction of cement in the concrete mix, degree of cement hydration, critical porosity (which corresponds to the percolation limit of the solid), and porosity of the ITZ parameters to determine the compressive strength of concrete. Prior research, however, did not account for both the specimen's and the ITZ's porosity combined. Additionally, the prior expressions did not take into account the percentage of coarse aggregate included in the concrete mixture or the coarse aggregate's crushing value. According to the research by Zhang et al. (2017), notably for RAC due to the adherent loose mortar layer to RCA, aggregate crushing value has a sig-

Table 7.8 Tests or methods performed to determine the parameters related to the proposed expression for compressive strength

Parameters	Test or Method
p	X-ray microtomography
r_m	X-ray microtomography
α	Thermogravimetric analysis
p_i	Image analysis of BSE images
t_i	Nanoindentation and image analysis of BSE images
C	Mix proportion of concrete
f_{ca}	Mix proportion of concrete
A_{cr}	Aggregate crushing test

nificant impact on the macro-mechanical properties of concrete. Additionally, as the ITZ is the weakest link of concrete, it is anticipated that the ITZ thickness and average modulus of elasticity will affect the macro-mechanical performance of concrete. Because different types of aggregate have varying physical and mechanical qualities, its thickness and micro-mechanical characteristics also depend on the kind of aggregate, cement content, w/c, mixing technique, and mix design procedure. In order to predict the compressive strength of the concrete, Pradhan et al. (2020a) took into account the porosity of the specimen (p), mean radius of the pores (r_m), fraction of cement content in the concrete mixture (C), degree of hydration of cement (α), porosity of ITZ parameters (p_i), coarse aggregate fraction in the concrete mixture (f_{ca}), aggregate crushing value (A_{cr}), and thickness of the ITZ (t_i) and represented in Eq. 7.10.1. As a result, the suggested expression was an improvement of the equation offered by Kondraivendhan and Bhattacharjee (2010) (Table 7.7) and shown in Eq. 7.10.2.

$$\sigma = f(p, r_m, C, \alpha, f_{ca}, A_{cr}, t_i, p_i) \tag{7.10.1}$$

$$\sigma = K\,(C\alpha)^x \left[\frac{(1-p)}{\sqrt{r_m}}\right]^y \left(\frac{f_{ca}}{A_{cr}}\right)^z \left(\frac{1}{p_i t_i}\right)^m \tag{7.10.2}$$

where K, x, y, z, and m are the constants. The iterative process is used to find the values of these constants, and as a result, the experimental data gathered were put to use by using various techniques to examine the microstructural properties. In this context, the performed experiments and adopted methods to determine the related parameters represented in Eq. 7.10.1 are shown in Table 7.8.

The final expression for the correlation between concrete's microstructural properties and compressive strength is given in Eq. 7.10.3, which was obtained by replacing the constants in Eq. 7.10.2. With an estimated R^2 value of 0.94, it was found that the proposed expression (Eq. 7.10.3) had a good correlation with the experimental findings. Its accuracy in predicting the compressive strength of both NAC and RAC can also be seen in Fig. 7.29, which exhibited a maximum error of 5.81%.

$$\sigma = 6.9\,(C\alpha)^{0.99}\left[\frac{(1-p)}{\sqrt{r_m}}\right]^{0.1}\left(\frac{f_{ca}}{A_{cr}}\right)^{0.25}\left(\frac{1}{p_i t_i}\right)^{0.025} \tag{7.10.3}$$

7.11 Closure

The properties of four different types of concrete mixes prepared by Pradhan et al. (2017) were explored at the micro- and meso-levels in the current chapter. In this context, TGA, nanoindentation, image analysis of BSE images, and X-ray tomography images methods were subjected. Following observations and propositions are the essence of this investigation:

- A method for estimating α from the observed chemically bound water content in the TGA test of hydrated cement paste was proposed by merging two earlier methods.
- Regardless of the type of aggregate, it was projected that PPM mix planned concrete had lower α values than conventional mix-designed concrete. For a certain type of concrete, the compressive strength parameter was found to be linearly related to α at various curing ages, however this connection was not supported for concrete with alternative aggregate types and mix design approaches.
- The distribution of different hydration components and voids in ITZ as well as the paste matrix may be visualized using the nanoindentation technique. Further, the variation in elastic modulus in the lateral direction from the aggregate boundary was used to calculate the thickness of the ITZ.
- Using the BSE images, the quantity of voids and unhydrated cement in the ITZ and paste matrix were measured. Additionally, the ITZ's thickness was calculated by examining the variation in void content across the ITZ in the lateral direction from the aggregate boundary.
- To forecast concrete's compressive strength at various curing ages, an equation was proposed. In this case, the thickness of the ITZ parameters as well as the cement content, the degree of hydration, the meso-level porosity, the mean radius of voids, the coarse aggregate percentage, the crushing value of the coarse aggregate, and the void content were taken into consideration.

References

Atzeni C, Massidda L, Sanna U (1987) Effect of pore size distribution on strength of hardened cement pastes. In: Proceedings of the first international RILEM congress on pore structure and material properties, pp 195–202

Balshin MY (1949) Relation of mechanical properties of powder metals and their porosity and the ultimate properties of porous-metal ceramic materials. Doklady Akademii Nauk SSSR 67(4):831–834

Bernard O, Ulm FJ, Lemarchand E (2003) A multiscale micromechanics-hydration model for the early-age elastic properties of cement-based materials. Cem Concr Res 33(9):1293–1309
Bhat PA, Debnath NC (2011) Theoretical and experimental study of structures and properties of cement paste: the nanostructural aspects of C – S – H. J Phys Chem Solids 72(8):920–933
Bhatty JI (1986) Hydration versus strength in a portland cement developed from domestic mineral wastes—a comparative study. Thermochimica Acta 106:93–103
Bhatty JI (1991) A review of the application of thermal analysis to cement-admixture systems. Thermochimica Acta 189(2):313–350
Chakradhara Rao M (2010) Characterisation and behaviour of recycled aggregate concrete. PhD thesis, IIT Kharagpur
Chen X, Wu S, Zhou J (2013) Influence of porosity on compressive and tensile strength of cement mortar. Constr Build Mater 40:869–874
Chindaprasirt P, Jaturapitakkul C, Chalee W, Rattanasak U (2009) Comparative study on the characteristics of fly ash and bottom ash geopolymers. Waste Manag 29:539–543
Choudhary HK, Anupama AV, Kumar R, Panzi ME, Matteppanavar S, Sherikar BN, Sahoo B (2015) Observation of phase transformations in cement during hydration. Constr Build Mater 101:122–129
Constantinides G, Ulm FJ (2004) The effect of two types of C-S-H on the elasticity of cement-based materials: results from nanoindentation and micromechanical modeling. Cem Concr Res 34(1):67–80
Copeland LE, Kantro DL, Verbeck GJ (1960) Chemistry of hydration of Portland cement. Citeseer
Deboucha W, Leklou N, Khelidj A, Oudjit MN (2017) Hydration development of mineral additives blended cement using thermogravimetric analysis (TGA): methodology of calculating the degree of hydration. Constr Build Mater 146:687–701
del Bosque IFS, Zhu W, Howind T, Matías A, de Rojas MIS, Medina C (2017) Properties of interfacial transition zones (ITZs) in concrete containing recycled mixed aggregate. Cem Concr Compos 81:25–34
Delgado AH, Paroli RM, Beaudoin JJ (1996) Comparison of IR techniques for the characterization of construction cement minerals and hydrated products. Appl Spectrosc 50(8):970–976
Diamond S (2001) Considerations in image analysis as applied to investigations of the ITZ in concrete. Cem Concr Compos 23(2–3):171–178
El-Jazairi B, Illston J (1980) The hydration of cement paste using the semi-isothermal method of derivative thermogravimetry. Cem Concr Res 10(3):361–366
Elsharief A, Cohen MD, Olek J (2003) Influence of aggregate size, water cement ratio and age on the microstructure of the interfacial transition zone. Cem Concr Res 33:1837–1849
Gao Y, Hu C, Zhang Y, Li Z, Pan J (2018) Characterisation of the interfacial transition zone in mortars by nanoindentation and scanning electron microscope. Mag Concr Res 70(18):965–972
Govin A, Peschard A, Guyonnet R (2006) Modification of cement hydration at early ages by natural and heated wood. Cem Concr Compos 28:12–20
Hasselman D (1969) Griffith flaws and the effect of porosity on tensile strength of brittle ceramics. J Am Ceram Soc 52(8):457–457
Hemalatha T, Mapa M, George N, Sasmal S (2016) Physico-chemical and mechanical characterization of high volume fly ash incorporated and engineered cement system towards developing greener cement. J Clean Prod 125:268–281
Huang J, Krabbenhoft K, Lyamin AV (2013) Statistical homogenization of elastic properties of cement paste based on X-ray microtomography images. Int J Solids Struct 50(5):699–709
IS: 10262 (2009) Concrete mix proportioning - guidelines
Kondraivendhan B, Bhattacharjee B (2010) Effect of age and water-cement ratio on size and dispersion of pores in ordinary Portland cement paste. ACI Mater J 107(2):147–154
Kong D, Lei T, Zheng J, Ma C, Jiang J, Jiang J (2010) Effect and mechanism of surface-coating pozzalanics materials around aggregate on properties and ITZ microstructure of recycled aggregate concrete. Constr Build Mater 24(5):701–708

Kumar R, Bhattacharjee B (2003) Porosity, pore size distribution and in situ strength of concrete. Cem Concr Res 33(1):155–164

Leite M, Monteiro P (2016) Microstructural analysis of recycled concrete using X-ray microtomography. Cem Concr Res 81:38–48

Li J, Xiao H, Zhou Y (2009) Influence of coating recycled aggregate surface with pozzolanic powder on properties of recycled aggregate concrete. Constr Build Mater 23(3):1287–1291

Li W, Xiao J, Sun Z, Kawashima S, Shah SP (2012) Interfacial transition zones in recycled aggregate concrete with different mixing approaches. Constr Build Mater 35:1045–1055

Lura P, Winnefeld F, Fang X (2017) A simple method for determining the total amount of physically and chemically bound water of different cements. J Thermal Anal Calorim 130(2):653–660

Mendes A, Gates WP, Sanjayan JG, Collins F (2011) NMR, XRD, IR and synchrotron NEXAFS spectroscopic studies of OPC and OPC/slag cement paste hydrates. Mater Struct 44:1773–1791

Mondal P, Shah SP, Marks L (2007) A reliable technique to determine the local mechanical properties at the nanoscale for cementitious materials. Cem Concr Res 37(10):1440–1444

Mondal P, Shah SP, Marks LD (2008) Nanoscale characterization of cementitious materials. ACI Mater J 105(2):174–179

Monteagudo SM, Moragues A, Gálvez JC, Casati MJ, Reyes E (2014) The degree of hydration assessment of blended cement pastes by differential thermal and thermogravimetric analysis. Morphological evolution of the solid phases. Thermochimica Acta 592:37–51

Mukharjee BB (2014) Behaviour of concrete incorporating recycled aggregates and nanosilica. PhD thesis, IIT Kharagpur

Mukharjee BB, Barai SV (2014) Influence of incorporation of nano-silica and recycled aggregates on compressive strength and microstructure of concrete. Constr Build Mater 71:570–578

Nasrazadani S, Eghtesad R, Sudoi E, Vupputuri S, Ramsey J, Ley M (2016) Application of Fourier transform infrared spectroscopy to study concrete degradation induced by biogenic sulfuric acid. Mater Struct 49(5):2025–2034

Oliver WC, Pharr GM (1992) An improved technique for determining hardness and elastic modulus using load and displacement sensing indentation experiments. J Mater Res 6(7):1564–1583

Ollivier JP, Maso JC, Bourdette B (1995) Interfacial transition zone in concrete. Adv Cem Based Mater 2(1):30–38

Pan Z-Y, Li G, Hong C-Y, Kuang H-L, Yu Y, Feng F-X, Liu D-M, Du H (2015) Modified recycled concrete aggregates for asphalt mixture using microbial calcite precipitation. R Soc Chem Adv 5(44):34854–34863

Pane I, Hansen W (2005) Investigation of blended cement hydration by isothermal calorimetry and thermal analysis. Cem Concr Res 35(6):1155–1164

Peyvandi A, Holmes D, Soroushian P, Balachandra AM (2014) Monitoring of sulfate attack in concrete by Al 27 and Si 29 MAS NMR spectroscopy. J Mater Civil Eng 27(8):04014226

Pradhan S, Kumar S, Barai SV (2017) Recycled aggregate concrete: particle packing method (PPM) of mix design approach. Constr Build Mater 152:269–284

Pradhan S, Kumar S, Barai SV (2020a) Multi-scale characterisation of recycled aggregate concrete and prediction of its performance. Cem Concr Compos 106:103480

Pradhan S, Kumar S, Barai SV (2020b) Understanding the behavior of recycled aggregate concrete by using thermogravimetric analysis. Front Struct Civil Eng 14(6):1561–1572

Rożek P, Król M, Mozgawa W (2018) Spectroscopic studies of fly ash-based geopolymers. Spectrochim Acta Part A: Mol Biomol Spectrosc 198:283–289

Ryshkewitch E (1953) Compression strength of porous sintered alumina and zirconia: 9th communication to ceramography. J Am Ceram Soc 36(2):65–68

Sanahuja J, Dormieux L, Chanvillard G (2007) Modelling elasticity of a hydrating cement paste. Cem Concr Res 37(10):1427–1439

Schiller KK (1971) Strength of porous materials. Cem Concr Res 1(4):419–422

Scrivener KL (2004) Backscattered electron imaging of cementitious microstructures: understanding and quantification. Cem Concr Compos 26(8):935–945

Scrivener KL, Bentur A, Pratt PL (1988) Quantitative characterization of the transition zone in high strength concretes. Adv Cem Res 1(4)

Scrivener KL, Crumbie AK, Laugesen P (2004) The interfacial transition zone (ITZ) between cement paste and aggregate in concrete. Interface Sci 12(4):411–421

Scrivener KL, Lothenbach B, De Belie N, Gruyaert E, Skibsted J, Snellings R, Vollpracht A (2015) TC 238-SCM: hydration and microstructure of concrete with SCMs: state of the art on methods to determine degree of reaction of SCMs. Mater Struct 48(4):835–862

Sidorova A, Vazquez-Ramonich E, Barra-Bizinotto M, Roa-Rovira JJ, Jimenez-Pique E (2014) Study of the recycled aggregates nature's influence on the aggregate-cement paste interface and ITZ. Constr Build Mater 68:677–684

Stefan L, Benboudjema F, Torrenti JM, Bissonnette B (2010) Prediction of elastic properties of cement pastes at early ages. Comput Mater Sci 47(3):775–784

Tam VWY, Gao XF, Tam CM (2005) Microstructural analysis of recycled aggregate concrete produced from two-stage mixing approach. Cem Concr Res 35(6):1195–1203

Tam VWY, Gao XF, Tam CM, Ng KM (2009) Physio-chemical reactions in recycle aggregate concrete. J Hazard Mater 163(2–3):823–828

Thomas C, de Brito J, Gil V, Sainz-Aja J, Cimentada A (2018) Multiple recycled aggregate properties analysed by X-ray microtomography. Constr Build Mater 166:171–180

Vedalakshmi R, Raj AS, Srinivasan S, Babu KG (2003) Quantification of hydrated cement products of blended cements in low and medium strength concrete using TG and DTA technique. Thermochimica Acta 407(1–2):49–60

Xiao J, Li W, Sun Z, Lange DA, Shah SP (2013) Properties of interfacial transition zones in recycled aggregate concrete tested by nanoindentation. Cem Concr Compos 37(1):276–292

Ye G, Liu X, De Schutter G, Poppe A-M, Taerwe L (2007) Influence of limestone powder used as filler in SCC on hydration and microstructure of cement pastes. Cem Concr Compos 29:94–102

Ylmén R, Jäglid U, Steenari B-M, Panas I (2009) Early hydration and setting of Portland cement monitored by IR, SEM and Vicat techniques. Cem Concr Res 39(5):433–439

Young JF, Hansen W (1986) Volume relationships for C-S-H formation based on hydration stoichiometries. MRS Proc 85:313

Zeng Q, Li K, Fen-Chong T, Dangla P (2012) Determination of cement hydration and pozzolanic reaction extents for fly-ash cement pastes. Constr Build Mater 27(1):560–569

Zhang H, Zhao Y (2015) Integrated interface parameters of recycled aggregate concrete. Constr Build Mater 101:861–877

Zhang H, Zhao Y, Meng T, Shah SP (2015) The modification effects of a nano-silica slurry on microstructure, strength, and strain development of recycled aggregate concrete applied in an enlarged structural test. Constr Build Mater 95:721–735

Zhang J, Scherer GW (2011) Comparison of methods for arresting hydration of cement. Cem Concr Res 41(10):1024–1036

Zhang Z, Wang H, Provis JL (2012) Quantitative study of the reactivity of fly ash in geopolymerization by FTIR. J Sustain Cem Based Mater 1(4):154–166

Zhang Z, Zhang Y, Yan C, Liu Y (2017) Influence of crushing index on properties of recycled aggregates pervious concrete. Constr Build Mater 135:112–118

Zuo Y, Qi B, Gao J, Li W (2018) Application of X-ray computed tomography in monitoring chloride ion penetration paths in recycled aggregate concrete. Adv Mater Sci Eng 2018:1–5

Part IV
Sustainability Assessment of Recycled Aggregate Concrete

Chapter 8
Life Cycle Assessment and Cost Analysis

8.1 Introduction

The researchers suggested various mix design techniques and mixing strategies to prepare RAC with acceptable mechanical characteristics. The DWR method, DVR method, EMV method (Fathifazl et al. 2009), and PPM (Pradhan et al. 2017) are the many mix design methodologies used for the preparation of RAC. For the purpose of preparing RAC, the performance indicators recommend EMV and PPM mix design approaches. However, due to the practical challenges of completely removing the connected mortar from RCA, the total replacement of NCA with RCA was not investigated in the EMV approach. In this sense, the PPM mix design satisfies both the performance criteria and the 100% utilization of RCA. A common procedure to evaluate a product's environmental effects is called a life cycle assessment (LCA). Understanding the potential of using RCA in place of NCA from the standpoint of environmental implications will therefore be helpful. While performing the LCA of RAC, it is crucial to take both the replacement percentage of RCA and cement content into consideration without compromising the performance of RAC with regard to NAC.

8.2 Life Cycle Assessment

Using LCA tools, it is possible to undertake a quantitative evaluation of a system's or product's environmental impact in accordance with ISO (ISO 14040 2006; ISO 14044 2006). The extraction of recycled aggregate (RA) from C&D waste as well as the preparation of RAC using RA need to be taken into account in the LCA research in order to compare the sustainability benefits of RAC with respect to conventional concrete. The environmental impact of processing natural aggregate (NA) and RA made from C&D waste was examined by Simion et al. (2013), who found that the

S. Pradhan et al., *Particle Packing Method for Recycled Aggregate Concrete*,
https://doi.org/10.1007/978-981-99-7516-7_8

latter scenario had about seven times lower CO_2 emissions. According to the study by Coelho and de Brito (2013a, b), replacing RA with virgin aggregate resulted in approximately 10 times and 8 times less CO_2 eq emissions and energy usage, respectively. The recycling facility's ability to produce and its transportation capabilities have a significant impact on the environment. It was recommended in this context by Coelho and de Brito (2013a, b) to build a recycling facility with a bigger capacity (it must not be over-sized) and find an alternative and more efficient form of transportation to replace the heavy-duty diesel-powered vehicles. By combining the RA and cementitious material extraction methods from C&D waste, Guignot et al. (2015) reported a large decrease in environmental consequences. According to Hossain et al. (2016), the manufacturing of RA conserves non-renewable resources by 58 percent and minimizes greenhouse gas emissions by 65%. According to the sensitivity study, the collection of C&D waste can be transported up to 20% farther or closer without having any effect on the environmental impact of RA production relative to NA processing (Hossain et al. 2016). According to Ghanbari et al. (2017), who took the Iranian context into account, there would be a 36% and a 30% reduction in annual CO_2 emissions and energy consumption, respectively. The research on LCA of the production of NA and RA in Brazil showed lower environmental impacts for the latter case (Rosado et al. 2017). Additionally, the sensitivity analysis shows that the recycling facility, which is up to 20 km away from the NA manufacturing facility, has lesser impact on the environment (Rosado et al. 2017).

According to Marinković et al. (2010), RAC preparation in Serbia is more affected by RA and cement production than other countries. As a result, while evaluating the environmental effects of RAC, the transit distance of RA relative to NA is crucial. According to Knoeri et al. (2013), RAC has an estimated 30% lower environmental effect. Moreover, additional cement content of 22–40 kg/m^3 in RAC or RA transportation distance of 15 km results in similar environmental as NAC production. Completely recyclable concrete (CRC) was the subject of a life cycle assessment (LCA) based on the cradle-to-cradle theory (De Schepper et al. 2014). The study found that the global warming potential for high-strength CRC and medium-strength CRC, respectively, was reduced by 66–70% and 7–35%. Jiménez et al. (2015) conducted a comparison of the LCA of RAC and NAC prepared using the EMV technique and the traditional mix design method. The components of the concrete system are responsible for at least 67% of each impact category. Better environmental performance was documented for the EMV technique for both NAC and RAC. According to the cradle-to-cradle theory, a closed-loop scenario of RAC production in China was examined by Ding et al. (2016). The study demonstrated that the use of cement and transportation are the two main sources of CO_2 emissions and energy use, which can be reduced by utilizing RCA.

The comparative analysis of the LCA of both RAC and NAC, which were prepared utilizing PPM mix design (Pradhan et al. 2017) and IS: 10262 (2009) approach, was the main emphasis of the work by Pradhan et al. (2019). According to Hossain et al. (2016); Mankelow et al. (2010), the transportation operations required for the manufacture or extraction of raw materials to the concrete batching plant account for around 40% of the overall energy consumption in the construction industry. As a result, sensitivity analysis was required, specifically taking into account various probable transport scenarios in India. The LCA study carried out by Pradhan et al. (2019) will be discussed in the sections that follow.

8.3 Methodology

The principles outlined in ISO 14040 (2006) and ISO 14044 (2006) serve as the foundation for conducting a LCA of a system or product to assess its environmental impacts over the course of its existence. The analysis was carried out in accordance with four clearly defined steps: (1) goal and scope definition, (2) creation of life cycle inventory (LCI), (3) assessment of the environmental impacts, and (4) interpretation of the results.

8.3.1 Goal and Scope Definition

Based on cradle-to-gate theory, Pradhan et al. (2019) sought to (a) establish an LCI for NCA and RCA production in India using data gathered from the basalt quarry and recycling plant, respectively, and (b) conduct a comparative LCA study to evaluate the environmental effects of NAC and RAC preparation. The impact of using RCA in place of NCA and using the PPM mix design methodology in place of the IS: 10262 (2009) method was compared using the LCA of the four different types of mixes.

Functional Unit

The production of 1 m^3 of NAC and RAC with a notional compressive strength of 30 MPa after 28 days of curing was the functional unit of the Pradhan et al. (2019) research assessment.

System Boundaries

The following processes are taken into account while preparing concrete inside the system boundary: (a) raw material extraction and production; (b) raw material transportation to the concrete mixing plant; and (c) concrete preparation at the concrete mixing plant. Figures 8.1 and 8.2, respectively, show the system boundaries for NAC and RAC preparation. Furthermore, the processing of NCA and RCA scenario in India is depicted in Figs. 8.3 and 8.4, respectively. The data used to evaluate the environmental impact during the preparation of NCA and RCA is shown in Appendix A.

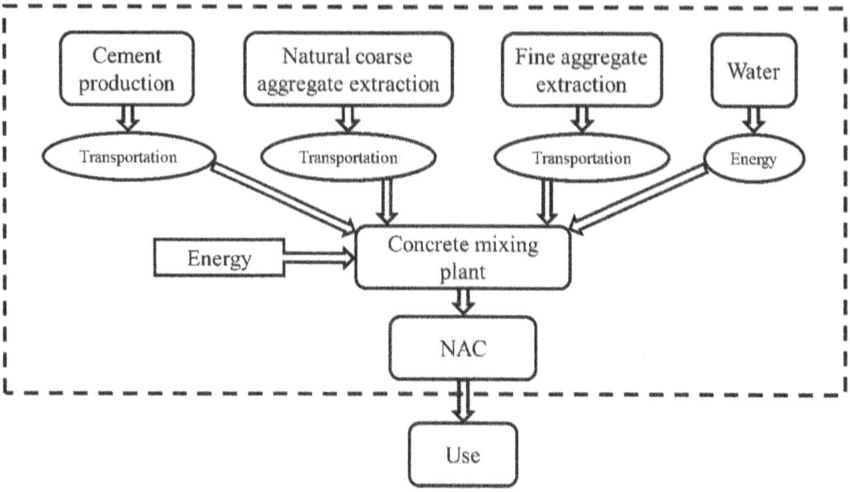

Fig. 8.1 System boundary of NAC (Pradhan et al. 2019)

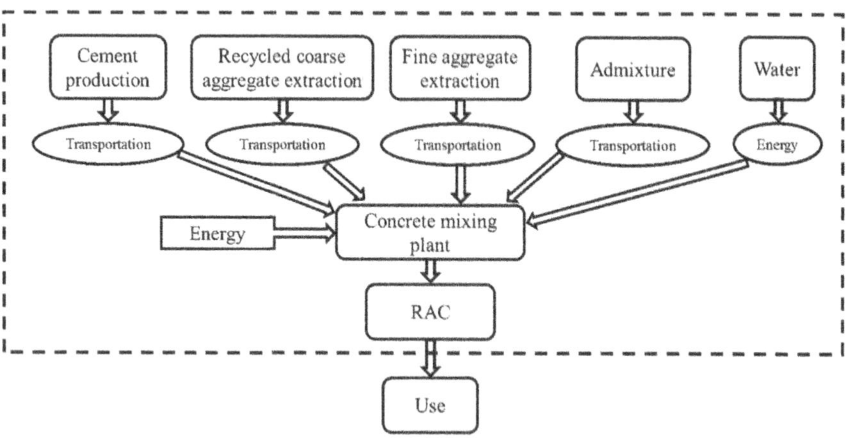

Fig. 8.2 System boundary of RAC (Pradhan et al. 2019)

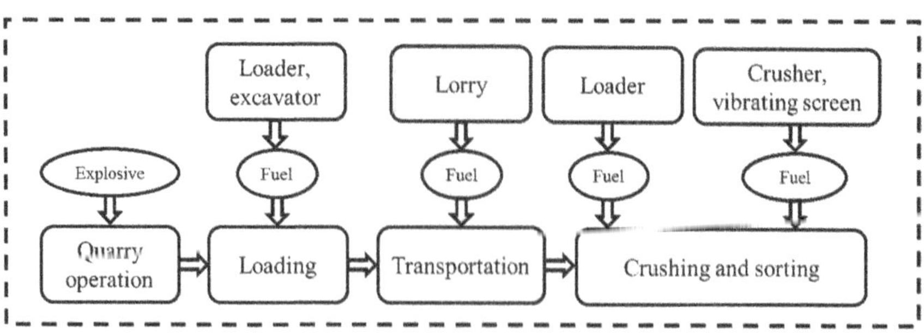

Fig. 8.3 Production process of NCA (Pradhan et al. 2019)

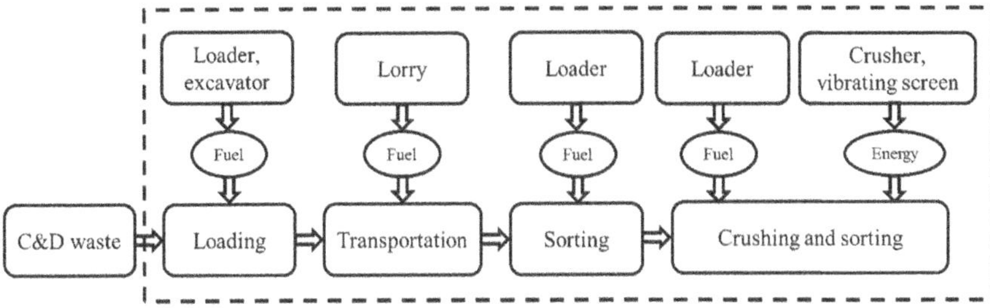

Fig. 8.4 Production process of RCA (Pradhan et al. 2019)

Assumptions

1. The sources of the C&D waste are the demolished structure, collapsed structure, and tested laboratory specimens. Both natural processes and anthropogenic activities can produce C&D waste. Therefore, the phases involved in the creation of C&D waste are not taken into account.
2. The recovery and recycling of steel scrap (reinforcement present in concrete) are not accounted.
3. Transportation within the NCA processing plant and recycling plant is not considered.
4. In waste concrete, the reclamation ratio (the proportion of recycled coarse aggregate to recycled fine aggregate) is 60:40, which is consistent with past findings (Jiménez et al. 2015; Marinković et al. 2010; Nagataki et al. 2004).
5. As little material is lost during the production of NCA and RCA, this burden is not taken into account.
6. Due to incomplete data provided by the relevant companies, the dust emission during the manufacturing of NCA and RCA is not taken into account. The environmental effects of these emissions are, however, less important (Braga et al. 2017).

8.3.2 Life Cycle Inventory (LCI) Data

The Life Cycle Inventory (LCI) is a list of the raw material, energy, and fuel inputs as well as the outputs as emissions to the air, water, and land within the system boundary for processing each functional unit (Marinković et al. 2010; Rosado et al. 2017). The required information from the crushing plant and recycling facility, respectively, was gathered to create the LCI for the manufacturing of NCA and RCA. On the other hand, the LCI for fuel (diesel), power (electricity), cement production, water consumption, loading operation (loader or excavator), and transportation (lorry) were extracted from the Ecoinvent 3.1 database.

NCA Production

Basalt was extracted from the quarry by using explosive. In this context, the pits were drilled to place the explosive. After the explosion, the loose basalt boulders were loaded into trucks using hydraulic excavators and driven to the processing facility. The production capacity of the basalt crushing plant is 180 t/h (1500 t/day). The trucks unloaded the basalt in the hopper of the crushing plant, from where it goes to the vibrating feeder. The conveyor belt carried the material from the vibrating feeder to the jaw crusher. The material then travelled on a conveyor belt to the cone crusher. The vibrating screen was used to separate the crushed material into various aggregate sizes. The over-sized aggregates were then transported by conveyor belt to the vertical shaft impact crusher before being discharged onto the vibrating screen. Appendix A contains information on the direct burdens (consumption of explosive, diesel, energy, and water) associated with the manufacturing of NCA.

RCA Production

The C&D garbage was sent to the recycling facility (IL&FS Environmental Infrastructure and Services Ltd., New Delhi), which covers a space of roughly 28,000 m^2 and has a production capacity of nearly 40 t/h (280 t/day). When the C&D debris arrived at the recycling facilities, waste concrete was manually separated from it by visual inspection. The waste concrete was then loaded into a wheel loader and transported to a recycling plant. To remove RCA of the necessary size, the waste concrete was broken in the jaw crusher after being transferred through the conveyor belt to the impact crusher. Natural magnets were used to extract the ferrous components from the crushed concrete. The suction bags and water sprinkler were employed close to the crusher to reduce dust emissions. The RCAs of various sizes were separated using vibrating screens. Finally, with the aid of an excavator, the prepared RCAs were gathered and stacked in open-air heaps. In Appendix A, the specifics of the direct burdens (consumption of diesel, energy, and water) incurred during the manufacturing of RCA are reported.

8.3.3 Life Cycle Impact Assessment (LCIA)

The Life Cycle Impact Assessment (LCIA) phase estimates the magnitude of the environmental impacts and resources used for the specified LCI phase. The three basic steps included in this phase are: (1) selection of the impact categories, (2) classification and characterization of the assigned LCI to the selected impact categories, and (3) conversion into indicator result by aggregating the LCI results (Marinković et al. 2010; Hossain et al. 2016). These steps are performed as specified in the ISO (ISO 14040 2006; ISO 14044 2006).

The two approaches for the impact assessment are: (1) problem-oriented mid-point approach and (2) damage-oriented end-point approach (Marinković et al. 2010; Hossain et al. 2016). The CML, ReCiPe, Impact2002+, and EDIP LCIA methods can be used to evaluate the problem-oriented mid-point approach, which characterizes the environmental impacts in terms of global warming, acidification, eutrophication, ozone layer depletion, abiotic depletion, photochemical oxidant creation, and human toxicity. The problem-oriented mid-point impacts are converted using the damage-oriented end-point approach, which quantifies impacts in terms of harm to people, resources, the environment, and climate change. The LCIA methods for this approach are ReCiPe, Ecoindicator 99, and Impact 2002+ (Hossain et al. 2016).

The SimaPro program, which includes the Ecoinvent 3.1 database, was used to conduct the impact analysis. It is important to note that the admixture utilized to prepare the RAC was not present in the Ecoinvent 3.1 database. This is why the LCIA of polycarboxylate-based admixture created by European Federation of Concrete Admixtures Associations Ltd. (EFCA) (2015) was utilized. Due to this limitation, the impact categories measured using the CML baseline technique are global warming potential, depletion of ozone layer, acidification potential, eutrophication potential, and formation of tropospheric ozone photochemical oxidants. Additionally, both renewable and non-renewable primary energy consumptions were estimated using the Cumulative Energy Demand technique.

8.4 Results and Discussion

The different transportation situations and distances (km) observed in the study of Pradhan et al. (2019) for different raw materials (NCA, RCA, cement, sand, and admixture) which were considered during the analysis are mentioned in Table 8.1.

Table 8.1 Actual transportation distances of different raw materials (Pradhan et al. 2019)

Material	From	To	Distance (km)	Vehicle type
Loose basalt	Mine	Crushing plant	5	16–32 t
NCA	Crushing plant	Concrete batching plant	50	7.5–16 t
C&D waste	Demolition site	Recycling plant	35	16–32 t
RCA	Recycling plant	Concrete batching plant	25	7.5–16 t
Sand	River	Concrete batching plant	30	7.5–16 t
Cement	Cement factory	Concrete batching plant	70	7.5–16 t

8.4.1 Interpretation of Results

Influence of Different Life Cycle Phases

The assessed environmental impacts for each category are shown in Fig. 8.5, where the percentage contributions of the various life cycle phases (cement, water, sand, transportation, blasting, admixture, diesel, and electricity) are shown for each impact category and each type of concrete. It is clear that cement makes up a significant portion (greater than 60%) of each impact category. Cement makes up at least 75% of all other impact categories considered, excluding ADP. Cement contributes more than 90% of the overall impact of the GWP, as well. In earlier investigations, comparable findings were also presented (Marinković et al. 2010; De Schepper et al. 2014; Jiménez et al. 2015; Ding et al. 2016; Serres et al. 2016; Braga et al. 2017). After cement, transportation was the second-largest contributor to each impact category, where it has a greater impact on ADP, ODP, and ADP (fossil fuels). The transportation-related activities were shown to have a similar effect, according to Marinković et al. (2010). The ADP is mainly affected by the use of sand. The majority of the impact categories affected by the employment of explosive to dismantle the

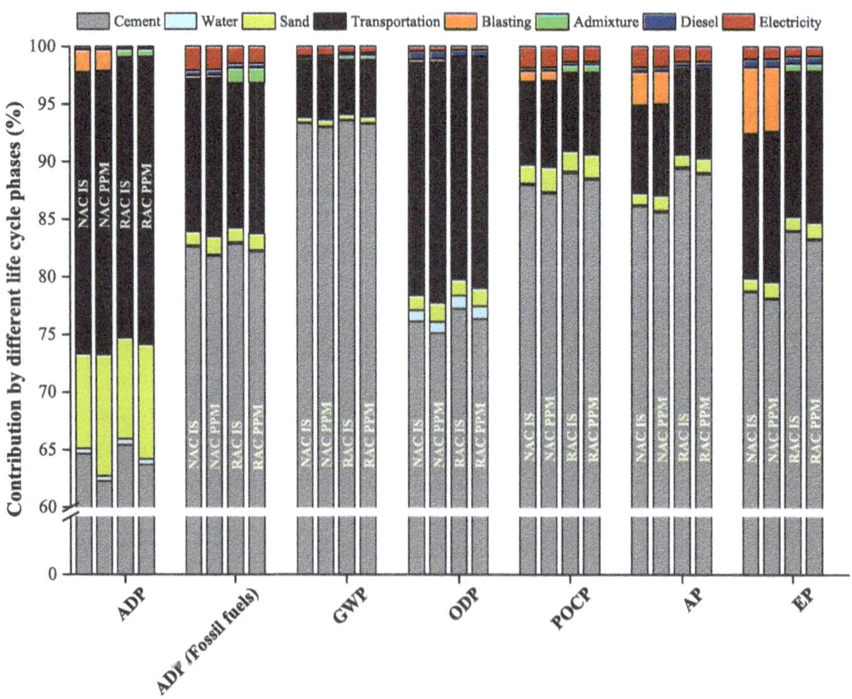

Fig. 8.5 Contribution by different life cycle phases to different impact categories (Pradhan et al. 2019)

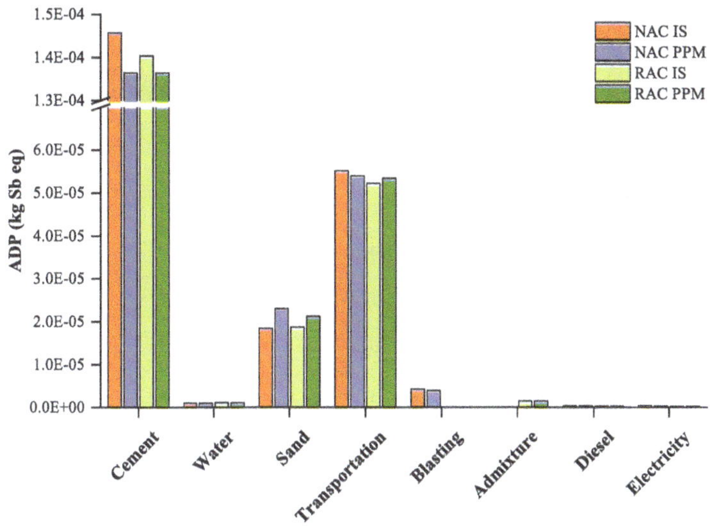

Fig. 8.6 Contribution by different life cycle phases to ADP (kg Sb eq) (Pradhan et al. 2019)

rock are EP, AP, ADP, and POCP (Fig. 8.5). The use of admixture for the preparation of RAC IS and RAC PPM concrete has more effect on ADP (fossil fuels) in comparison to other impact categories (Figs. 8.7 and 8.5). Additionally, the use of admixture in RAC results in a modest impact on ADP, AP, and POCP. The use of electric energy has little impact on the impact categories for ADP (fossil fuels), POCP, AP, EP, and GWP. The life cycle phases, such as diesel and water, have negligible contributions in the discussed impact categories. The absolute contributions of various life cycle phases to various effect categories are shown in Figs. 8.6, 8.7, 8.8, 8.9, 8.10, 8.11, and 8.12.

Influence of Aggregate Type

The estimated environmental consequences for various concrete types were normal ized in relation to the highest value obtained, and the resulting normalized values are shown in Fig. 8.13. It has been noted that NAC IS has the greatest influence for each impact category. This is mostly due to the use of maximum cement content, which can be supported by Fig. 8.6, 8.7, 8.8, 8.9, 8.10, 8.11, and 8.12. Additionally, RAC IS concrete had greater impacts than NAC PPM concrete in the categories of ADP (fossil fuels), GWP, and POCP. The usage of more cement content is mostly to blame for this. NAC PPM concrete, however, demonstrated higher impacts than RAC IS concrete in the ADP, AP, and EP impact categories. This shows that when comparing concrete mixtures with various types of aggregate (NCA and RCA), cement is not the determining factor for ADP, AP, and EP. Figures 8.6, 8.11, and 8.12 suggest that, transportation and blasting are the critical contributors in the aforementioned

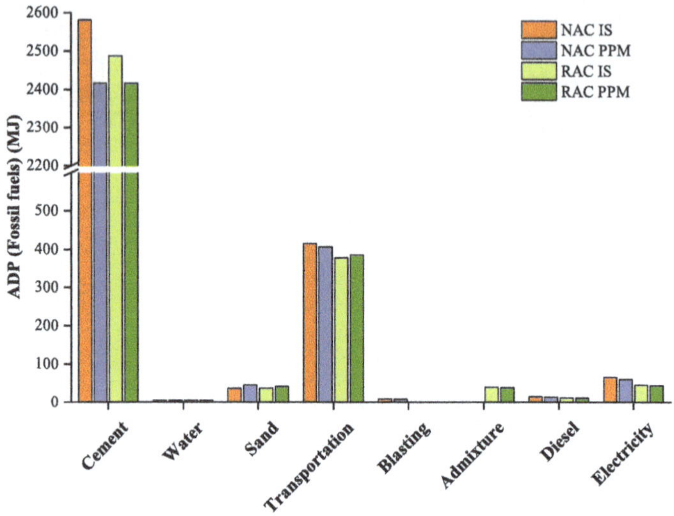

Fig. 8.7 Contribution by different life cycle phases to ADP (fossil fuels) (MJ) (Pradhan et al. 2019)

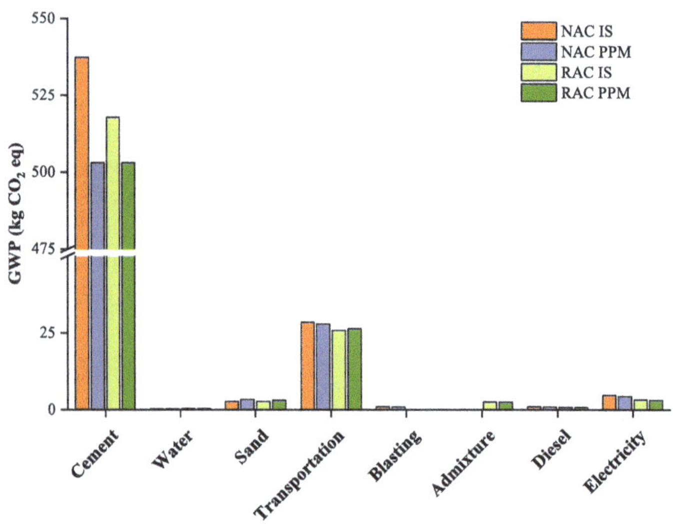

Fig. 8.8 Contribution by different life cycle phases to GWP (kg CO_2 eq) (Pradhan et al. 2019)

impact categories. Additionally, the use of more sand in NAC PPM concrete adds to the burden on ADP, AP, and EP. It can be shown that RAC PPM concrete has the lowest value for each effect category. This is a result of the additive effects of RAC PPM concrete's lower cement content, shorter transportation distance, and lack of explosive use.

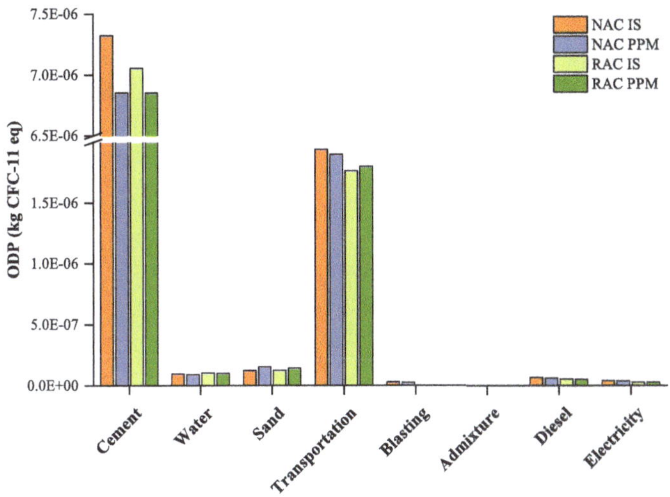

Fig. 8.9 Contribution by different life cycle phases to ODP (kg CFC-11 eq) (Pradhan et al. 2019)

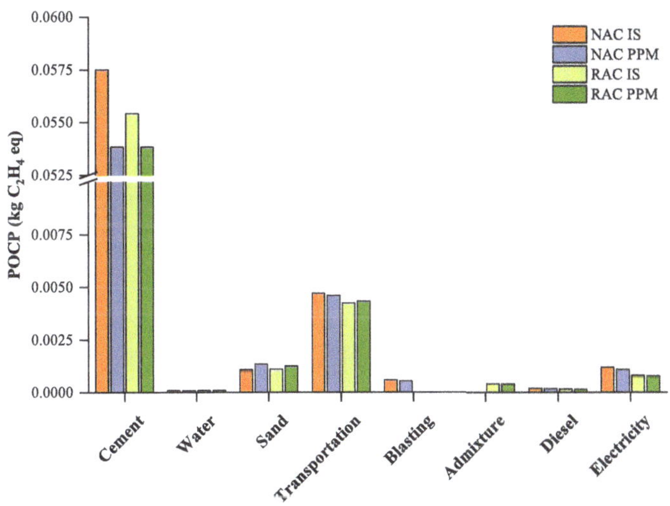

Fig. 8.10 Contribution by different life cycle phases to POCP (kg C_2H_4 eq) (Pradhan et al. 2019)

Influence of Mix Design Method

While employing the same aggregate type (NCA or RCA), the conventionally prepared concrete (NAC IS and RAC IS) showed larger environmental consequences for each category than the PPM mix-designed concrete (NAC PPM and RAC PPM)

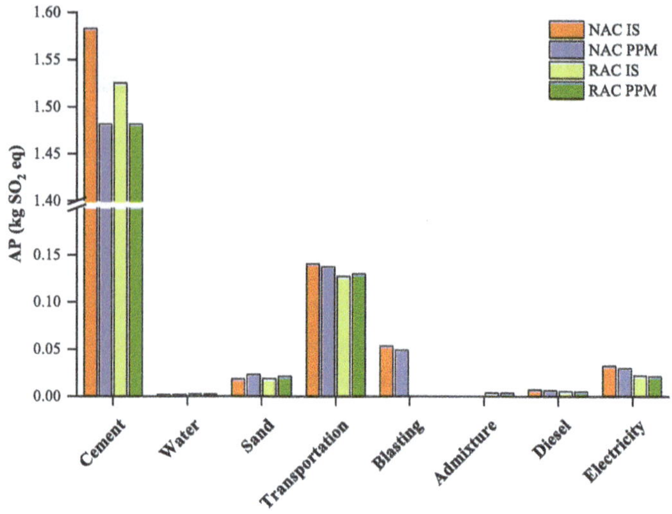

Fig. 8.11 Contribution by different life cycle phases to AP (kg SO_2 eq) (Pradhan et al. 2019)

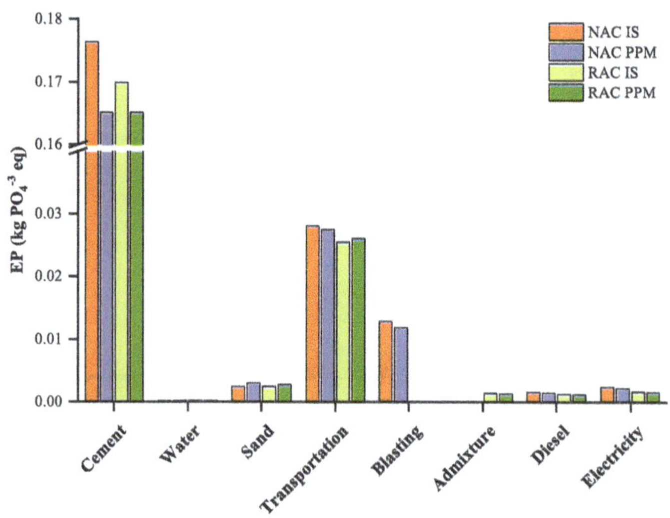

Fig. 8.12 Contribution by different life cycle phases to EP (kg PO_4^{-3} eq) (Pradhan et al. 2019)

(Fig. 8.13). Such a trend is predominantly due to the higher cement requirement in case of IS code method (Figs. 8.6, 8.7, 8.8, 8.9, 8.10, 8.11, and 8.12) although the contributions by the sand and transportation were higher for PPM mix-designed concrete (Figs. 8.6, 8.7, 8.8, 8.9, 8.10, 8.11, and 8.12).

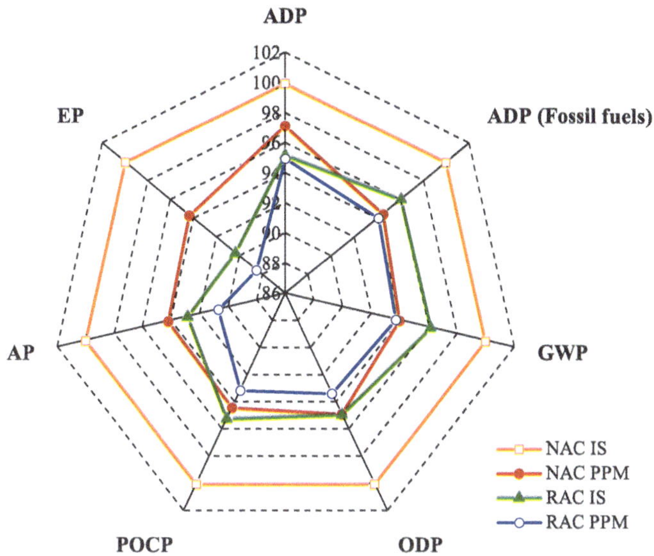

Fig. 8.13 Contribution by different life cycle phases to different impact categories (Pradhan et al. 2019)

8.5 Sensitivity Analysis

The environmental impacts were estimated for the fixed transportation scenario specific to the study of Pradhan et al. (2017). However, this restricts the study's use, and it might not be acceptable to evaluate the effects on the environment of various situations where the transport distance of aggregates is greater or smaller. Hence, a sensitivity analysis of transport distances of crushing plant to concrete batching plant (D_1') for NCA and demolition site to recycling plant (D_2) and recycling plant to concrete batching plant (D_2') for RCA in the maximum range of 500 km was studied by keeping all other life cycle phases constant (Table 8.2). According to Pradhan et al. (2017) study, the actual distance between the crushing plant and the batching plant was a maximum of 250 km. This could be lessened if the batching factory is situated inside the city limits. This distance is maintained at 5 km, however, in accordance with the prevailing belief that the crushing plant is close to the quarry. The location of the recycling plant from the demolition site was considered in three different distances, i.e. 5 km—when the location is very close, 35 km—when the location is within the city limits and the higher distance of 100 km can be justified for India, where there are very few recycling plants operating and the recycling of C&D waste is still not widely accepted. The calculated ADP (fossil fuels) in MJ, GWP in kg CO_2 eq, AP in kg SO_2 eq, and EP in kg PO_4^{-3} eq for various transportation instances are shown in Table 8.2. For most of the cases, NAC IS exhibited the maximum impacts irrespective of the category over the considered transport distance, D_1'. As a result, the maximum distance between the aggregate preparation plant and the

Table 8.2 Different transport distance of NCA and RCA considered for sensitivity analysis

Type of aggregate	Transportation scenario	Transport distance (km)			Type of vehicle
		Case1	Case 2	Case 3	
NCA	Mine to crushing plant (D_1)	5	5	5	16–32 t
	Crushing plant to concrete batching plant (D_1')	0, 50, 100, 150, 200, 250, 300, 400, and 500	0, 50, 100, 150, 200, 250, 300, and 400	0, 50, 100, 150, 200, 250, 300, and 400	7.5–16 t
RCA	Demolition site to recycling plant (D_2)	5	35	100	16–32 t
	Recycling plant to concrete batching plant (D_2')	0, 50, 100, 150, 200, 250, 300, 400, and 500	0, 50, 100, 150, 200, 250, 300, and 400	0, 50, 100, 150, 200, 250, 300, and 400	7.5–16 t

concrete batching plant for NAC PPM, RAC IS, and RAC PPM was established in relation to NAC IS, which denotes the maximum distance below which the environmental effects of the corresponding aggregate are less than NAC IS. Furthermore, the limit transport distance for the environmental impacts corresponds to 0 km and 250 km transport distance of NCA (D_1') for the preparation of one functional unit of NAC IS.

8.5.1 *Case 1*

NAC IS and RAC PPM show the highest and least environmental consequences, respectively, regardless of the impact category (Fig. 8.14). Additionally, as the transport distance of NCA increases for the preparation of NAC IS, the gap between the limit transport distance of RCA and the transport distance increases. The limit transport distance is observed to be minimum for ADP (Fossil fuels) impact category (Table 8.3).

8.5.2 *Case 2*

Irrespective of the impact category, NAC IS exhibits maximum environmental impacts. For AP and EP impact categories, RAC PPM exhibits minimum values over the considered distance D_2'. However, in ADP (fossil fuels) and GWP impact categories, NAC PPM concrete imparts lowest environmental impacts up to certain

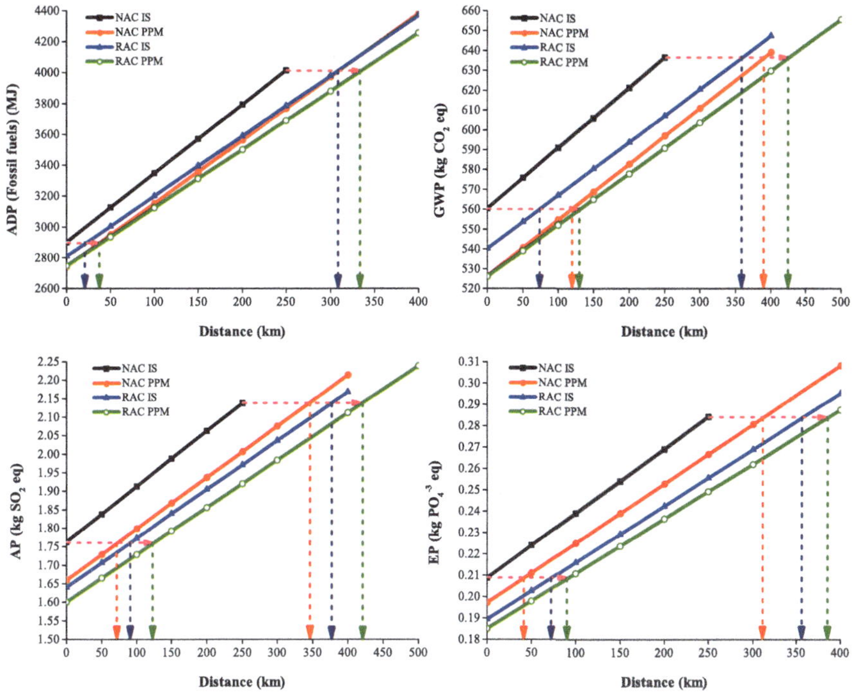

Fig. 8.14 Influence of different concrete on ADP, GWP, AP, and EP for Case 1 (Pradhan et al. 2019)

Table 8.3 Limit transport distance of NCA and RCA for Case 1

Type of concrete	Limit transport distance (km)							
	ADP (Fossil fuels)		GWP		AP		EP	
NAC IS	**0**	**250**	**0**	**250**	**0**	**250**	**0**	**250**
NAC PPM	40	310	120	395	70	350	45	315
RAC IS	20	310	75	360	90	375	75	360
RAC PPM	40	340	130	425	125	425	90	390

distance $D_1^{'}$ (300 km and 275 km, respectively) and further increase in $D_1^{'}$ results in higher ADP (fossil fuels) and GWP than RAC PPM (Fig. 8.15). The limit transport distance of RCA and the transport distance of NCA diverge, similar to Case 1, and this discrepancy grows as the transport distance of NCA increases for the preparation of NAC IS concrete. The limit transport distance is observed to be minimum for ADP (fossil fuels) impact category (Table 8.4).

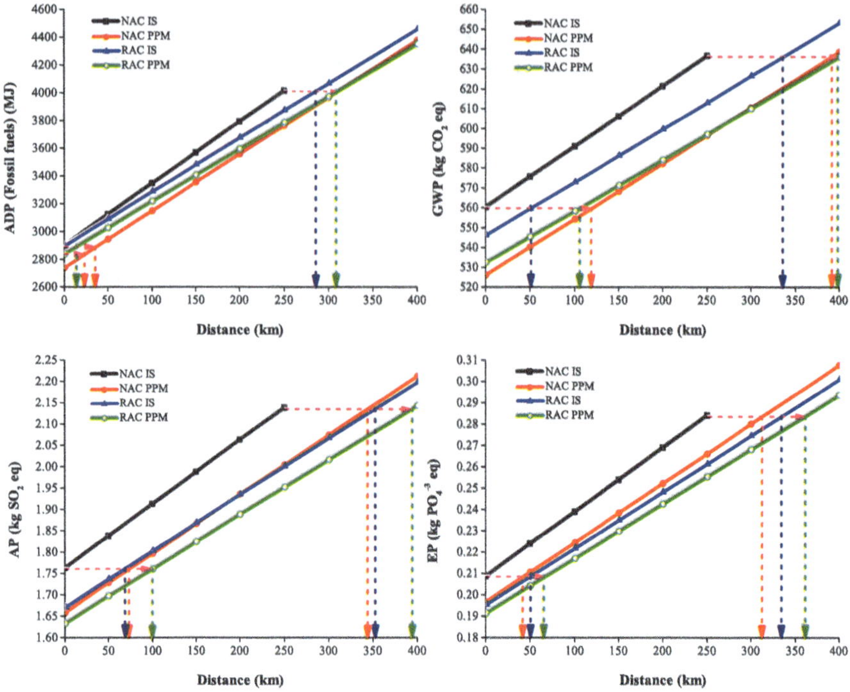

Fig. 8.15 Influence of different concrete on ADP, GWP, AP, and EP for Case 2 (Pradhan et al. 2019)

Table 8.4 Limit transport distance of NCA and RCA for Case 2

Type of concrete	Limit transport distance (km)							
	ADP (Fossil fuels)		GWP		AP		EP	
NAC IS	**0**	**250**	**0**	**250**	**0**	**250**	**0**	**250**
NAC PPM	35	310	120	390	75	345	40	315
RAC IS	0	285	50	335	70	355	50	335
RAC PPM	15	310	105	400	100	395	70	360

8.5.3 Case 3

For GWP, AP, and EP impact categories, NAC IS exhibits maximum impacts over the considered D_1', whereas for ADP (fossil fuels) impact category RAC IS has maximum impact (Fig. 8.16). Apart from the ADP (fossil fuels) impact category, the limit transport distance of RAC IS is lower than the considered distance D_1' for

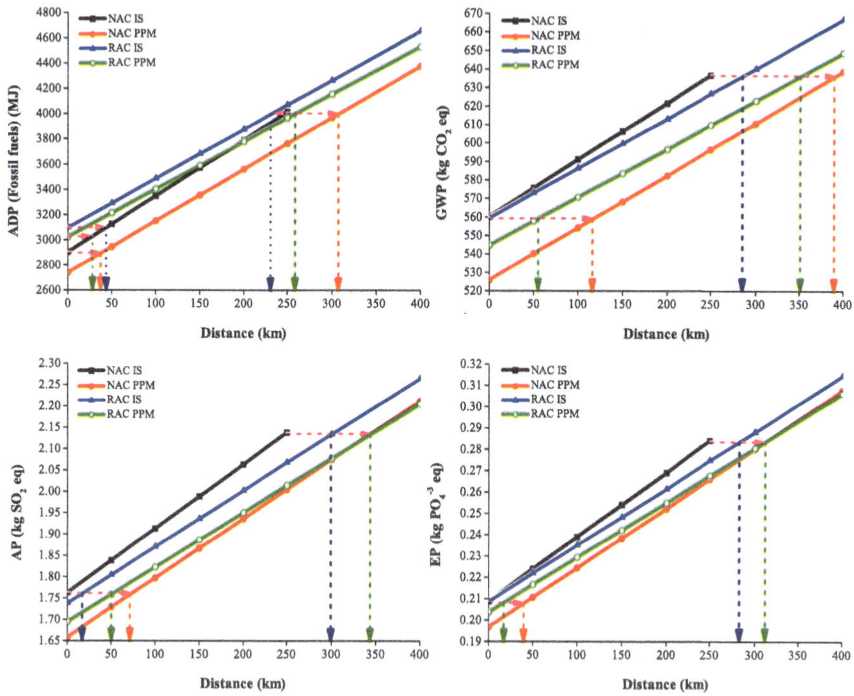

Fig. 8.16 Influence of different concrete on ADP, GWP, AP, and EP for Case 3 (Pradhan et al. 2019)

Table 8.5 Limit transport distance of NCA and RCA for Case 3

Type of concrete	Limit transport distance (km)							
	ADP (Fossil fuels)		GWP		AP		EP	
NAC IS	**0**	**250**	**0**	**250**	**0**	**250**	**0**	**250**
NAC PPM	35	305	120	390	70	345	40	310
RAC IS	–45	235	0	285	20	300	0	285
RAC PPM	–30	260	55	350	50	345	20	310

NAC IS, whereas for RAC PPM concrete the limit transport distance is lower up to 190 km of $D_1^{'}$ for NAC IS (Table 8.5). The environmental impact for each category is higher for RAC IS than NAC PPM and RAC PPM over the considered $D_1^{'}$ and $D_2^{'}$. Similar to the earlier cases, the limit transport distance is minimum for ADP (fossil fuels) impact category.

Table 8.6 Limit collection distance of C&D waste for RAC with respect to different D_1' of NAC IS

Type of concrete	Limit collection distance of C&D waste	D_1' in case of NAC IS (km)					
		0	50	100	150	200	250
RAC IS	D_2 (km)	34.5	44	53	62	70	79
RAC PPM		55	67	79	91	103	115

8.5.4 Critical Analysis

The limit transport distance of NCA and RCA as reported in Tables 8.3 to 8.5 indicates that, D_1' (for NAC PPM) and D_2' (RAC IS and RAC PPM) are higher than D_1' of NAC IS, while considering GWP, AP, and EP impact categories (except RAC IS in Case 3 in EP impact category). However, for ADP (fossil fuels) impact category, D_2' of RAC IS in Case 2 and both RAC IS and RAC PPM in Case 3 are lower than D_1' of NCA in NAC IS. From the three different cases, it can be inferred that D_2 of C&D waste influences D_2' of processed RCA. Moreover, after certain D_2 of C&D waste, D_2' is adversely affected irrespective of the mix design method of concrete. In this context, the limiting value of D_2 is determined for different D_1' of NAC IS (D_2' is same as D_1') and represented in Table 8.6. It can be observed that, the scope for collection of C&D waste from a larger distance to the recycling plant increases with the increase in D_2'. The increment in D_2 is approximately 9 km (13%–26%) and 12 km (12%–22%) for RAC IS and RAC PPM, respectively, for the increment of 50 km of D_2'. For D_2' within 250 km, D_2 in case of RAC PPM is about 45%–59% higher than RAC IS. Moreover, for D_2' beyond 200 km, C&D waste can be collected from more than 100 km distance in case of RAC PPM. Additionally, the largest limit transport distance recorded for the same impact category shows that the usage of RAC PPM has the least contribution to GWP. The whole analysis shows the benefits of the PPM mix design methodology over the traditional IS code mix design method.

8.6 Cost Analysis

To examine the viability of various types of concrete from an economic standpoint, the costs of the four types of concrete were estimated. As a result, the price of preparation of 1 m^3 of concrete was evaluated. The three main components of the cost analysis of concrete preparation, namely, (a) material cost, (b) labour cost, and (c) additional costs, are taken into account in this context.

The materials used in the study of Pradhan et al. (2017) include cement, sand, aggregate (NCA, RCA), and admixture. The cost of particulars used to prepare different types of concrete are presented in Table 8.7. The number of skilled, semi-

Table 8.7 Cost of the materials used for concrete preparation

Material	Cost (INR/kg)
Cement	6
Sand	1.1
NCA	0.8
RCA	0.6
Admixture	150

Table 8.8 Cost of 1 m^3 of concrete of different types

Type of concrete	Cost (INR/m^3)
NAC IS	5907
NAC PPM	5817
RAC IS	5450
RAC PPM	5445

skilled, and unskilled labours required to prepare 1 m^3 of concrete was determined as per the provisions specified in IS: 7272 (Part I) (1974). During the process of preparing concrete, a certain amount of labour charge was added for the mixing, lifting, carrying, and compacting operations. The labour information are (a) mixing of concrete = 3 h/m^3, (b) lifting and carrying concrete = 1.20 h/m^3, (c) compacting concrete = 0.80 h/m^3, and (d) levelling surface of concrete = 0.10 h/m^3. The other information regarding concrete production includes (a) skilled labour cost = INR 380/day, (b) semi-skilled labour cost = 320 INR/day, (c) unskilled labour cost = INR 280/day, and (d) overhead charges = extra 12%.

The estimated costs of NAC IS, NAC PPM, RAC IS, and RAC PPM are shown in Table 8.8. According to the study, RAC PPM concrete production is 8.5% more cost-effective than NAC IS. PPM mix design, however, has very little impact on lowering the cost of producing concrete; a maximum saving of 1.5% was calculated for NAC PPM compared to NAC IS.

8.7 Closure

The study evaluated the environmental effects of concrete using the CML baseline technique, and a comparison analysis was done by taking into account two types of coarse aggregates (NCA and RCA) and two mix design methodologies (IS: 10262 (2009) and PPM). Four distinct concrete mix costs were examined, along with their preparation costs. The important findings are listed below:

- Because the processing step of aggregate has a relatively lesser environmental impact than NCA, using RCA is preferred for a particular mix design approach. With the same aggregate type, the PPM mix design approach, however, had a noticeably reduced environmental impact because it used less cement.

- The sensitivity study showed that the reduced cement demand of RAC PPM permits the transportation of RCA across greater distances, which is also reflected in its lowest GWP contribution.
- For higher collection distance of C&D waste to recycling plant, the limit transport distance for RAC IS and RAC PPM was observed to be lower than NAC IS in ADP (fossil fuels) category. The maximum collection distance for C&D waste was therefore assessed in relation to ADP (fossil fuels) for various supply distances of processed RCA as NCA.
- The cost of preparation of RAC PPM is the lowest. The PPM mix design method has very little effect on lowering the cost of concrete preparation.

References

Braga AM, Silvestre JD, de Brito J (2017) Compared environmental and economic impact from cradle to gate of concrete with natural and recycled coarse aggregates. J Clean Prod 162:529–543

Coelho A, de Brito J (2013) Environmental analysis of a construction and demolition waste recycling plant in Portugal-Part I: Energy consumption and CO2emissions. Waste Manag 33(5):1258–1267

Coelho A, de Brito J (2013) Environmental analysis of a construction and demolition waste recycling plant in Portugal-Part II: Environmental sensitivity analysis. Waste Manag 33(5):147–161

De Schepper M, Van den Heede P, Van Driessche I, De Belie N (2014) Life cycle assessment of completely recyclable concrete. Materials 7(8):6010–6027

Ding T, Xiao J, Tam VW (2016) A closed-loop life cycle assessment of recycled aggregate concrete utilization in China. Waste Manag 56:367–375

European Federation of Concrete Admixtures Associations Ltd. (EFCA) (2015) Concrete admixtures–plasticisers and superplasticisers. Technical report

Fathifazl G, Abbas A, Razaqpur AG, Isgor OB, Fournier B, Foo S (2009) New mixture proportioning method for concrete made with coarse recycled concrete aggregate. J Mater Civil Eng 21(10):601–611

Ghanbari M, Abbasi AM, Ravanshadnia M (2017) Production of natural and recycled aggregates: the environmental impacts of energy consumption and CO2 emissions. J Mater Cycles Waste Manag 20(2):810–822

Guignot S, Touzé S, Von der Weid F, Ménard Y, Villeneuve J (2015) Recycling construction and demolition wastes as building materials: a life cycle assessment. J Ind Ecol 19(6):1030–1043

Hossain MU, Poon CS, Lo IM, Cheng JC (2016) Comparative environmental evaluation of aggregate production from recycled waste materials and virgin sources by LCA. Resour Conserv Recycl 109:67–77

IS: 10262 (2009). Concrete mix proportioning–guidelines

IS: 7272 (Part I) (1974) Recommendation for labour output constants for building work

ISO 14040 (2006) Environmental management-life cycle assessment-principles and framework. ISO, Geneva

ISO 14044 (2006) Environmental management-life cycle assessments-requirements and guidelines. ISO, Geneva

Jiménez C, Barra M, Josa A, Valls S (2015) LCA of recycled and conventional concretes designed using the equivalent mortar volume and classic methods. Const Build Mater 84:245–252

Knoeri C, Sanyé-Mengual E, Althaus HJ (2013) Comparative LCA of recycled and conventional concrete for structural applications. Int J Life Cycle Assess 18(5):909–918

Mankelow JM, Oyo-Ita D, Birkin M (2010) Assessing the cabon footprint of transporting primary aggregates. In: Scott P, Walton G (eds) Proceedings of the 15th extractive industry geology conference. EIG conferences Ltd, pp 41–45

Marinković S, Radonjanin V, Malešev M, Ignjatović I (2010) Comparative environmental assessment of natural and recycled aggregate concrete. Waste Manag 30(11):2255–2264

Nagataki S, Gokce A, Saeki T, Hisada M (2004) Assessment of recycling process induced damage sensitivity of recycled concrete aggregates. Cement Concr Res 34(6):965–971

Pradhan S, Kumar S, Barai SV (2017) Recycled aggregate concrete: particle packing method (PPM) of mix design approach. Constr Build Mater 152:269–284

Pradhan S, Tiwari BR, Kumar S, Barai SV (2019) Comparative LCA of recycled and natural aggregate concrete using particle packing method and conventional method of design mix. J Clean Prod 228:679–691

Rosado LP, Vitale P, Penteado CSG, Arena U (2017) Life cycle assessment of natural and mixed recycled aggregate production in Brazil. J Clean Prod 151:634–642

Serres N, Braymand S, Feugeas F (2016) Environmental evaluation of concrete made from recycled concrete aggregate implementing life cycle assessment. J Build Eng 5:24–33

Simion IM, Fortuna ME, Bonoli A, Gavrilescu M (2013) Comparing environmental impacts of natural inert and recycled construction and demolition waste processing using LCA. J Environ Eng Land Manag 21(4):273–287

Part V
Structural Applications

Chapter 9
Structural Applications: Beam

9.1 Introduction

The fundamental characteristics of RCA and how it is used to create RAC were covered in prior chapters. In this context, the influence of the PPM mix design technique was explored. This approach, coupled with the TSMA, increased the performance of concrete and reduced the detrimental impact of RCA on RAC. This encourages the use of concrete prepared using the PPM mix design approach and TSMA to investigate the behaviour of reinforced NAC and RAC structural elements. Therefore, the present chapter investigates the shear and flexure behaviour of NAC and RAC beams constructed with PPM mix at various longitudinal reinforcement contents. The appropriateness of the current expressions to forecast the load carrying capacity of RAC beams with and without transverse reinforcement is also verified.

9.2 Reinforced RAC Beams

9.2.1 Flexure Behaviour

The flexural behaviour of reinforced concrete (RC) beam with different RCA substitution levels and reinforcement ratios were adopted during the experimental investigations. According to Etxeberria et al. (2007) research, there is hardly any difference in the ultimate load between RAC beams and conventional RC beams for three different replacement ratios of 25, 50, and 100%. A strong agreement between the experimental results of RAC beams and the currently accepted theoretical approaches was proven by the comparative analysis. The usage of RCA up to a 25% replacement ratio was finally advised for structural application. Using 50, 70, and 100% replacement levels of RCA, Bai and Sun (2010) evaluated the flexural behaviour of RAC beams and discovered no differences in the cracking pattern or mode of failure of

S. Pradhan et al., *Particle Packing Method for Recycled Aggregate Concrete*,
https://doi.org/10.1007/978-981-99-7516-7_9

these flexure-critical RAC beams. There was no discernible difference between the cracking moment and the final moment, despite the fact that the number of cracks and ultimate deflection were on the greater side for RAC beams. The experimental study by Knaack and Kurama (Knaack 2013; Knaack and Kurama 2013, 2014, 2015) at a replacement ratio of 50 and 100%, identified no significant difference in nonlinear behaviour or mode of failure in flexure critical reinforced RAC beams as compared to conventional RC beams. Although RAC beams showed a reduction in initial stiffness, the difference in ultimate moment was insignificant. For RAC beams, they have reported greater ultimate deflection. Finally, they have proposed that reinforced RAC beams can be constructed using the existing design guidelines and analytical models for conventional RC beams. Even when using 100% RCA, Arezoumandi et al. (2015b) found that the fracture morphology and crack advancement for RAC beams were extremely comparable to those for standard concrete beams. Although RAC beams displayed lower cracking moment, there was no discernible difference in yielding moment between RAC and NAC beams. Higher ultimate deflection than typical RC beams was produced as a result of the RAC beams' reduced rigidity following an early first crack.

Experimental research was done on RAC beams with 100% RCA and varied tensile reinforcement contents (0.84 and 1.5%), and it was discovered that there was no change in failure mechanism compared to NAC beams (Ajdukiewicz and Kliszczewicz 2007). The ultimate load was reduced by an average of 3.5% for the RAC beams, while greater deformation was observed for the same. The experiment conducted by Fathifazl et al. (2009b, 2010) using the Equivalent Mortar Volume (EMV) method of mix design (63.5 and 74.3%) with tensile reinforcement ratio ranged from 0.49 to 3.31% revealed that the crack spacing and cracking moment for both singly reinforced and doubly reinforced RAC beams were less than those of conventional concrete beams. However, the maximum midspan deflection observed for RAC beams was within the permissible limit recommended by ACI 318 (2008) and Eurocode 2 (2004). According to Ignjatović et al. (2012), who evaluated the flexural performance of RAC beams with various longitudinal reinforcement contents (0.28, 1.46, and 2.54%), RAC beams did not significantly differ from NAC beams in terms of crack width, crack spacing, or load–deflection behaviour for the same longitudinal reinforcement content, regardless of the replacement ratio of RCA (50 and 100%). The size of the failure surface and the severity of the damage to the concrete, however, were negatively impacted by the RCA's quality and replacement ratio. There was no exception found in the flexural performance studied by Choi et al. (2012) on doubly reinforced sections with 100% replacement of RCA. The current design provisions and allowable limits specified in the ACI code for conventional RC beams are also suitable for the RAC beams. In experiments, the flexural performance of RAC beams with tensile reinforcement ratios of 0.5, 1, 1.5, and 1.8% for replacement ratios of 15, 30, and 50% were examined (Kang et al. 2014). Although there were more cracks on the RAC beams, the use of RCA had no effect on the crack patterns or serviceability traits. However, the weakening of the bond between the aggregate interface and mortar caused by the increase in replacement ratio negatively impacted the flexural capacity of RAC beams. The flexural design of RAC beams with a maximum replacement ratio of 30% is compatible with the current design codes.

9.2.2 *Shear Behaviour*

González-Fonteboa and Martínez-Abella (2007) conducted the shear test of reinforced RAC beams consisting of 50% of RCA and overall longitudinal reinforcement content of 2.98%. Although premature cracking and horizontal splitting tension fractures along the tension reinforcement were seen, the ultimate load and deflection of RAC beams showed no appreciable variation from the typical concrete beam. More significantly, they came to the conclusion that the RC beam shear design provisions currently in use can be easily applied to RAC beams without modification. In this context, Ji et al. (2008) advised that the shear design of RAC beams does not require further revision due to the existing codes' conservative approach, with the exception of Eurocode 2 (2004). According to the experiment performed by Etxeberria et al. (2007) on RC beams having 25, 50, and 100% of RCA, the reduction in cracking load is inevitable when RCA is used. Additionally, the use of current design standards was advised for RAC beams with a 25% RCA replacement ratio, and it was noted that more research was required to determine how well these beams will function in shear under greater RCA replacement ratios (Etxeberria et al. 2007). Kim et al. (2013) investigated how the size of RAC beams affected shear behaviour with 50 and 100% RCA utilization. The experimental findings demonstrated that the shear stress dropped with the same increase in beam depth, while the shear force of RAC beams increased. Additionally, the design of RAC beams was recommended using the current design standards because identical behaviour of RAC beams was seen up to 600 mm depth. The good quality of the RCA employed in the study, according to the scientists, may have contributed to the RAC beams' satisfactory performance. The experimental analysis carried out by Knaack and Kurama (2014) showed that the addition of RCA had no effect on the performance of RC beams. Although it was recommended that RAC beams utilize the same design criteria as conventional beams, Knaack and Kurama (2014) argued that more research utilizing RCA of different qualities be done before its assured adoption. At a 50% replacement of RCA, Sadati et al. (2016) observed a 16% reduction in the shear strength of RAC beams and reported that ACI 318 (2008) expression, JSCE (2007) expression, fracture mechanics approaches, and modified compression field theory (MCFT) conservatively predict the shear strength of RAC beams. Ignjatović et al. (2017) investigated the shear performance of RAC beams reinforced both with and without transverse reinforcement. The trials carried out for two distinct replacement ratios (50 and 100%) did not show any appreciable differences in fracture pattern or manner of failure. Again, there was very little fluctuation in the service load deflection. The whole investigation came to the conclusion that the RAC beam design does not require any further change due to the inherent conservatism of the current design requirements. Rahal and Alrefaei (2017) noted a 13–18% loss in shear strength of RAC beams at 15% replacement of NCA. Rahal and Alrefaei (2017) verified the shear strength of RAC beams at 5, 10, 16, 23, 35, 50, 75, and 100% substitution of RCA. Additionally, it was determined whether it was appropriate to forecast the shear strength of RAC beams using their compressive strength, and it was advised to do so by applying a reduction factor (λ_r) of 20%. Prior to the exemption of any restrictions on the replacement level of RCA, it has been advised by Rahal and Alrefaei (2017) to conduct additional experimental research on the shear behaviour of RAC beams.

The shear behaviour of RAC beams without transverse reinforcement was studied by Sogo et al. (2004) and Sato et al. (2007) for varying tensile reinforcement contents (2.39, 4.03, and 4.22%) and employing 100% RCA. According to their observations, the fracture pattern and failure mode of RAC beams were quite comparable to those of conventional beams. For the RAC beams, however, a drop of 10–20% was noted. González-Fonteboa and Martínez-Abella (2004) noted that the performance of RAC beams under shear was not noticeably improved by the addition of 8% silica fume. According to Fathifazl et al. (2009a, 2011), using the Equivalent Mortar Volume (EMV) mix proportioning method instead of the traditional mix design approach improved the shear strength of RAC beams. However, the shear strength of the RAC beams made using the EMV approach was lower than that of the NAC beams, suggesting that the aggregate interlocking mechanism may have played a less significant part in this. Due to its innate conservatism, RAC beams can be designed reliably using the current shear design procedures for ordinary concrete beams in Canadian standards, ACI 318 (2008) and Eurocode 2 (2004). Using varied shear span-to-depth ratios (1.5, 2.5, and 3.25), Choi et al. (2010) evaluated the effects of RCA (30, 50, and 100% replacement ratio) on the shear performance of RC beams with varying longitudinal reinforcement content (0.53, 0.81, and 1.61%). Both NAC and RAC beams showed a shear tension mode of failure in the examined beam specimens, and the ultimate shear strength of RAC beams was found to be lower than that of NAC beams. In addition to this, the reduction in shear strength with the increase in shear span-to-depth ratio was more significant in RAC beams. The existing design criteria for RAC beams are effective at providing enough shear strength due of their conservatism, according to Choi et al. (2010). Arezoumandi et al. (2014) conducted an experimental research on RAC beams by completely substituting NCA with longitudinal reinforcement ratios of 1.27, 2.03, and 2.73%. He found that the fracture pattern and load–deflection relationship were identical to those of the NAC beams. The findings of the tested beams fell within the 95

9.3 Impact of PPM Mix Design Approach

9.3.1 Experimental Program

Pradhan et al. (2018a, b) studied the shear and flexure behaviour of RAC beams using 100% RCA and different longitudinal reinforcement content. Both NAC and RAC were prepared using PPM mix design approach and the corresponding mix

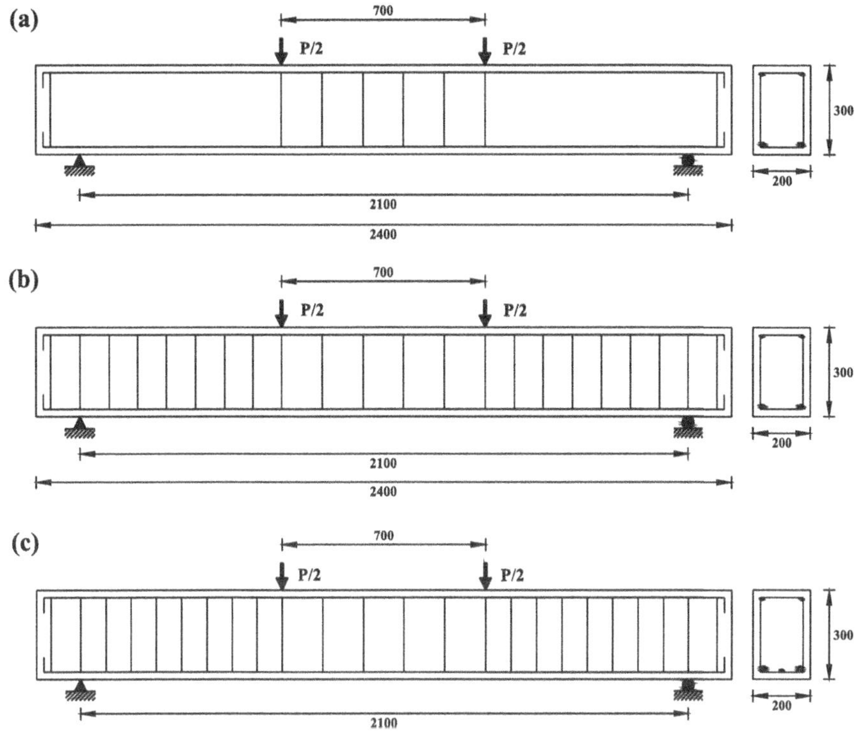

Fig. 9.1 **a** Reinforcement details of beams for shear test having $\rho = 1.31\%$ and $\rho = 0.75\%$, **b** Reinforcement details of beams for flexure test having $\rho = 0.42\%$ and $\rho = 0.75\%$, and **c** Reinforcement details of beams for flexure test having $\rho = 1.31\%$ and $\rho = 1.61\%$ (Pradhan et al. 2018a)

proportion is represented in Table 4.1 in chapter "Particle Packing Method of Mix Proportioning and Modified Mixing Approach". To examine the performance and behaviour in shear, six reinforced NAC and RAC beams (three NACs and three RACs) that lacked transverse reinforcement were evaluated. A total of fourteen beams (seven NAC beams and seven RAC beams) were used to study the flexural behaviour of reinforced NAC and RAC beams. All of the tested beams were 2400 mm in length and had a rectangular cross section with dimensions of 200 mm in width, 300 mm in total depth, and 265 mm in effective depth (d). For all the beams, a fixed shear span-to-depth ratio (a/d) of 2.6 was used. By conducting experiments on beams with longitudinal reinforcement ratios (rho) of 0.75% (NAC 0.75B1-S, NAC 0.75B2-S, RAC 0.75B1-S, and RAC 0.75B2-S) and 1.31% (NAC 1.31B-S and RAC 1.31B-S), the shear performance of RAC beams was compared to that of NAC beams. The shear spans of these shear critical beams did not have the transverse reinforcing. The reinforcement details of the beam specimens without transverse reinforcement and with transverse reinforcement are illustrated in Figs. 9.1 and 9.2.

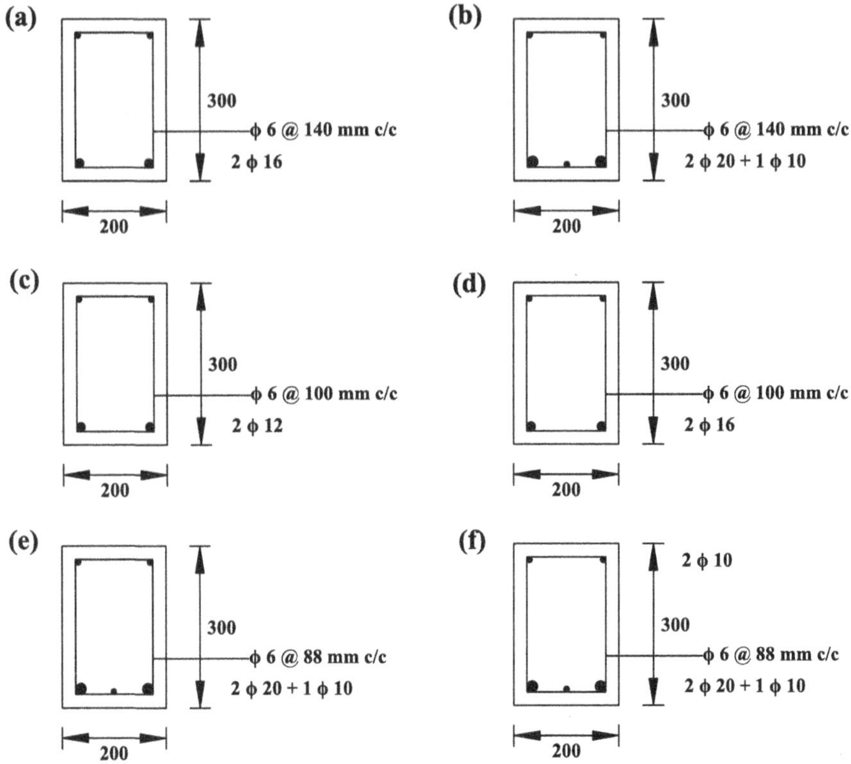

Fig. 9.2 **a** Cross-sectional view of midspan of beams for shear test having $\rho = 0.75\%$, **b** Cross-sectional view of midspan of beams for shear test having $\rho = 1.31\%$, **c** Cross-sectional view of shear span of beams for flexure test having $\rho = 0.42\%$, **d** Cross-sectional view of shear span of beams for flexure test having $\rho = 0.75\%$, **e** Cross-sectional view of shear span of beams for flexure test having $\rho = 1.31\%$, **f** Cross-sectional view of shear span of beams for flexure test having $\rho = 1.61\%$ (Pradhan et al. 2018a)

The performance of PPM mix designed beams in flexure was studied using longitudinal reinforcement ratio (ρ) of 0.42%[1] (NAC 0.42B1-F, NAC 0.42B2-F, RAC 0.42B1-F, and RAC 0.42B2-F), 0.75%[2] (NAC 0.75B1-F, NAC 0.75B2-F, RAC 0.75B1-F, and RAC 0.75B2-F), 1.31%[3] (NAC 1.31B-F, and RAC 1.31B-F),

[1] For $\rho = 0.42\%$, tensile reinforcement consisted of two 12 mm diameter bars, transverse reinforcement consisted of 6 mm diameter bar @ 140 mm c/c at the mid-span (6 stirrups), and 100 mm c/c at the shear span (8 stirrups).

[2] For $\rho = 0.75\%$, tensile reinforcement consisted of two 16 mm diameter bars, transverse reinforcement consisted of 6 mm diameter bar @ 140 mm c/c at the midspan (6 stirrups) and 100 mm c/c at the shear span (8 stirrups).

[3] For $\rho = 1.31\%$, tensile reinforcement consisted of two 20 mm diameter bars and one 10 mm diameter bar, transverse reinforcement consisted of 6 mm diameter bar @ 140 mm c/c at the midspan (6 stirrups) and 88 mm c/c at the shear span (9 stirrups).

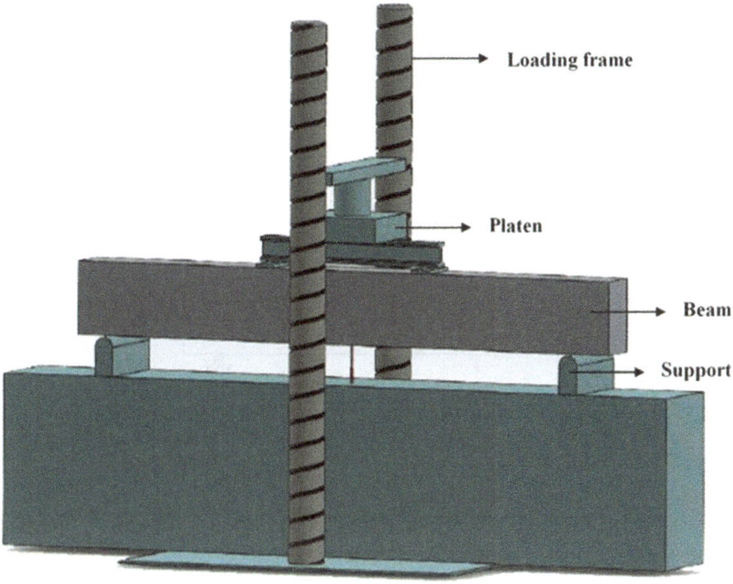

Fig. 9.3 Test setup for the four-point bending test of beams (Authors')

and 1.61%[4] (NAC 1.61B1-F, NAC 1.61B2-F, RAC 1.61B1-F, and RAC 1.61B2-F). Transverse reinforcements were provided as per the requirements of IS: 456 (2000) to preclude the shear failure. The details of the four types of beams are illustrated in Figs. 9.1 and 9.2.

9.3.2 Test Setup and Procedure

The beams were held in place 150 mm from either end by a roller and a pin support. The load was applied symmetrically at a distance of 700 mm from each support with the assistance of a servo-controlled hydraulic actuator, resulting in a clear span between the two supports of 2100 mm. Using a four-point bending setup, the beam specimens were readily supported and tested. With a maximum displacement of 90 mm, the load was applied at a displacement-controlled rate of 1 mm/min (Fig. 9.3).

The deflection at the mid-span and two loading sites, respectively, were measured using an LVDT (linear variable differential transducer) and two dial gauges. Electrical strain gauges were put in place in the beam's side face and tensile reinforcements

[4] For $\rho = 1.61\%$, tensile reinforcement consisted of two 20 mm diameter bars and one 10 mm diameter bar and compression reinforcement consisted of two 10 mm diameter bars, transverse reinforcement consisted of 6 mm dia bar @ 140 mm c/c at the midspan (6 stirrups) and 88 mm c/c at the shear span (9 stirrups).

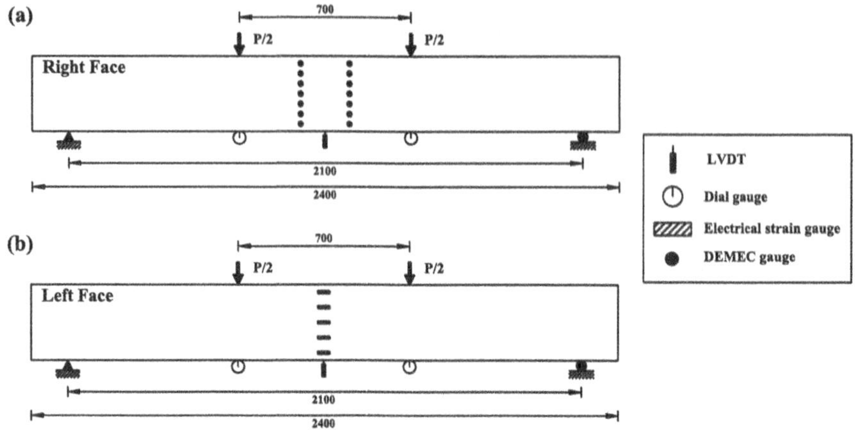

Fig. 9.4 Instrumentation details of the tested beam specimens (Pradhan et al. 2018a)

to measure the strain in the steel and concrete, respectively. Concrete strain was also measured using the demountable mechanical strain gauges (DEMEC gauges). The DEMEC gauge readings and strain gauge readings were recorded every 10 kN. Figure 9.4 shows the specifics of the test arrangement.

9.4 Discussion on Shear and Flexure Performance

The discussion of longitudinal reinforcement and concrete for beams with and without transverse reinforcement includes the crack pattern, failure mode, load–deflection relationship, and load–strain relationship and covers in the subsequent subsections.

9.4.1 Shear Performance

Crack Pattern and Failure Mode

Both NAC and RAC beams had identical crack morphologies and crack propagation, however, RAC beams had substantially closer crack spacing (Fig. 9.5). For all of the beams, the first flexural crack was discovered closer to the mid-span, or the region of maximum bending moment. With the steady rise in applied stress, additional flexural cracks were seen between the loading point and the support. The inclined flexure-shear cracks were discovered in the shear span after an additional increase in applied stress. The flexure-shear cracks spread both horizontally along the tensile reinforcement towards the support of the beams and inclinedly towards the loading plate. The beams then failed in a brittle manner through the main diagonal flexure-

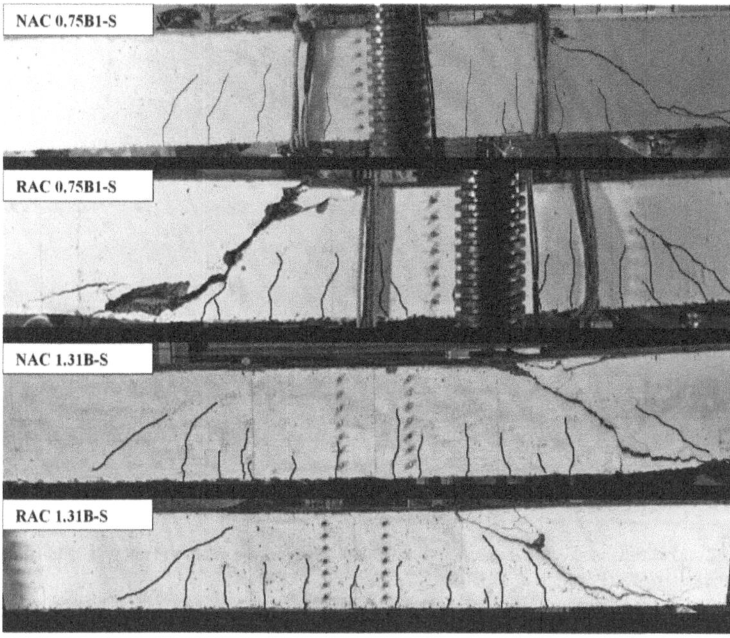

Fig. 9.5 Final condition of beam specimens without transverse reinforcement (Pradhan et al. 2018b)

shear crack, which was also accompanied by the crushing of concrete beneath the loading plate and local bond failure of tensile reinforcement close to the supports. Because there was no transverse reinforcement in the shear span, the tension failure was not noticed in any of the tested beams. For both NAC and RAC beams, the main flexure-shear crack was slanted at a roughly 45° inclination. The post-failure cracking pattern of the tested NAC and RAC beam specimens (Fig. 9.5) without transverse reinforcement clearly demonstrates the evidence of the horizontal crack along the tensile reinforcement merging with the flexure-shear crack. This suggests that the shear stress mode of failure was shared by the NAC and RAC beams.

Load–Deflection Relationship

The critical observations of the test results are summarized in Table 9.1 and the load versus mid-span deflection curves for both NAC and RAC beams without shear reinforcement are shown in Figs. 9.6 and 9.7. The shear strength of both NAC and RAC beams increases as the longitudinal reinforcement content increases. Similar behaviour was also reported by Choi et al. (2010) and Kim et al. (2013). This is attributed to the increment in the compression zone because of the downward shifting of the neutral axis and additional contribution by the dowel action in shear resistance mechanisms (Arezoumandi et al. 2014, 2015a). The load versus mid-span

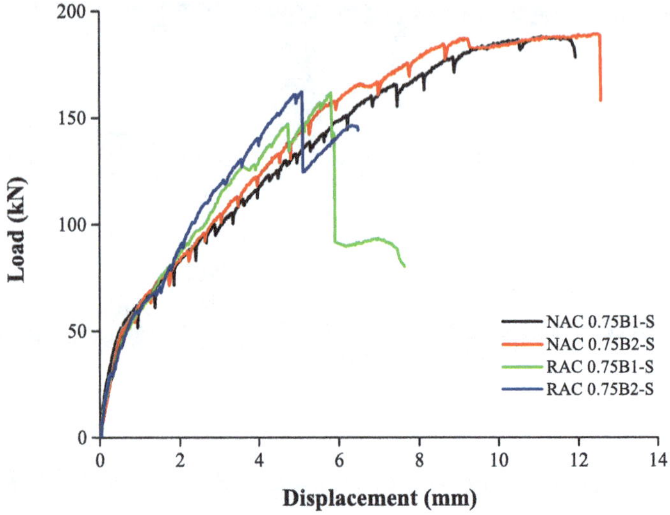

Fig. 9.6 Load versus mid-span deflection relationship for beams without transverse reinforcement and with $\rho = 0.75\%$ (Pradhan et al. 2018b)

deflection relationship of tested beams without shear reinforcement (Figs. 9.6 and 9.7) indicates that the performance of RAC beam specimens is adversely affected by the inclusion of RCA. A significant drop in cracking load as well as the ultimate shear strength of RAC beam specimens was witnessed (Table 9.1). The RAC beams exhibited a 17–22% reduction in initial cracking load, while a drop of 14% was observed in the ultimate load as compared to NAC beams. The earlier studies on the shear behaviour of RAC beams without transverse reinforcement also reported a similar trend (González-Fonteboa and Martínez-Abella 2004, 2007; Etxeberria et al. 2007; Choi et al. 2010; Kim et al. 2013; Arezoumandi et al. 2014; Knaack and Kurama 2014; Sadati et al. 2016; Katkhuda and Shatarat 2016; Ignjatović et al. 2017; Rahal and Alrefaei 2017; González-Fonteboa et al. 2009; Michaud 2015). The lower tensile strength value of RAC due to the presence of two ITZs attributes to the initiation of flexural cracks at a lower load in RAC beams (Choi et al. 2010; Rahal and Alrefaei 2017; Arezoumandi et al. 2015b). Apart from this, the aggregate shape of RCA (angularity of the aggregate) and higher early shrinkage (Knaack and Kurama 2014, 2015) of RAC also impart a negative effect on the flexural tensile strength of RAC beams. It is important to note that the effectiveness of a reinforced concrete beam under shear is influenced by factors such as uncracked concrete, stirrups, dowel action, and aggregate interlocking. The uncracked concrete, dowel action, and aggregate interlocking support the system's shear force in the absence of stirrups. These three shear-resisting mechanisms are related to the mechanical characteristics of aggregate and concrete. As a result, the shear strength of RAC beams is found to be lower than that of conventional concrete beams, indicating that RCA and RAC have a lesser contribution in shear resistance mechanisms than NCA and NAC.

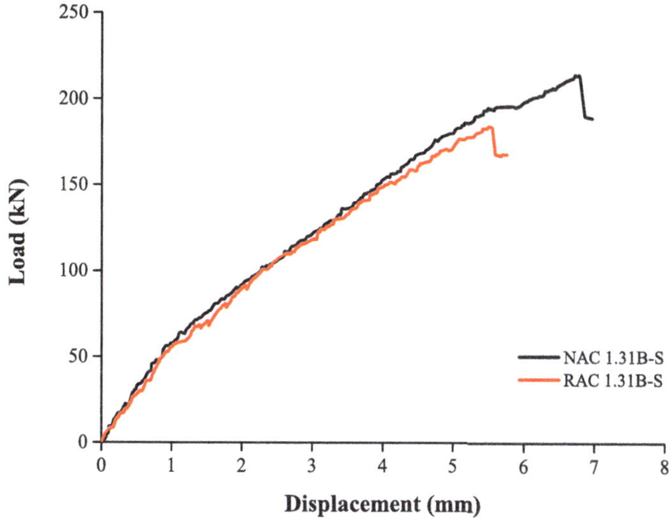

Fig. 9.7 Load versus mid-span deflection relationship for beams without transverse reinforcement and with $\rho = 1.31\%$ (Pradhan et al. 2018b)

Table 9.1 Summary of test results of beams without transverse reinforcement

Beam identifier	ρ (%)		V_{cr} (kN)	V_y (kN)	V_u (kN)	δ_{cr} (mm)	δ_y (mm)	δ_u (mm)	Failure mode
	ρ_t	ρ_c							
NAC 0.75B1-S	0.75	–	25.81	–	94.14	0.55	–	11.07	Shear tension
NAC 0.75B2-S			25.72	–	94.93	0.58	–	12.45	Shear tension
RAC 0.75B1-S	0.75	–	20.64		81.11	0.43	–	5.80	Shear tension
RAC 0.75B2-S			19.48	–	81.32	0.46	–	5.05	Shear tension
NAC 1.31B-S	1.31	–	35.79	–	107.24	1.37	–	6.77	Shear tension
RAC 1.31B-S	1.31	–	29.74	–	92.29	1.20	–	5.50	Shear tension

The mid-span deflections of the tested beams at cracking load (δ_{cr}) and ultimate load (δ_u) were determined from the load–deflection relationship and represented in Table 9.1. The δ_{cr} of both NAC and RAC beams increases as the longitudinal reinforcement content increases, whereas δ_u decreases for both NAC and RAC beams as the longitudinal reinforcement content increases. Choi et al. (2010) and Kim et al. (2013) also reported similar trend of δ_u for RAC beams with 100% RCA. The tested RAC beams without transverse reinforcement exhibited about 12.4–20.7% reduction in δ_{cr}. The δ_u of RAC beams without transverse reinforcement exhibited a reduction of 18.8% at $\rho = 1.31\%$, while a huge drop of 53.9% was recorded for $\rho = 0.75\%$ as compared to NAC beams. Similar observations were also reported by Choi et al. (2010), Rahal and Alrefaei (2017), Arezoumandi et al. (2015a), and Sadati et al. (2016). However, Knaack and Kurama (2014) and González-Fonteboa et al. (2009)

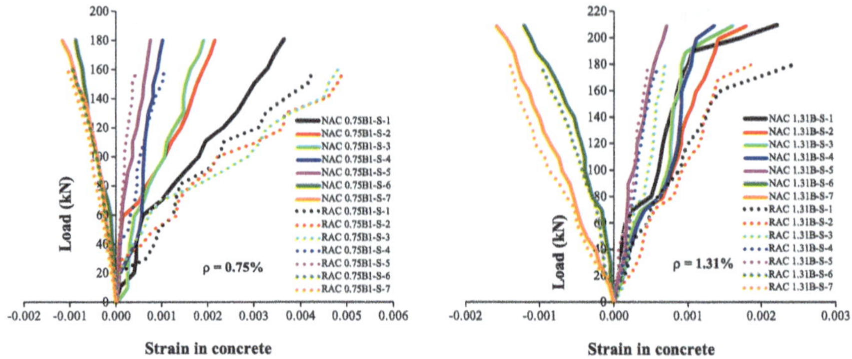

Fig. 9.8 Load versus strain in concrete relationship for beams without transverse reinforcement (Authors')

observed larger deflection in RAC beams in comparison to the conventional concrete beams.

The absence of shear reinforcement adversely affects the ductility of RAC beams. The load–deflection curves (Figs. 9.6 and 9.7) show linear behaviour of both NAC and RAC beams prior to the appearance of initial crack. More importantly, lower pre-cracked stiffness was reported for RAC beams, which is similar to the observations of other authors (Fathifazl et al. 2009a; Knaack and Kurama 2014; Rahal and Alrefaei 2017). The RAC beams without transverse reinforcement showed 1.2 and 5.2% lower pre-cracked stiffness than the NAC beams with $\rho = 0.75\%$ and $\rho = 1.31\%$, respectively (Fig. 9.8).

Load–Strain Relationship

Regardless of the kind of concrete, the concrete strain in the tension region rapidly rose following the initial flexural crack. When compared to NAC beams, the increase in RAC beams, however, was very significant. For RAC beams, it was found that the compressive strain in concrete was higher above neutral axis, particularly near the initial flexural crack's source. This response is explained by the faster rate of crack development in RAC beams, which led to a faster rate of neutral axis shift towards the compression zone (Kang et al. 2014). The lower bond strength of RAC (Butler et al. 2011, 2014) is possibly the reason for higher crack growth rate and this may be attributed to the inferior strength of RCA (Butler et al. 2014), lower aggregate interlocking effect and higher rate of bond deterioration between RCA and mortar interface (Butler et al. 2011; Kang et al. 2014).

The pattern of load versus strain relationship in tensile steel (Fig. 9.9) was not influenced by the incorporation of RCA and such observation has been reported by Etxeberria et al. (2007). The strain in the tensile reinforcement of RAC beams was found to be higher than that of NAC beams before the initial flexural crack,

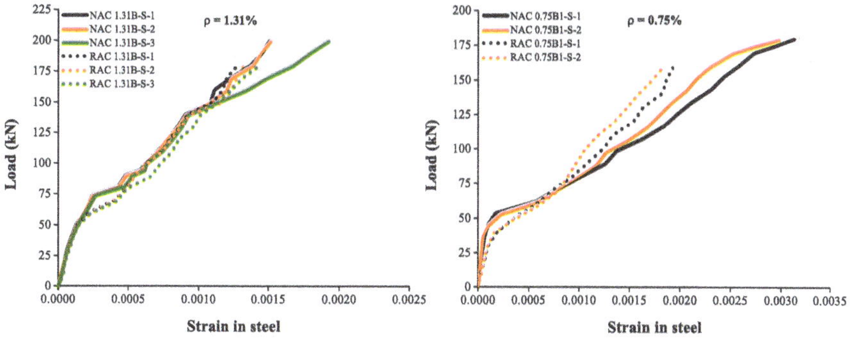

Fig. 9.9 Load versus strain in steel relationship for beams without transverse reinforcement (Authors')

similar to the strain in concrete for the same longitudinal reinforcement content. Higher deflection of RAC beams than NAC beams at the same load is caused due to the lower bond stiffness of RAC (Sato et al. 2007; Kang et al. 2014). The same behaviour was seen in beams without transverse reinforcement as well. Higher steel strain was visible in RAC beams because of the greater deflection before the initial flexural crack.

The strain in steel was found to be lower for RAC beams at the point of final failure of the specimens. The maximum strain in steel at failure was reduced on average by 18 and 41% for shear critical RAC beams with $rho = 1.31\%$ and $rho = 0.75\%$, respectively. Similar reduction in strain of longitudinal reinforcement is also reported in other studies (Etxeberria et al. 2007; Arezoumandi et al. 2014, 2015a; Ignjatović et al. 2017). This behaviour can be attributed to the notable lower failure load for RAC beams. Owing to the lower bond strength between RAC and longitudinal reinforcement (Butler et al. 2011, 2014; Xiao and Falkner 2007) the contribution by dowel action in shear resistance mechanisms is possibly less in RAC beams than that of NAC beams. This could be a contributing factor to the RAC beams' reduced steel strain and lower ultimate shear strength.

9.4.2 *Flexure Performance*

Crack Pattern and Failure Mode

The first flexural crack was spotted near the mid-span of all the beams, i.e. in the region of maximum bending moment. The formation of more flexural cracks was observed between loading point and support with the gradual increase in applied load which propagated towards the compression zone and the inclined flexure-shear cracks were detected. Subsequently, the tensile reinforcement yielded and followed by rapid deformation of the beams. Finally, the failure of the beam specimens occurred as a result of crushing of concrete in the compression fibre or due to the propagation of a number of flexure cracks to the compression zone. All the NAC beams and RAC beams with longitudinal reinforcement content (ρ) of 0.75 and 0.42% failed due to the propagation of flexure cracks to the compression zone. However, RAC beams with ρ of 1.31 and 1.61% failed immediately through a major diagonal crack. The typical flexural cracks and final condition of the beam specimens after failure are shown in Fig. 9.10. Typical flexural tension mode of failure was observed for both NAC and RAC beams (except for RAC beams having reinforcement ratios of 1.31 and 1.61%) with definite yielding of tensile reinforcement. The RAC beams with longitudianl reinforcement content of 1.31 and 1.61% failed in diagonal tension mode. A similar behaviour was also observed in other studies (Ajdukiewicz and Kliszczewicz 2007; Fathifazl et al. 2009b; Ignjatović et al. 2012), in which the RAC beams with transverse reinforcement failed uncharacteristically in diagonal tension mode. The crack morphology and their progression were similar for both NAC and RAC beams, whereas the spacing between cracks was much closer in case of RAC beams. This is similar to the observations reported in other studies (Maruyama et al. 2004; Fathifazl et al. 2009b; Kang et al. 2014; Arezoumandi et al. 2015b). This difference in cracking and failure behaviour may be due to different tension stiffening behaviour. Moreover, the number of cracks increases for both NAC and RAC beams as the longitudinal reinforcement ratio increases and a similar behaviour was also observed by Ignjatović et al. (2012).

Load–Deflection Relationship

The experimental results of Pradhan et al. (2018a) are summarized in Table 9.2. The load–deflection curves of both NAC and RAC beams having longitudinal reinforcement ratios of 1.61, 1.31, 0.75, and 0.42% are shown in Figs. 9.11, 9.12, 9.13, and 9.14, respectively. The cracking load (V_{cr}), yield load (V_y), ultimate load (V_u), and corresponding mid-span displacement are the important responses of flexure critical beam, which are determined from the load–deflection relationship of the tested beams. The reported V_{cr}, V_y, and V_u in Table 9.2 are half of the applied load ($P/2$).

It was found that RAC beams had a lower initial cracking load than NAC beams. Initial cracking load drops as longitudinal reinforcement ratio decreases, and as longitudinal reinforcement ratio decreases, so does the initial cracking load reduction percentage for RAC beams relative to NAC beams. Kang et al. (2014) and Arezoumandi et al. (2015b) both noted the same pattern. The initial cracking load of RAC beams having ρ of 1.61, 1.31, 0.75, and 0.42% are 8.2, 3.8, 9.8%, and 34.3% lower than that of NAC beams, respectively. This indicates that the reduction in cracking load of RAC beams is more significant with the reduction in longitudinal reinforcement content. The reduction in initial cracking load may be attributed to the aggregate shape (lower angularity of the RCA) and higher early age shrinkage of RAC (Knaack and Kurama 2014, 2015). In addition to this, the lower tensile strength of RAC due to the presence of two ITZs, the crack initiates earlier in RAC beams (Arezoumandi et al. 2015b; Ignjatović et al. 2017).

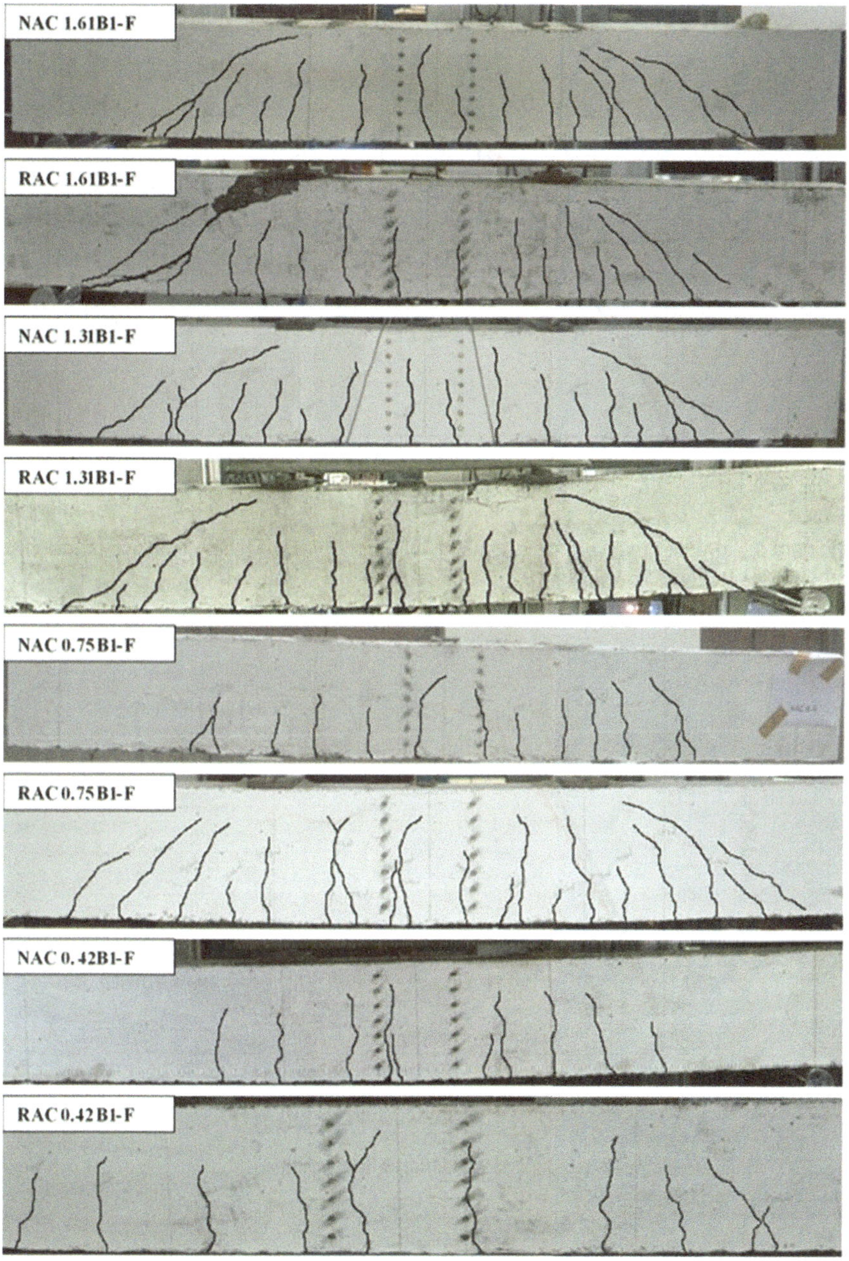

Fig. 9.10 Final condition of beam specimens with transverse reinforcement (Pradhan et al. 2018a)

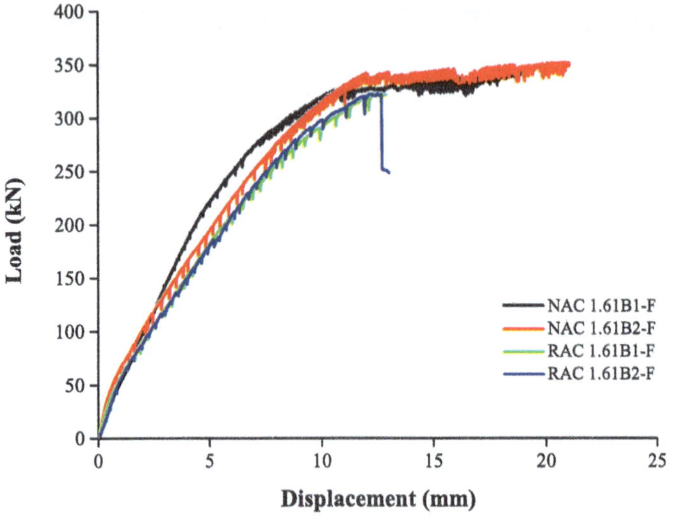

Fig. 9.11 Load versus mid-span deflection relationship for beams with transverse reinforcement and $\rho = 1.61\%$ (Pradhan et al. 2018a)

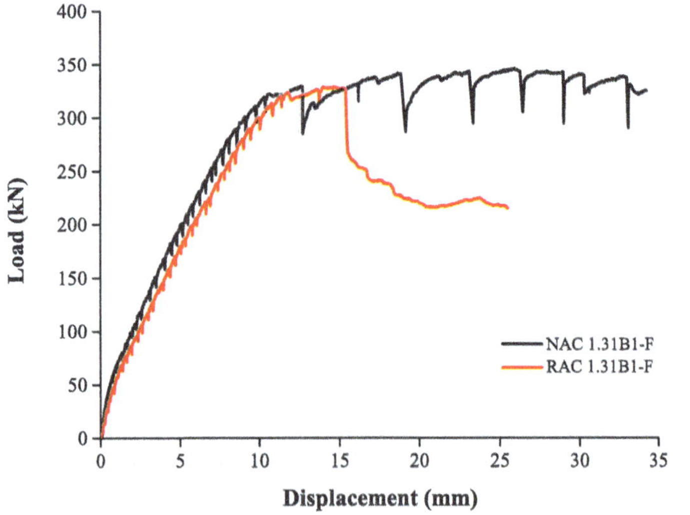

Fig. 9.12 Load versus mid-span deflection relationship for beams with transverse reinforcement and $\rho = 1.31\%$ (Pradhan et al. 2018a)

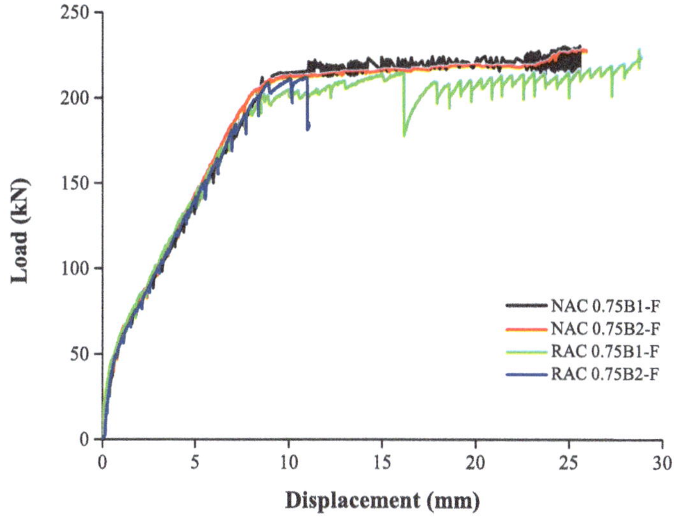

Fig. 9.13 Load versus mid-span deflection relationship for beams with transverse reinforcement and $\rho = 0.75\%$ (Pradhan et al. 2018a)

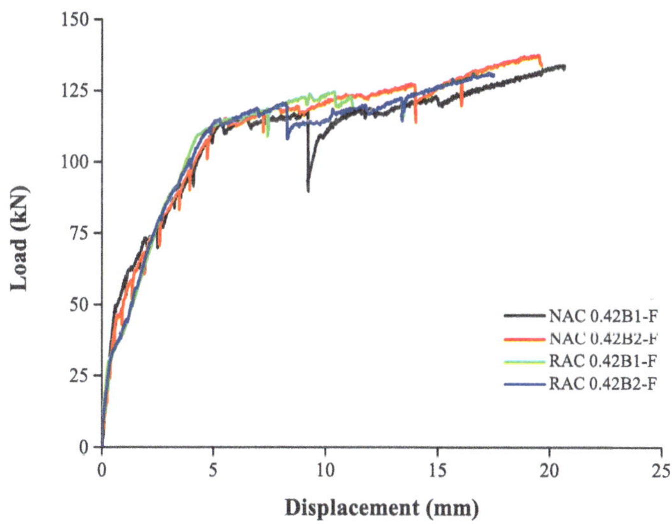

Fig. 9.14 Load versus mid-span deflection relationship for beams with transverse reinforcement and $\rho = 0.42\%$ (Pradhan et al. 2018a)

Table 9.2 Summary of test results of beams with transverse reinforcement

Beam identifier	ρ (%)		V_{cr} (kN)	V_y (kN)	V_u (kN)	δ_{cr} (mm)	δ_y (mm)	δ_u (mm)	Failure mode
	ρ_t	ρ_c							
NAC 1.61B1-F	1.31	0.30	25.76	162.85	171.10	1.01	11.50	18.85	Flexure
NAC 1.61B2-F			26.23	168.64	176.65	0.67	12.12	20.73	Flexure
RAC 1.61B1-F	1.31	0.30	23.91	–	161.97	0.81	–	12.48	Shear tension
RAC 1.61B2-F			23.82	–	162.11	0.74	–	12.24	Shear tension
NAC 1.31B1-F	1.31	–	33.02	160.43	173.00	0.98	11.23	25.91	Flexure
RAC 1.31B1-F	1.31	–	31.78	159.30	165.04	1.20	11.19	14.65	Flexure
NAC 0.75B1-F	0.75	–	29.57	103.90	115.33	0.94	8.56	25.56	Flexure
NAC 0.75B2-F			29.51	102.61	114.77	0.88	8.26	25.74	Flexure
RAC 0.75B1-F	0.75	–	26.52	100.49	114.75	0.69	8.74	28.71	Flexure
RAC 0.75B2-F			26.76	101.09	114.40	0.84	8.43	26.73	Flexure
NAC 0.42B1-F	0.42	–	24.69	56.82	67.04	0.67	5.38	20.56	Flexure
NAC 0.42B2-F			23.24	57.03	69.07	0.72	5.34	19.48	Flexure
RAC 0.42B1-F	0.42	–	15.37	54.60	65.05	0.31	4.22	16.45	Flexure
RAC 0.42B2-F			16.11	55.43	65.89	0.40	4.76	17.26	Flexure

As the longitudinal reinforcement ratio decreases, so does the yield load for NAC and RAC beams. Furthermore, for longitudinal reinforcement ratios of 1.31, 0.75, and 0.42%, respectively, the yield loads of RAC beams were 0.70, 2.40, and 3.36% lower than those of NAC beams. The difference between the yield loads of NAC and RAC beams, however, was not as great as the difference between the cracking loads. Similar conduct was seen in other studies, including those by Ignjatović et al. (2012), Kang et al. (2014), Arezoumandi et al. (2015b).

The comparison of initial cracking load and ultimate load shows that, the initial cracking load was 15.0, 19.1, 25.7, and 35.2% of ultimate load of NAC beams and 14.7, 19.3, 23.3, and 24.0% of RAC beams for the ρ value of 1.61, 1.31, 0.75, and 0.42%, respectively. This comparison shows that, for both NAC and RAC beams, the difference between the initial cracking load and ultimate load increases as the reinforcement ratio decreases. With the exception of the beams with a reinforcement ratio of 0.42%, a comparable amount of difference between ultimate load and cracking load was seen for both NAC beams and RAC beams. This suggests that the RAC beams not only have a reduced cracking load but also a lower ultimate load. For reinforcement ratios of 1.61, 1.31, 0.75, and 0.42%, respectively, the reduction in ultimate load for RAC beams was 6.8, 4.6, 0.4, and 3.8%. In other investigations, similar findings were also reported (Etxeberria et al. 2007; Ajdukiewicz and Kliszczewicz 2007; Ignjatović et al. 2012; Choi et al. 2012; Kang et al. 2014; Arezoumandi et al. 2015b). As the longitudinal reinforcement content decreases, the difference between the ultimate loads of RAC and NAC beams continuously reduces, unlike the cracking load. This suggests that the lower tensile strength of RAC affects the first cracking load of RAC beams, while the integration of RCA has a relatively small impact on the ultimate load.

The load–deflection curves were used to determine the deflection of the beams at the first cracking load (δ_{cr}), yield load (δ_y), and ultimate load (δ_u). Reduced tensile reinforcement content causes RAC beams' deflection at cracking loads to decrease, and this behaviour is consistent with findings from earlier investigations (Ignjatović et al. 2012; Choi et al. 2012; Kang et al. 2014). The RAC beams showed lesser δ_{cr} in comparison to the NAC beams for the same reinforcement content except for $\rho = 1.31\%$. RAC beams exhibited 7.6, 15.9, and 48.9% lower δ_{cr} for longitudinal reinforcement ratio of 1.61, 0.75, and 0.42%, respectively, whereas, for $\rho = 1.31\%$, the RAC beam showed 22.5% higher δ_{cr} with respect to the NAC beam. A similar trend was also observed in δ_y. However, the reduction in case of RAC beams was very minimal (0.4, 1.2, and 16.2% for ρ of 1.31, 0.75, and 0.42%, respectively) as compared to δ_{cr}. The ultimate deflection of RAC beams was found to be lower as compared to NAC beams for $\rho = 1.61\%$, $\rho = 1.31\%$, and $\rho = 0.42\%$ (37.5, 43.5, and 15.8%, respectively), whereas, for $\rho = 0.42\%$ interestingly, δ_u was increased by 79% for RAC beams having $\rho = 0.75\%$. The investigation shows that RAC beams' flexural behaviour was unaffected by reduced longitudinal reinforcement ratios despite having lower δ_u.

The cracking, yielding, and ultimate failure loads as well as the associated mid-span displacement can be readily seen for both NAC and RAC beams with ρ of 1.31, 0.75, and 0.42%. These three are the distinguishing characteristics of the load–deflection relationship in flexural failure of a beam. The load–deflection relationship suggests that the RAC beams (RAC 1.61B1-F and RAC 1.61B2-F) failed prior to the yielding of tensile reinforcement in NAC beams with $\rho = 1.61\%$, which showed longitudinal reinforcement yielding and the typical flexure tension mode failure. Similar load–deflection relationships were found for beams with longitudinal reinforcement ratio of 2.54% in the experimental study conducted by Ignjatović et al. (2012). Moreover, Fathifazl et al. (2009b) also reported shear compression mode failure for transversely reinforced RAC beams with longitudinal reinforcement ratio of 4%. The crack pattern of tested beams (Fig. 9.10) suggests that the mode of failure for RAC beams with ρ of 1.31 and 1.61% was diagonal tension. Due to insufficient transverse reinforcing or the shear resistance of uncracked concrete, the diagonal tension mode of failure may occur. Compression reinforcement strengthens the compression zone and decreases its depth in a double-reinforced concrete beam. By enabling the tensile reinforcement to yield before the concrete is crushed allows the beams to fail in tension. Since the use of compression steel enhances the ultimate load carrying capacity of the beam, the shear critical zone needs to be strengthened with shear reinforcement to resist increased shear stress. In contrast, the same amount of shear reinforcement was found to be insufficient for RAC beams with ρ of 1.31 and 1.61% and failed in the diagonal tension mode. The shear reinforcement (0.323%) provided for NAC beams with $\rho = 1.61\%$ (NAC 1.61B1-F and NAC 1.61B2-F) was sufficient for the desired flexure tension mode failure. Studies on the behaviour of RAC beams without transverse reinforcement demonstrate that RAC has weaker shear resistance mechanisms (Sadati et al. 2016; Rahal and Alrefaei 2017; Choi et al. 2010; Arezoumandi et al. 2014, 2015a; Pradhan et al. 2018b).

Table 9.3 Variation in stiffness of tested beams

ρ (%)	K_{uncr}		K_{cr}	
	NAC	RAC	NAC	RAC
1.61	32.33	30.81	12.75	–
1.31	33.69	26.48	12.43	12.76
0.75	32.50	35.15	9.52	9.40
0.42	34.56	44.93	7.07	9.53

The uncracked stiffness (K_{uncr}), the post crack stiffness (K_{cr}) of both NAC and RAC beams for different longitudinal reinforcement ratio were calculated using Eq. 9.4.1 and 9.4.2, respectively, and summarized in Table 9.3. The K_{uncr} of RAC beams increases as the longitudinal reinforcement ratio decreases, whereas, for NAC beams, nearly similar K_{uncr} was observed irrespective of the longitudinal reinforcement content. However, K_{cr} of both NAC and RAC beams were nearly similar, apart from the beams with longitudinal reinforcement ratio of 0.42%. In other studies also, similar behaviour in K_{cr} of NAC and RAC beams were observed (Choi et al. 2012; Kang et al. 2014).

$$K_{uncr} = \frac{V_{cr}}{\delta_{cr}} \tag{9.4.1}$$

$$K_{cr} = \frac{V_y - V_{cr}}{\delta_y - \delta_{cr}} \tag{9.4.2}$$

Ductility Ratio

The ductility ratio is expressed as the ratio of δ_u to δ_y. The ductility ratio increases as the longitudinal reinforcement ratio reduces, irrespective of the type of concrete (NAC or RAC) used (Table 9.4). Similar behaviour was also reported by Kang et al. (2014). The ductility ratio of RAC beams with $\rho = 1.31\%$ was 43% lower than the NAC beams, whereas only marginal difference is observed in the ductility ratio of NAC and RAC beams with longitudinal reinforcement ratio of 0.75 and 0.42%. Choi et al. (2012) also observed negligible difference in the ductility ratio of NAC and RAC beams. However, Kang et al. (2014) and Maruyama et al. (2004) reported lower ductility ratio of RAC beams. The study by Pradhan et al. (2018a) indicates that the ductility of RAC beams is adversely affected at higher longitudinal reinforcement content.

All NAC and RAC beams failed in the flexure tension mode, with the exception of the RAC beams with ρ of 1.31 and 1.61%. The failure of the RAC beams occurred before the longitudinal reinforcement yielded in an unfavourable diagonal tension mode for increased longitudinal reinforcement content. The maximum

Table 9.4 Ductility ratio of tested beam specimens

ρ (%)	Ductility ratio	
	NAC beams	RAC beams
1.61	1.68	–
1.31	2.31	1.31
0.75	3.05	3.13
0.42	3.74	3.77

allowed content of longitudinal reinforcement for RAC beam can be limited, or the shear reinforcement content can be altered, to overcome this issue. The maximum permitted longitudinal reinforcement ratio for the RAC beams can be established using the ductility ratio. In order to calculate the minimum and maximum requirements of flexural reinforcement ratio to assure flexure tension mode failure of RAC beams, it is necessary to study the impact of the size of the RAC beams on ductility ratio. Further experimental analysis of the flexural behaviour of RAC beams with increased longitudinal reinforcement content is necessary in this regard.

Load–Strain Relationship

Electrical strain gauges and DEMEC gauges were used to measure the strain in the concrete at the midpoint of the beam. Regardless of the type of beam, the concrete strain in the stress zone increased significantly as anticipated following the initial flexural crack. Comparing RAC and NAC beams, Fig. 9.15 shows that the RAC beams exhibited more strain at a given load in the tension zone. Also noted to be higher for RAC beams was the compression zone strain. Ajdukiewicz and Kliszczewicz (2007) and Kang et al. (2014) also reported similar behaviour of RAC beams. According to Kang et al. (2014), the higher compressive strain in RAC beams was caused by the greater upward shifting of neutral axis caused by the faster rate of crack propagation. The presence of two ITZ (Kang et al. 2014; Butler et al. 2011) confirms the lower bond strength of RAC and the greater crack growth rate by causing a lower aggregate interlocking effect in RAC and a higher rate of bond deterioration between RCA and mortar interface (Butler et al. 2011, 2014).

Electrical strain gauges were positioned in the middle of the tensile reinforcement to measure the strain in steel. The introduction of RCA does not change the pattern of load versus strain in the tensile steel relationship (Fig. 9.16), and comparable behaviour was also documented in previous experiments (Etxeberria et al. 2007; Kang et al. 2014; Arezoumandi et al. 2015b). But regardless of the longitudinal reinforcement ratio, it was found that the strain in steel at specimen failure was lower for RAC beams, which is consistent with the findings of the studies by Etxeberria et al. (2007) and Kang et al. (2014). Maximum tensile reinforcement strains were

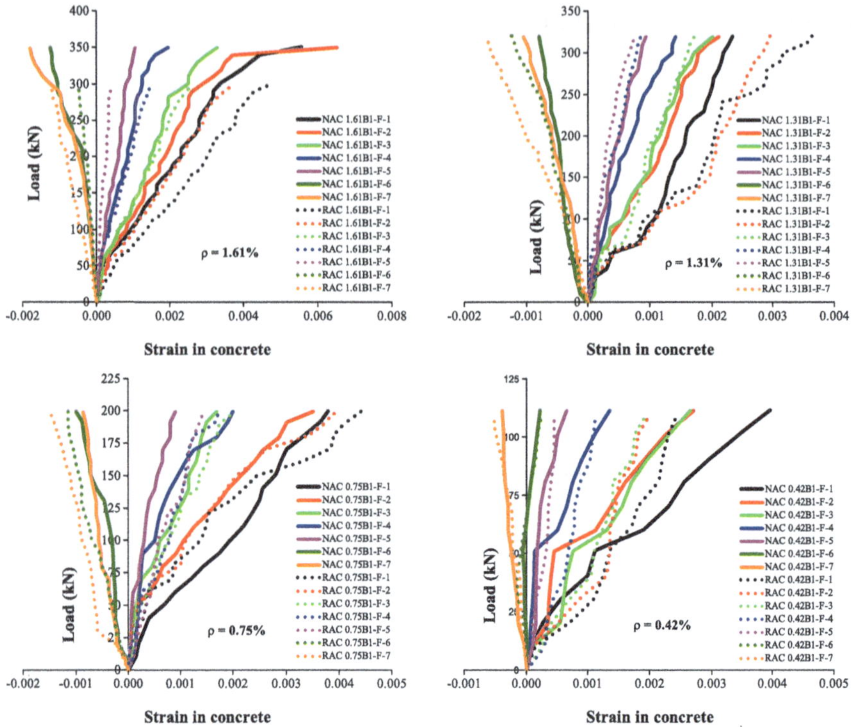

Fig. 9.15 Load versus strain in concrete relationship for beams with transverse reinforcement (Authors')

reduced by as much as 28% for RAC beams, on average by 21%. The lower failure load for RAC beams can be blamed for this tendency.

Shear Behaviour of Beams with Transverse Reinforcement

The investigations done on RAC beams without any transverse reinforcement proved that RAC is inferior in shear transfer mechanisms. Additionally, transversely reinforced RAC beams with longitudinal reinforcement ratios of 1.31 and 1.61% showed the diagonal tension mode of failure (Fig. 9.10).

The RAC beams having ρ of 1.31 and 1.61% failed suddenly through the major diagonal flexure-shear crack (Fig. 9.10). Furthermore, the connection between load and strain in steel (Fig. 9.16) indicates that RAC beams with $\rho = 1.61\%$ failed before the longitudinal reinforcement gave way. The RAC beams with ρ of 1.31 and 1.61% failed in the diagonal tension mode as opposed to the flexure tension mode, according to the evidence at hand. As a result, the RAC beams performed differently from the NAC beams at larger longitudinal reinforcement contents, exposing their weaknesses

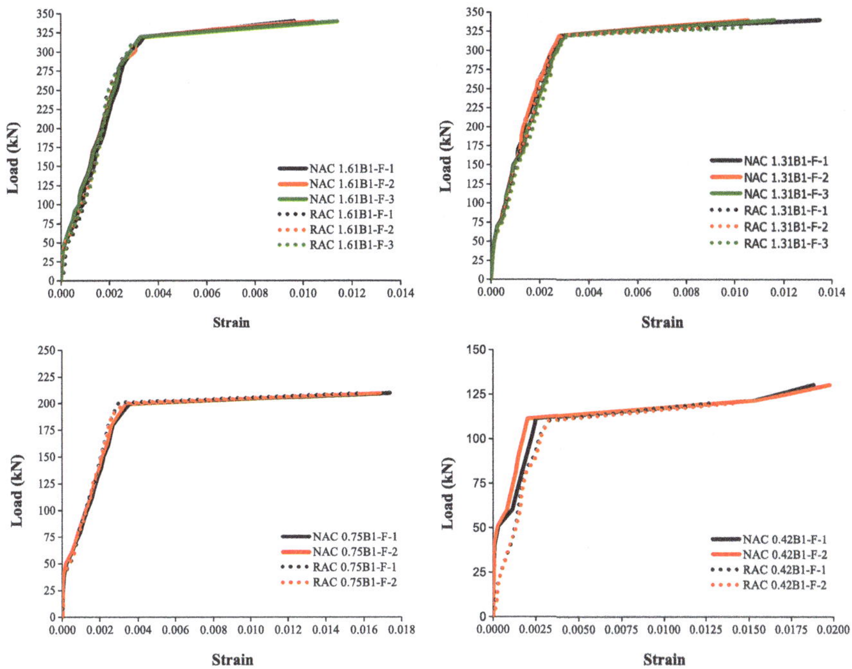

Fig. 9.16 Relationship between load and strain in steel for beams with transverse reinforcement (Authors')

in shear resistance. The provided transverse reinforcement content was found to be sufficient to cause the flexure tension mode of failure in NAC beams, whereas, for the same transverse reinforcement content, the RAC beams are failing in the unfavourable diagonal tension mode. Such behaviour might be explained by RAC's diminished contribution to shear resistance processes. The lower shear strength of shear critical RAC beams provides evidence of the RAC's decreased involvement in shear resistance processes. Under similar design and testing conditions, RAC beams behaved in an uncharacteristic manner for ρ 1.31 and 1.61% with respect to NAC beams, necessitating a revision of existing shear design provisions to make them applicable to RAC beams and prevent the undesirable diagonal tension mode of failure.

There was little to no difference between the compressive strengths of NAC and RAC, according to Pradhan et al. (2017), while RAC's flexural tensile strength and split tensile strength both decreased by 6.7 and 17.7%. It is important to note that once the tensile strength limit is achieved in a reinforced concrete beam, the flexural crack begins at the bottom fibre. Additionally, in the shear span, the main tensile stress causes flexural cracks to spread diagonally until they reach the level of tensile reinforcement. As a result, the resistance of a reinforced concrete beam is significantly influenced by the tensile strength of concrete. One of the main causes of RAC's lower shear strength when compared to conventional concrete beams is its lower tensile strength.

9.5 Shear Strength Assessment of RAC Beams

9.5.1 Comparative Study on Shear Strength Prediction of RAC Beams Without Stirrups

The experimental study of the shear behaviour of both RAC and NAC beams revealed that the addition of RCA had a negative impact on the shear performance of reinforced concrete beams. In order to validate the applicability of the existing equations for predicting the shear strength of RC beams found in various codes and literature (Table 9.5), an exercise was conducted to compare experimental results with them.

The expressions available in IS: 456 (2000), ACI 318 (2008), fib MC (2010), Eurocode 2 (2004), BS 8110-1 (1997), and New Zealand code (1984) and proposed equations by Zsutty (1968, 1971), Niwa et al. (1986), Kim and Park (1996), Rebeiz (1999), and Russo et al. (2005) consider the compressive strength of concrete and longitudinal reinforcement ratio to predict the shear strength of RC beams. In addition to this, ACI 318 (2008), fib MC (2010), Zsutty (1968, 1971), Niwa et al. (1986), Kim and Park (1996), Rebeiz (1999), and Russo et al. (2005) also consider the shear span-to-depth ratio. Eurocode 2 (2004), BS 8110-1 (1997), and New Zealand code (1984) account for the effect of depth of the beam along with the compressive strength of concrete and longitudinal reinforcement ratio. The equations proposed by Bažant and Kim (1984) and Bažant and Sun (1987) are governed by the fracture mechanics approach, where the effect of maximum size of the aggregate (d_0) is accounted in addition to the compressive strength of concrete, longitudinal reinforcement ratio, shear span-to-depth ratio, and depth of the beam. In Gastebled and May (2001) and Bažant and Yu (2005a, b) expressions, the effect of maximum size of the aggregate is not accounted. Additionally, to predict the diagonal cracking strength of RC beams without transverse reinforcement, the equation put forth by Bažant and Yu (2005a, b) takes into account the compressive strength of concrete, longitudinal reinforcement ratio, and depth of the beam characteristics.

The experimental data from the available literature attests to the appropriateness of the aforementioned formulas to forecast the diagonal cracking strength of RAC beams without shear reinforcement. The experimental parameters and test outcomes of the RAC beams without stirrups were compiled into a database, which is provided in Appendix B. The measured values and predicted values of the existing expressions were compared using this database. The average values and coefficient of variation (CoV) for the ratio of measured to anticipated values of diagonal cracking strength ($v_{c,meas}/v_{c,pred}$) are shown in Fig. 9.17. The expressions of Eurocode 2 (2004), New Zealand code (1984), Zsutty (1968, 1971), Bažant and Sun (1987), Niwa et al. (1986), Kim and Park (1996), and Russo et al. (2005) overestimate the diagonal cracking

Table 9.5 Existing equations to predict v_c of RC beams without shear reinforcement

Codes and authors	Expressions
IS: 456 (2000)	$v_c = \frac{0.85\sqrt{f_{ck,cu}}\left(\sqrt{1+5\beta}-1\right)}{6\beta}$ where $\beta = \frac{0.8 f_{ck,cu}}{6.89\rho} \nless 1$
ACI 318 (2008)	$v_c = \frac{1}{7}\left(\sqrt{f_c} + 120\frac{\rho}{a/d}\right)$
fib MC (2010)	$v_c = 0.15\left(1+\sqrt{\frac{200}{d}}\right)\left(\frac{3d}{a}\right)^{\frac{1}{3}}(\rho f_c)^{\frac{1}{3}}$
Eurocode 2 (2004)	$v_c = 0.18\left(1+\sqrt{\frac{200}{d}}\right)(\rho f_c)^{\frac{1}{3}}$
BS 8110-1 (1997)	$v_c = \frac{0.79}{\gamma_m}\left(\frac{\rho A_s}{bd}\right)^{\frac{1}{3}}\left(\frac{400}{d}\right)^{\frac{1}{4}}\left(\frac{f_c}{25}\right)^{\frac{1}{3}}$
New Zealand code (1984)	$v_c = (0.07 + 10\rho)\sqrt{f_c}$
Zsutty (1968, 1971)	$v_c = 2.21\left(f_c\rho\frac{d}{a}\right)^{\frac{1}{3}}$, for $\frac{a}{d} > 2.5$ $v_c = 2.21\left(f_c\rho\frac{d}{a}\right)^{\frac{1}{3}}\left(\frac{2.5d}{a}\right)$, for $\frac{a}{d} \leq 2.5$
Niwa et al. (1986)	$v_c = 1.125\frac{\rho^{\frac{1}{3}}}{d^{\frac{1}{4}}} f_c^{\frac{1}{3}}\left(0.75 + \frac{1.4}{a/d}\right)$
Kim and Park (1996)	$v_c = 3.5 f_c^{\frac{1}{3}}\rho^{\frac{3}{8}}\left(0.4 + \frac{d}{a}\right)\lambda(d)$ where $\lambda(d)$ = Shape adjustment factor
Rebeiz (1999)	$v_c = 0.4 + \sqrt{f_c\rho\frac{d}{a}}\,(2.7 - 0.4A_d)$ where $A_d = \frac{a}{d}$ for $1 < \frac{a}{d} < 2.5$ and 2.5 for $\frac{a}{d} \geq 2.5$
Russo et al. (2005)	$v_c = 1.13\frac{1+\sqrt{\frac{5.08}{d_0}}}{\sqrt{1+\frac{d}{25d_0}}}\left[\rho^{0.4} f_c^{0.39} + 0.5\rho^{0.83} f_y^{0.89}\left(\frac{a}{d}\right)^{-1.2-0.45\left(\frac{a}{d}\right)}\right]$
Bažant and Kim (1984)	$v_c = \frac{0.831\rho^{\frac{1}{3}}}{1+\frac{d}{25d_0}}\left[\sqrt{f_c} + 249\sqrt{\frac{\rho}{(a/d)^5}}\right]$
Bažant and Sun (1987)	$v_c = 0.54\rho^{\frac{1}{3}}\frac{1+\sqrt{\frac{5.08}{d_0}}}{\sqrt{1+\frac{d}{25d_0}}}\left[\sqrt{f_c} + 249.2\sqrt{\frac{\rho}{(a/d)^5}}\right]$
Gastebled and May (2001)	$v_c = 0.15\frac{37.41}{\sqrt{d}}\left(\frac{3d}{a}\right)^{\frac{1}{3}}(100\rho)^{\frac{1}{6}}\left(1-\sqrt{\rho}\right)^{\frac{2}{3}} f_c^{0.35}$
Bažant and Yu (2005a, b)	$v_c = 3.5\sqrt{f_c\frac{7\rho^{\frac{2}{3}}}{d}}$

where f_c = compressive strength of concrete cylinder (MPa), $f_{ck,cu}$ = compressive strength of cube (MPa), ρ = longitudinal reinforcement ratio, a/d = shear span-to-depth ratio, f_y = yield strength of longitudinal reinforcement (MPa), b = width of the beam (mm), d = effective depth of the beam (mm), d_0 = maximum aggregate size (mm)

strength of RAC beams without stirrups, which can be observed from Fig. 9.17. However, the remaining expressions (IS: 456 2000; ACI 318 2008; fib MC 2010; BS 8110-1 1997; Rebeiz 1999; Bažant and Kim 1984; Bažant and Sun 1987; Bažant and Yu 2005a, b; Gastebled and May 2001) discussed here, underestimate the v_c

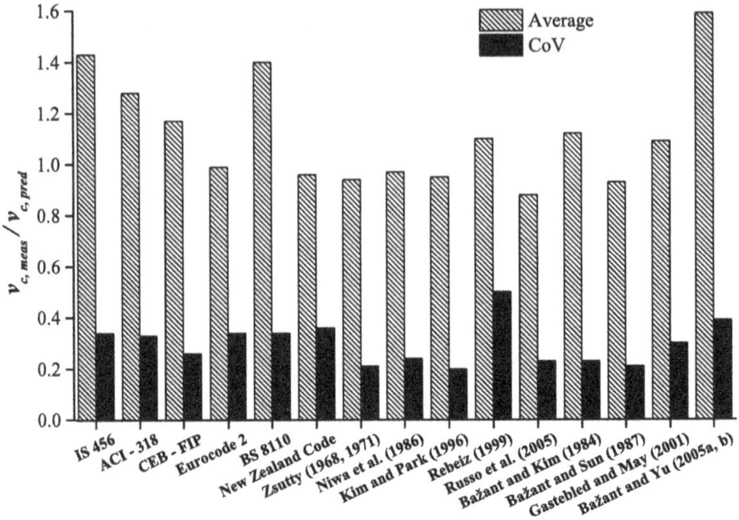

Fig. 9.17 Experimental and predicted diagonal tension cracking strength of RAC beams (Pradhan et al. 2018b)

of RAC beams without stirrups and the applicability of these expressions for RAC beams is verified using the average and CoV values obtained for $v_{c,meas}/v_{c,pred}$. In this regard, the average and CoV values of IS: 456 (2000), ACI 318 (2008), fib MC (2010), BS 8110-1 (1997), Rebeiz (1999), Bažant and Kim (1984), Gastebled and May (2001), and Bažant and Yu (2005a, b) are observed to be 1.43 (CoV = 0.34), 1.28 (CoV = 0.33), 1.17 (CoV = 0.26), 1.40 (CoV = 0.34), 1.10 (CoV = 0.50), 1.12 (CoV = 0.23), 1.09 (CoV = 0.30), and 1.59 (CoV = 0.39), respectively. In comparison to the other existing formulas for v_c of the RAC beams without stirrups, the equation presented by Bažant and Kim (1984) displays better prediction when average and CoV values are taken into account. Out of 91 tested beam outcomes, 29 unconservative predictions are found using this equation. Because of this, it was more acceptable to use a different equation to forecast the v_c of RAC beams. In this case, the available test results of RAC beams without shear reinforcement (Appendix B) were used.

9.5.2 Comparative Study on Shear Strength Prediction of RAC Beams with Stirrups

The ultimate load carrying capability of RAC beams with stirrups was examined through an investigation to see whether the current expression was appropriate. In this regard, the expressions available in ACI 318 (2008), New Zealand code (1984), Eurocode 2 (2004), fib MC (2010), Zsutty (1971), Bažant and Kim (1984), Arslan (2008), and Russo et al. (2013) as represented in Table 9.6 were adopted to predict the ultimate load of RAC beams with stirrups and compared with the experimental

Table 9.6 Existing expressions to predict ultimate load of RC beams

Codes and authors	Expressions
ACI 318 (2008)	$v_u = \frac{1}{7}\left(\sqrt{f_c} + 120\frac{\rho}{a/d}\right) + \rho_w f_{yw}$
New Zealand code (1984)	$v_u = (0.07 + 10\rho)\sqrt{f_c} + \rho_w f_{yw}$
Eurocode 2 (2004)	$V_{Rd} = V_{Rd,c} + V_{Rd,s}$
	$V_{Rd,c} = \left[C_{Rd,c}k\left(100\rho f_{ck}\right)^{1/3}\right]bd$
	$V_{Rd,s} = \frac{A_{sv}}{s_v} z f_{yw} \cot\theta$
fib MC (2010)	$V_{Rd} = V_{Rd,c} + V_{Rd,s}$
	$V_{Rd,c} = k_v \frac{\sqrt{f_{ck}}}{\gamma_c} bz$
	$V_{Rd,s} = \frac{A_{sv}}{s_v} z f_{yw} \cot\theta$
Zsutty (1971)	$v_u = 2.21\left(f_c\rho\frac{d}{a}\right)^{\frac{1}{3}}\left(\frac{2.5}{a/d}\right) + \rho_w f_{yw}$ (for, $\frac{a}{d} < 2.5$)
	$v_u = 2.21\left(f_c\rho\frac{d}{a}\right)^{\frac{1}{3}} + \rho_w f_{yw}$ (for, $\frac{a}{d} \geq 2.5$)
Bažant and Kim (1984)	$v_u = \frac{0.831\rho^{1/3}}{1+\frac{d}{25d_a}}\left(\sqrt{f_c} + 249\sqrt{\frac{\rho}{(a/d)^5}}\right) + \rho_w f_{yw}$
Arslan (2008)	$v_u = \left(0.15 f_c^{0.5} + 0.02 f_c^{0.65}\right)\left(\frac{2.5}{a/d}\right) + \rho_w f_{yw}$ (for, $\frac{a}{d} < 2.5$)
	$v_u = \left(0.15 f_c^{0.5} + 0.02 f_c^{0.65}\right) + \rho_w f_{yw}$ (for, $\frac{a}{d} \geq 2.5$)
Russo et al. (2013)	$v_u = 0.72\frac{1+\sqrt{\frac{5.08}{d_a}}}{\sqrt{1+\frac{d}{25d_a}}}\left[\rho^{0.4} f_c^{0.39} + 0.5\rho^{0.83} f_y^{0.89}\left(\frac{a}{d}\right)^{-1.2-0.45\frac{a}{d}}\right]$ $+0.17 f_c^{0.5}\left(\rho_w f_{yw}\right)^{0.7}$

where v_u = ultimate shear strength of RC beam with stirrups (MPa), V_{Rd} = shear resistance of RC beam (N), $V_{Rd,c}$ = contribution of concrete in shear resistance (N), $V_{Rd,s}$ = contribution of stirrups in shear resistance (N), ρ = longitudinal reinforcement ratio, f_y = yield strength of longitudinal reinforcement (MPa), f_c = compressive strength of concrete (MPa), b = width of the beam (mm), d = effective depth of the beam (mm), d_a = maximum size of the aggregate (mm), a/d = shear span-to-depth ratio, ρ_w = transverse reinforcement ratio = A_{sv}/bs_v (where A_{sv} = area of transverse reinforcement (mm^2), s_v = spacing of transverse reinforcement (mm)), and f_{yw} = yield strength of transverse reinforcement (MPa), $C_{Rd,c} = 0.18/\gamma_c$, γ_c = partial factor of safety for concrete = 1.5, $k = 1 + \sqrt{\frac{200}{d}}$, z = lever arm = $0.9d(mm)$, $k_v = \frac{0.4}{1+1500\epsilon_x}\left[1 - \frac{V_{E,test}}{V_{Rd,max}}\right] \geq 0$, $\theta_{min} = 20^\circ + 10000\epsilon_x$, $\epsilon_x = \frac{V_{E,test}\left[\frac{d}{z}\left(\frac{a}{d}-1\right)+1\right]}{2E_s A_{sl}}$, $V_{R,max} = k_\epsilon \eta_{fc} f_c z b_w \sin\theta\cos\theta$, $\eta_{fc} = \left(\frac{f_c}{30}\right)^{1/3} \leq 1.0$, $k_\epsilon = \frac{1}{1.2+55\epsilon_1} \leq 0.65$, $\epsilon_1 = \epsilon_x + (\epsilon_x + 0.002)\cot^2\theta$, E_s = modulus of elasticity of longitudinal reinforcement (MPa), ϵ_x = longitudinal strain at mid-depth of the beam, A_{sl} = area of longitudinal reinforcement (mm^2)

results. A database was prepared by collecting the experimental results of RAC beams with transverse reinforcement from the available literature (Etxeberria 2004; Etxeberria et al. 2007; Ajdukiewicz and Kliszczewicz 2007; Fathifazl et al. 2009b, 2010; Bai and Sun 2010; Ignjatović et al. 2012; Choi et al. 2012; Knaack 2013; Knaack and Kurama 2013, 2014; Kang et al. 2014; Arezoumandi et al. 2015b; González-Fonteboa and Martínez-Abella 2004, 2007) and also from the reported data by Tošić et al. (2016) and represented in Appendix C. This database was used to calculate the ratio of measured to predicted ultimate load ($v_{u,meas}/v_{u,pred}$) by employing the expressions represented in Table 9.6.

The variation in $v_{u,meas}/v_{u,pred}$ with respect to the corresponding $\rho_w f_{yw}$ value is represented in Fig. 9.18 for each expression. It is observed from Fig. 9.18 that, for $\rho_w f_{yw}$ value above 1.5 the existing expressions overestimate more than 90% of the test results. Using $v_{u,meas}/v_{u,pred}$ for each expression, the mean, coefficient of variation (CoV), and number of unconservative predictions are evaluated. The results are compiled in Table 9.7. The lower mean values show that the ultimate load of transversely reinforced RAC beams is overestimated by the available expressions. Furthermore, the current expressions record at least 60% unconservative forecasts. The database of RAC beams reported in Appendix C was used to create a separate expression to predict the ultimate load of RAC beams with transverse reinforcement while accounting for the RCA replacement ratio, mechanical properties of RAC, and longitudinal and transverse reinforcement content parameters.

9.5.3 *Critical Analysis of Design Parameters*

Without Transverse Reinforcement

The expressions covered in Sect. 9.5.1 take into account the compressive strength of concrete, longitudinal reinforcement ratio, shear span-to-depth ratio, effective depth of the beam, and the ratio of maximum aggregate size and effective depth of the beam to account for the size effect of the beam parameters and predict v_c of RAC beam without transverse reinforcement. The impact of the RCA replacement ratio on the shear behaviour of RAC beams is also taken into consideration in addition to these parameters (Pradhan et al. 2018b). The experimental findings presented in the prior research were used to examine the impact of these key design factors on the diagonal stress cracking strength of RAC beams. Figure 9.19 illustrates the change in $v_{c,RAC}/v_{c,NAC}$ in relation to the previously listed parameters and represents the study's summary in Table 9.8.

Figure 9.19 shows the diagonal cracking strength of RAC beams subjected to various RCA replacement ratios. When the replacement ratio is less than 100%, the shear strength of RAC beams is negatively impacted, as shown by the mean and coefficient of variation (CoV) values of 0.92 and 0.10, respectively. Additionally, with a replacement level of 100%, the mean value of 0.93 (CoV = 0.12) indicates a negative impact on the shear strength of RAC beams. Similar to this, the influence of RAC compressive strength on RAC beam diagonal tension cracking strength was examined in two groups of RAC compressive strength, namely, up to 30 MPa and above 30 MPa, and the corresponding mean value of $v_{c,RAC}/v_{c,NAC}$ was observed to be 0.88 (CoV = 0.14) and 0.94 (CoV = 0.10), respectively. This suggests that the shear behaviour of RAC beams is adversely affected by the compressive strength of RAC up to 30 MPa. Additionally, the d_a/d ratio was used to analyse the change in shear resistance of RAC beams in three groups: less than 0.08, from 0.08 to 0.1, and above 0.1. The group with the lowest d_a/d ratio was seen to have a mean value of $v_{c,RAC}/v_{c,NAC}$ of 0.89 (CoV = 0.11), followed by the group with the highest d_a/d

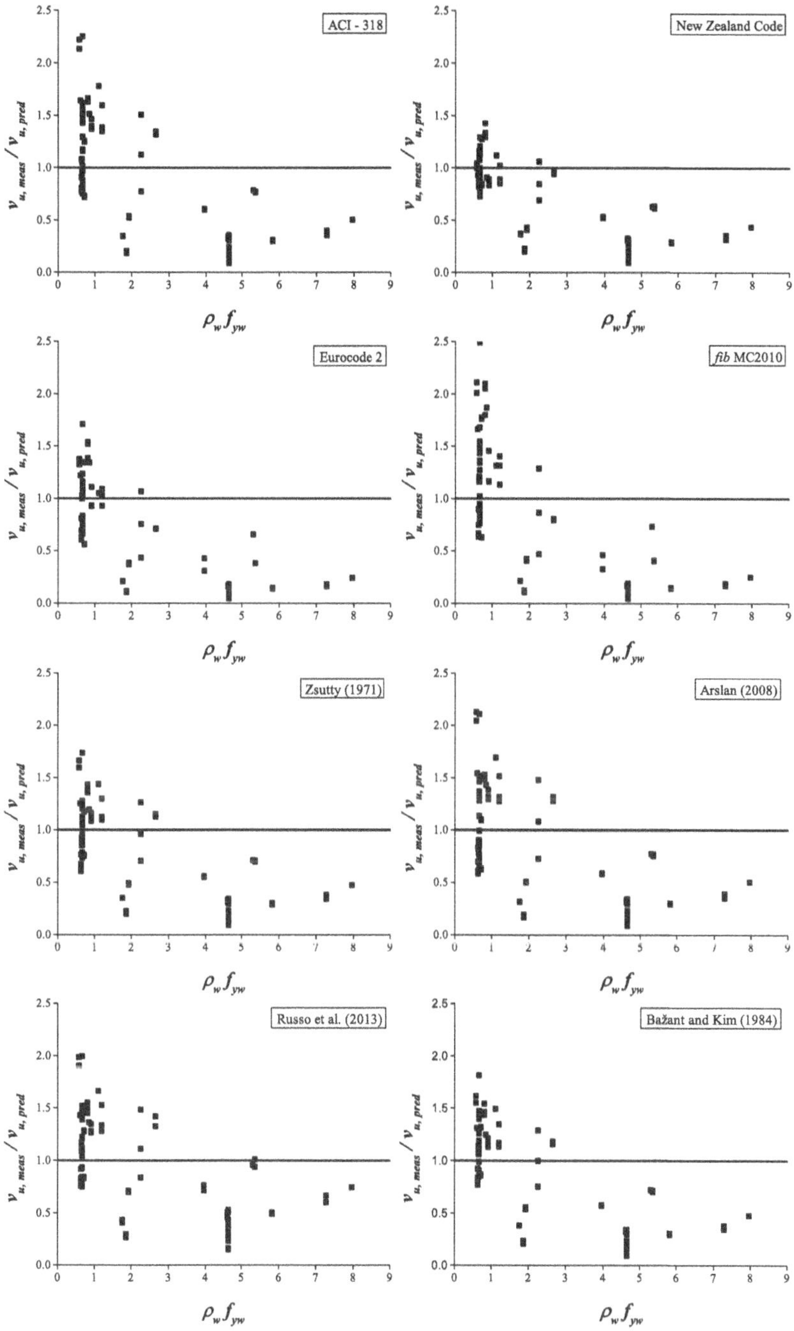

Fig. 9.18 Measured to predicted ultimate load ratio of RAC beams versus $\rho_w f_{yw}$ (Pradhan et al. 2018a)

Table 9.7 Comparison of existing expressions to predict the ultimate strength of RAC beams with stirrups

Expressions	Mean ($v_{u,meas}/v_{u,pred}$)	CoV	Number of unconservative predictions
ACI 318 (2008)	0.75	0.71	55
New Zealand code (1984)	0.62	0.61	67
Eurocode 2 (2004)	0.55	0.80	62
fib MC (2010)	0.67	0.89	58
Zsutty (1971)	0.65	0.65	61
Bažant and Kim (1984)	0.71	0.65	53
Arslan (2008)	0.68	0.72	60
Russo et al. (2013)	0.81	0.57	53

Table 9.8 Summary of the effect of critical parameters on $v_{c,RAC}/v_{c,NAC}$

Parameter	Limit	Mean ($v_{c,RAC}/v_{c,NAC}$)	CoV ($v_{c,RAC}/v_{c,NAC}$)
r (%)	< 100	0.92	0.10
	$= 100$	0.93	0.12
$f_{c,RAC}$ (MPa)	≤ 30	0.88	0.14
	> 30	0.94	0.10
d_a/d	< 0.08	0.89	0.11
	$0.08 \geq d_a/d \leq 0.10$	0.98	0.11
	> 0.10	1.02	0.06
d (mm)	≤ 200	1.05	0.10
	$200 > d \leq 350$	0.87	0.12
	> 350	0.90	0.12
ρ (%)	≤ 1.4	0.95	0.11
	$1.4 > \rho \leq 2.5$	0.91	0.14
	> 2.5	0.92	0.11
a/d	≤ 2.5	0.92	0.11
	$2.5 > a/d \leq 3.5$	0.87	0.12
	> 3.5	1.02	0.07

ratio of 0.98 (CoV = 0.11) and the lowest d_a/d ratio of 0.08 (CoV = 0.11). This implies that when the d_a/d ratio rises, so does the shear resistance of RAC beams. Shear strength was negatively impacted for RAC beams with d_a/d ratio up to 0.08. In three groups, the impact of the beam's effective depth on the strength of diagonal tension cracks was investigated. The first set of beams has an effective depth of up to 200 mm, the second group have an effective depth of 200–350 mm, and the third group have an effective depth of more than 350 mm. For the first, second, and third groups,

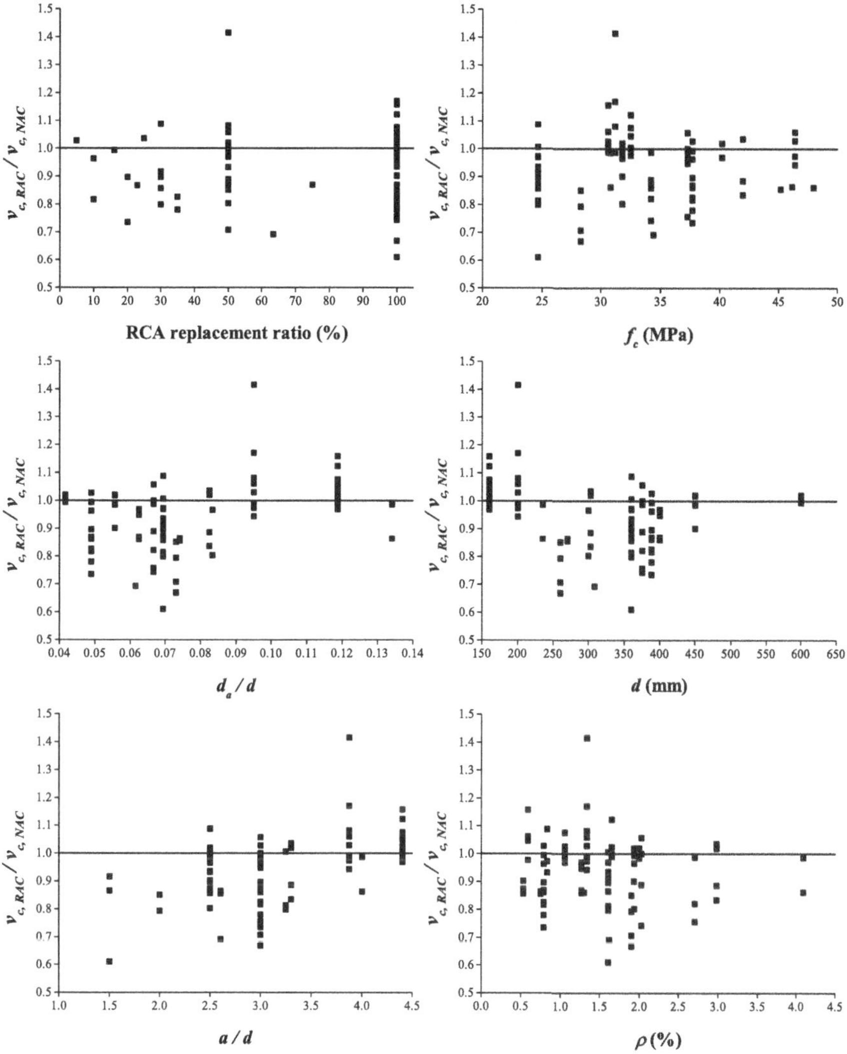

Fig. 9.19 Comparison of the effect of different design parameters on $v_{c,RAC}/v_{c,NAC}$ (Pradhan et al. 2018b)

respectively, the mean values were determined to be 1.05 (CoV = 0.10), 0.87 (CoV = 0.12), and 0.90 (CoV = 0.12). Therefore, the shear behaviour of RAC beams is adversely affected by effective depths above 200 mm. The diagonal cracking strength is negatively impacted by the longitudinal reinforcement content above 1.4%, as shown in Fig. 9.19. Three groups, including $\rho \leq 1.4$, $1.4 > \rho \leq 2.5$, and $\rho > 2.5$, were examined for the effect of ρ on the shear strength of RAC beams. The shear behaviour of RAC beams is more adversely affected by longitudinal reinforcement ratios over 1.4% as indicated by the lower mean values of 0.91 (CoV = 0.14) and 0.92 (CoV = 0.11) for the groups $1.4 > \rho \leq 2.5$ and $\rho > 2.5$ compared to 0.95 (CoV = 0.11) for $\rho \leq 1.4$. This tendency could be explained by RAC's inferior tensile and bonding strength than NAC. Additionally, the intrinsic poor quality of RCA and faster rate of bond deterioration between RCA and mortar interface are impacting the role played by concrete in shear resistance mechanisms in the case of RAC beams. The shear span-to-depth ratio (a/d) has an impact on the shear strength of RAC beams, as seen in Fig. 9.19. The $v_{c,RAC}/v_{c,NAC}$ shows mean values of 0.92 (CoV = 0.11) and 0.87 (CoV = 0.12), respectively, for $a/d \leq 2.5$ and $2.5 > a/d \leq 3.5$, whereas the mean value is 1.02 (CoV = 0.07) for $a/d > 3.5$. As a result, it can be inferred that the shear span-to-depth ratio, $a/d \leq 3.5$, has a detrimental effect on the shear resistance of RAC beams. This means that the inferior shear performance of RAC beams cannot be specifically attributed to the lower compressive strength of RAC. However, none of the design parameter has a singular significant negative effect on the shear behaviour of RAC beams. It is possible that the combined influence of all of the previously discussed characteristics is what causes the inferior shear behaviour of RAC beams. Additionally, each design parameter's critical region was investigated. For the shear strength prediction of RAC beams without shear reinforcement, a separate equation must be formulated that takes the replacement ratio parameter into account along with other factors like the tensile strength of the concrete, the effective depth of the beam, the shear span-to-depth ratio, and the longitudinal reinforcement ratio. In order to do this, the test data of RAC beams without transverse reinforcement were employed.

With Transverse Reinforcement

The existing expressions to predict the ultimate load of reinforced concrete beams with transverse reinforcement account for the compressive strength of concrete (f_c), geometric ratio of longitudinal reinforcement content (ρ), shear span-to-depth ratio (a/d), effective depth of the beam (d), maximum size of the aggregate to effective depth of the beam ratio (d_a/d), yield strength of longitudinal reinforcement (f_y), geometric ratio of transverse reinforcement (ρ_w), and yield strength of transverse reinforcement (f_{yw}). In addition to these parameters, the effect of RCA replacement ratio on v_u was accounted forPradhan et al. (2018a). The effect of these major design parameters on v_u was studied using the prepared database. The variation in the ratio of ultimate strength of RAC beams and the companion NAC beams ($v_{u,RAC}/v_{u,NAC}$) are analysed with respect to the aforementioned parameters (Fig. 9.20).

For various boundary conditions of the discussed parameters, Table 9.9 summarizes the mean value and CoV of $v_{u,RAC}/v_{u,NAC}$. The study demonstrates that none of the design factors have a particularly unfavourable impact on the flexural behaviour of RAC beams. In order to formulate a formula for the ultimate strength of a transversely reinforced RAC beam, each of the discussed parameters was taken into account and the database of RAC beams was used.

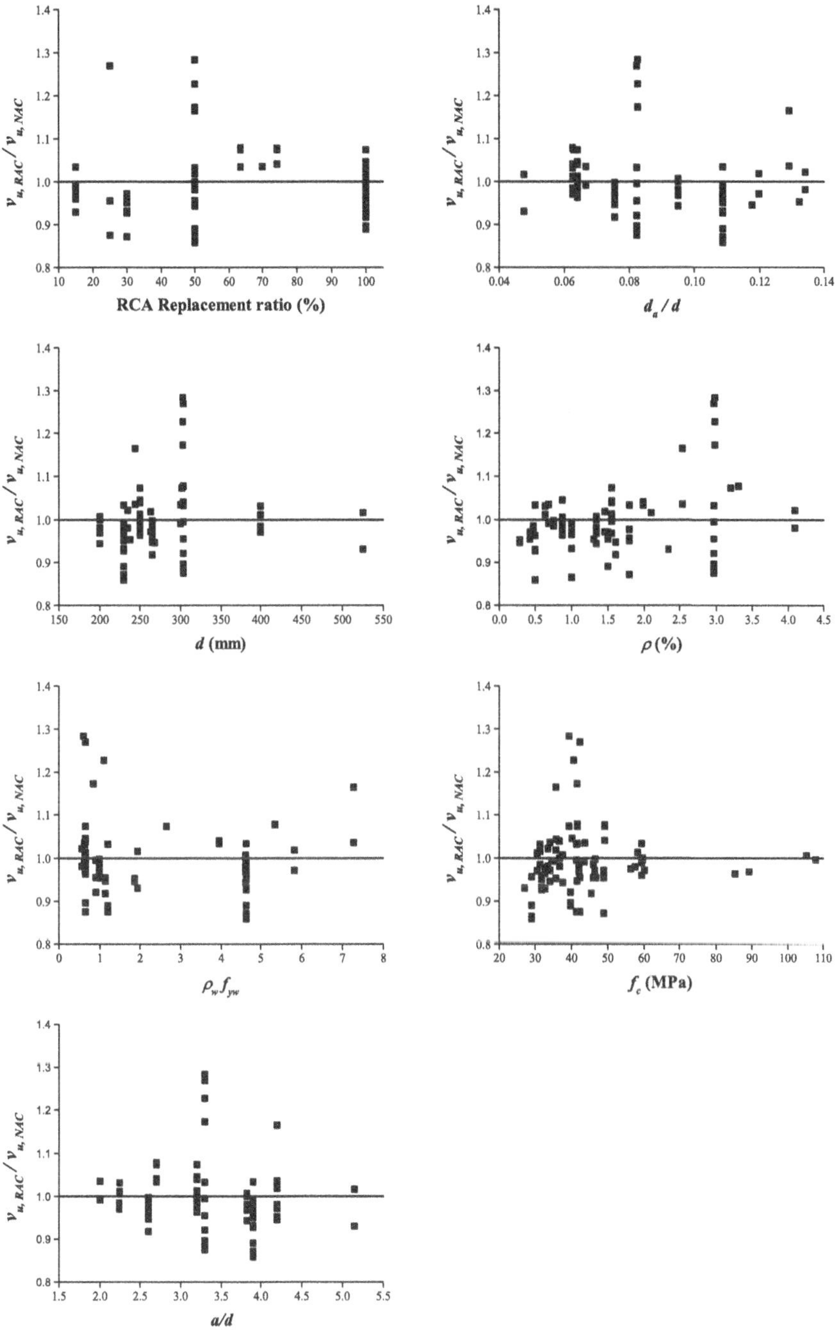

Fig. 9.20 Effect of different parameters on the ultimate strength of RAC beams (Pradhan et al. 2018b)

Table 9.9 Summary of the effect of critical parameters on $v_{u,RAC}/v_{u,NAC}$

Parameter	Limit	Mean ($v_{u,RAC}/v_{u,NAC}$)	CoV ($v_{u,RAC}/v_{u,NAC}$)
r (%)	< 100	1.00	0.10
	$= 100$	0.98	0.04
d_a/d	≤ 0.1	1.00	0.08
	> 0.1	0.97	0.06
d (mm)	≤ 250	0.98	0.05
	> 250	1.01	0.10
ρ (%)	≤ 1.4	0.97	0.04
	> 1.4	1.00	0.10
$\rho_w f_{yw}$	≤ 1.2	1.00	0.09
	> 1.2	0.98	0.06
$f_{c,RAC}$ (MPa)	≤ 40	0.98	0.08
	> 40	1.00	0.08
a/d	≤ 2.5	1.00	0.02
	> 2.5	0.99	0.08

9.6 Expression for the Shear Strength of RAC Beams

9.6.1 *Without Transverse Reinforcement*

In general, the longitudinal reinforcement ratio (ρ), shear span-to-depth ratio (a/d), and concrete's tensile strength (f_{ct}) influence a reinforced concrete (RC) beam's diagonal tension cracking strength (v_c) in the absence of transverse reinforcement. Additionally, f_{ct} can be visualized as a function of concrete's compressive strength (f_c). The RCA replacement ratio (r) parameter was also taken into account for RAC beams. To take the size effect of the concrete beam into account, a straightforward d_a/d function was added. As a result, it is believed that the following expression (Eq. 9.6.1) accurately predicts v_c for RAC beams lacking shear reinforcement.

$$v_c = K r^x f_c^y \rho^z \left(\frac{d_a}{d}\right)^m \left(\frac{a}{d}\right)^n \tag{9.6.1}$$

where K, x, y, z, m, and n are constants. The values of these constants were determined by an iterative process to attain the uniformity in the prediction of v_c for RAC beams and in this process the shear database (Appendix B) was operated. The final expression for v_c of RAC beams without shear reinforcement is represented in Eq. 9.6.2. According to the estimated $v_{c,meas}/v_{c,pred}$, which is shown in Fig. 9.21, the average value is 1.08 with a CoV of 0.17. Out of 91 test results for RAC beams without shear reinforcement, 23 unconservative predictions were found using Eq. 9.6.2. The formula of Pradhan et al. (2018b), however, is more appropriate and shows

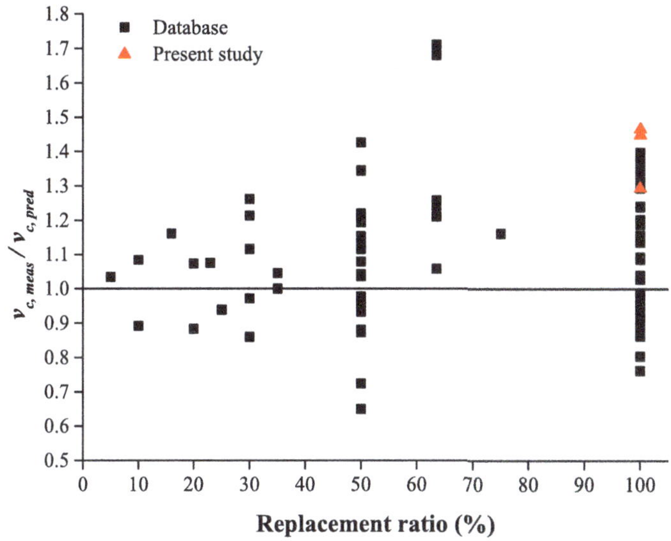

Fig. 9.21 Calculated $v_{c,meas}/v_{c,pred}$ values for the tested RAC beams (Pradhan et al. 2018b)

better prediction of the v_c for RAC beams without stirrups in contrast to the other expressions while considering the average value, CoV, and number of unconservative forecasts.

$$v_c = 1.6r^{-0.1} f_c^{0.6} \rho^{0.45} \left(\frac{d_a}{d}\right)^{0.48} \left(\frac{a}{d}\right)^{-0.91} \tag{9.6.2}$$

9.6.2 *With Transverse Reinforcement*

The reinforced concrete (RC) beam's transverse reinforcement boosts its shear resistance and prevents the catastrophic failure. The following expression (Eq. 9.6.3) can be used to represent the four shear resistance mechanisms that contribute to a reinforced concrete beam's ability to resist transverse shear stress (v_u: (1) the shear resistance of the uncracked concrete (v_c), (2) dowel action (v_d), (3) the vertical component of aggregate interlocking (v_a), and (4) the shear resistance carried by the transverse reinforcement (v_s) (Park and Paulay 1975).

$$v_u = v_c + v_d + v_a + v_s \tag{9.6.3}$$

Along with these mechanisms, other influencing factors that affect the relative contributions of various mechanisms include the material and geometric qualities, loading stage, and amount of cracking. Prior to the commencement of a flexural crack, the uncracked concrete completely resists the shear force applied to the system. Redistribution of stresses begins when flexural cracks occur. In this case, the uncracked concrete, dowel action, and aggregate interlocking mechanisms all work together to withstand the vertical shear stress. Therefore, as the diagonal tension crack formation occurs, the stirrups assist in shear resistance mechanisms. Even after a diagonal tension crack appears, the shear-reinforced reinforced concrete beam maintains its shear resistance, carrying shear loads until the stirrups yield in tension. The process of a fracture expanding is accelerated by the shear reinforcement yielding. Finally, flexural compressive stress and shear stress work together in a synergistic manner to cause the failure.

The contributions made by concrete and stirrup to the mechanics of shear resistance are not separate from one another in reality. By enhancing the dowel action, concrete tooth capacity, and aggregate interlock, transverse reinforcement improves an RC beam's shear resistance mechanisms (Park and Paulay 1975). The shear resistance mechanisms of RC beams are influenced by the quantity and distribution of stirrup (Russo et al. 2013; Russo and Puleri 1997; Pendyala and Mendis 2000). It is feasible to ensure the highest explicable shear stress provided by the stirrups ($\rho_w f_{yw}$) and it enhances the contribution of concrete (v_c) in shear resistance mechanisms in RC beams, where the beam action governs the shear resistance mechanisms (Russo and Puleri 1997). In this regard, Russo and Puleri (1997) developed the term v_{si} to account for the increase in shear strength of a transversely reinforced beam over an RC beam without stirrups. The v_{si} term takes into consideration both the irregularity of the truss action and the increase in v_c caused by beam action. Additionally, v_{si} makes sure that the contributions made by concrete and stirrup in shear resistance mechanisms interact with one another. In the light of this, Eq. 9.6.4 can be used to represent the ultimate shear stress experienced in an RC beam with stirrups (Russo et al. 2013; Russo and Puleri 1997).

$$v_u = v_c + v_{si} \tag{9.6.4}$$

$$\Rightarrow v_{si} = v_u - v_c \tag{9.6.5}$$

Equation 9.6.5 can be used to calculate the increase in shear strength of an RC beam with stirrups. Using the experimental results of paired RC beams with and without transverse reinforcement, Russo et al. (2013) calculated v_u and v_c, respectively. The accuracy of v_{si} to forecast the shear strength contribution by stirrups, however, may be impacted by the sparse amount of experimental results on paired RAC beams that are currently available. As a result, the formulas for predicting v_u and v_c were derived using the experimental data for RAC beams with stirrups and without stirrups, respectively. The v_c expression was derived from the test results of RAC beams without transverse reinforcement (Appendix B), and the result is shown in Eq. 9.6.2.

Equation 9.6.5 was used to derive v_{si} using the experimental v_u and calculated v_c values. Plotted with respect to $\rho_w f_{yw}$, the ratios of estimated v_{si} and empirically obtained v_u can be seen to vary from 6 to 65% (Fig. 9.22). The introduction of

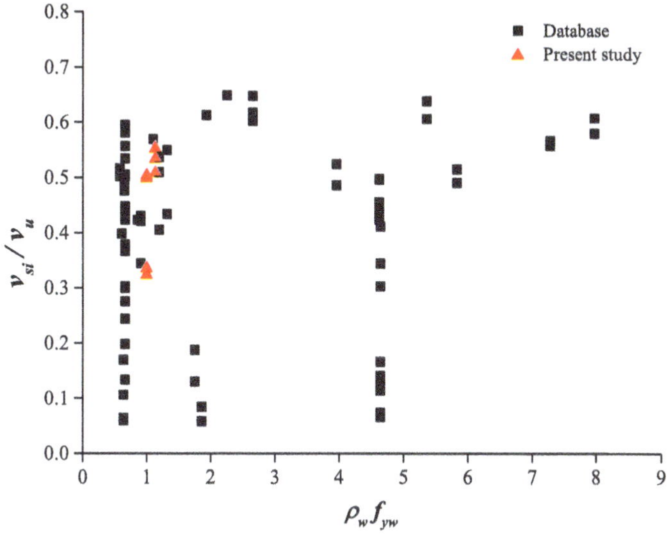

Fig. 9.22 Ratio of the increment in shear strength due to the inclusion of stirrups to the ultimate shear strength of RAC beams with stirrups (Pradhan et al. 2018a)

transverse reinforcement may have a varying impact on the shear strength of RAC beams, according to this. The stirrup's nominal shear strength ($\rho_w f_{yw}$) is expressed using the formula in ACI 318 (2008), which denotes the stirrups' contribution to the increase in the shear strength of the RC beam. The dependency of v_{si} in RAC beams on nominal shear strength of stirrups (v_s) is verified by the ratio of v_{si} and v_s (where $v_s = \rho_w f_{yw}$) and plotted with respect to $\rho_w f_{yw}$ (Fig. 9.23). The horizontal line with ordinate 1 in Fig. 9.23 represents the ideal correlation between v_{si} and v_s. As can be seen from Fig. 9.23, ACI 318 (2008) formulation for v_s does not evaluate the increase in shear strength of RAC beams caused by stirrups evenly for different nominal shear strengths of stirrup. Additionally, the correlation between v_{si} and v_s was unsatisfactory for stirrups with higher nominal shear strengths ($\rho_w f_{yw} > 3$) since the ratio of v_{si} to v_s was less than 0.5 for each test result (Fig. 9.23). The average value of v_{si}/v_s was found to be 0.51 while the CoV was 0.92. The high value of CoV suggests that contrary to what is assumed in ACI 318 (2008), the increase in shear strength of RAC beams brought on by the addition of stirrups is not only determined by the nominal shear strength of the stirrup. Further study is required to develop an equation to forecast the $v_s i$ of RAC beams more uniformly due to the variability in v_{si} determination for RAC beams. In this situation, an expression for the v_u of RAC beams with shear reinforcement was obtained. Eq. 9.6.5 can be used to calculate the contribution of stirrups in shear strength increment once v_u and v_c are known.

The three main factors that contribute to the shear-resisting mechanisms of an RC beam with stirrups, such as the action of the stirrup, the action of the longitudinal reinforcing dowels, and the action of the diagonal concrete struts, can be compared to

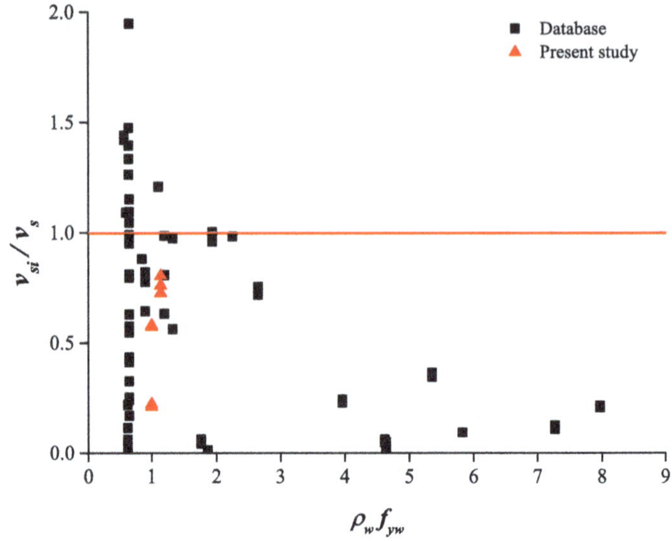

Fig. 9.23 Ratio of measured shear strength increment to the calculated increment in shear strength as per ACI code for RAC beams with stirrups (Pradhan et al. 2018a)

those of a statically determined truss. Concrete strut action in shear resisting mechanisms is controlled by the compressive strength of concrete (f_c), RCA replacement ratio (r), and the ratio of maximum aggregate size to effective depth of the beam (d_a/d). The longitudinal reinforcing content (ρ) in dowel action affects v_u. The geometric ratio of the transverse reinforcement (ρ_w) and the yield strength of the stirrup (f_{yw}) determine v_u in the stirrup action. Therefore, v_u's controlling parameters can be written as

$$v_u = f(r, f_c, d_a/d, \rho, f_y, a/d, \rho_w, f_{yw}) \quad (9.6.6)$$

Similar to the expression suggested by Russo et al. (2013) to predict v_u, the following equation was considered for v_u by adjudging the governing parameters as represented in Eq. 9.6.6:

$$v_u = \left(\frac{d_a}{d}\right)^x \left[K_1 r^m f_c^n \rho^p + K_2 \rho^y f_y^z \left(\frac{a}{d}\right)^a\right] + K_3 r^b f_c^c \rho^i (\rho_w f_{yw})^j \quad (9.6.7)$$

where K_1, K_2, K_3, x, y, z, m, n, p, a, b, c, i and j are constants. The iterative technique was used to determine the values of these constants. The experimental findings of RAC beams with transverse reinforcement (Appendix C) were utilized in this context. There are 96 results for transversely reinforced RAC beams in the database. However, it was noted that 82 beams had failed in the flexure tension mode and the remaining 14 had failed in the diagonal tension mode. The constants of Eq. 9.6.7 were calculated using the parameters of the beams that showed the flexure tension mode of failure. The values of the constants obtained are; $K_1 = 0.025$, $K_2 = 0.001$, $K_3 =$

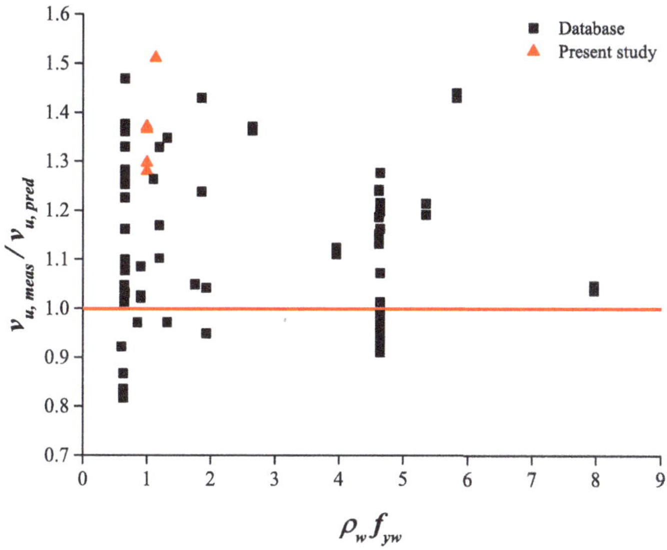

Fig. 9.24 Ratio of measured and predicted ultimate strength of RAC beams (Pradhan et al. 2018a)

0.001, $x = -0.45$, $y = 0.75$, $z = 1.2$, $m = -0.1$, $n = 0.4$, $p = 0.5$, $a = -1.5$, $b = -0.1$, $c = 0.48$, $i = 0.92$ and $j = 0.7$. Consequently, the expression for v_u can be written as

$$v_u = \left(\frac{d_a}{d}\right)^{-0.45}\left[0.025r^{-0.1}f_c^{0.4}\rho^{0.5} + 0.001\rho^{0.75}f_y^{1.2}\left(\frac{a}{d}\right)^{-1.5}\right] + 0.001r^{-0.1}f_c^{0.48}\rho^{0.92}(\rho_w f_{yw})^{0.7} \tag{9.6.8}$$

The $v_{u,meas}/v_{u,pred}$ was calculated for RAC beams with transverse reinforcement using Eq. 9.6.8. The variation in $v_{u,meas}/v_{u,pred}$ with respect to $\rho_w f_{yw}$ is represented in Fig. 9.24. The average value and CoV observed for calculated $v_{u,meas}/v_{u,pred}$ are 1.14 and 0.14, respectively. The average value indicates that the v_u of RAC beams with transverse reinforcement is slightly underestimated by the expression suggested by Pradhan et al. (2018a). Furthermore, the recommended expression showed 15 unconservative predictions out of 82 test findings. In comparison to the other available formulas (Table 9.6), Eq. 9.6.8 displays better prediction of v_u for RAC beams with transverse reinforcement when considering the average value, CoV, and quantity of unconservative predictions.

The proposed expressions (Pradhan et al. 2018a, b) for v_u and v_c were incorporated in Eq. 9.6.5 to estimate the contribution of stirrups in shear strength increment (v_{si}) of RAC beams with transverse reinforcement. The ratio of measured v_{si} to the calculated v_{si} ($v_{si,meas}/v_{si,pred}$) was estimated and represented in Fig. 9.25. The calculated v_{si} using Eq. 9.6.5 exhibits better correlation in comparison to the expression recommended by ACI 318 (2008) (Fig. 9.23) with the measured value of v_{si}.

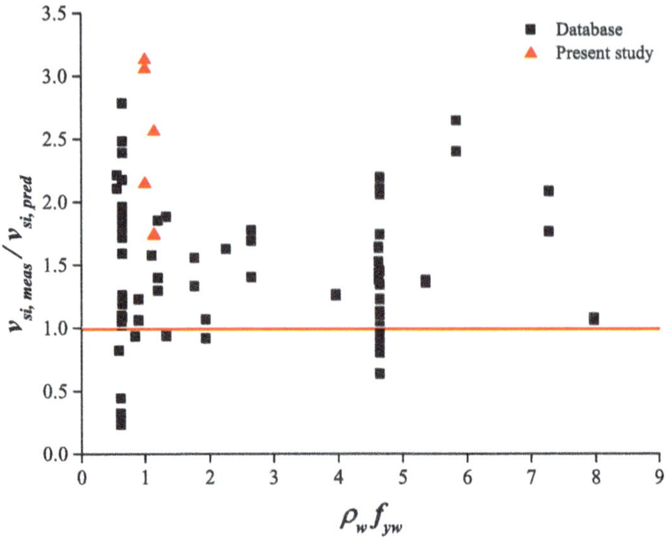

Fig. 9.25 Ratio of measured shear strength increment to the calculated increment in shear strength as per the derived expression for RAC beams with stirrups (Pradhan et al. 2018b)

The average value of $v_{si,meas}/v_{si,pred}$ was found to be 1.44 with a CoV of 0.39. The computed v_{si} significantly understates the role stirrups play in increasing the shear strength of RAC beams, yet the prediction of v_{si} was consistent for stirrups with varying nominal shear strengths.

9.6.3 Effectiveness of Stirrups in Shear Resistance

Transverse reinforcement in RC beams improves dowel action, concrete tooth capacity, and aggregate interlock, which in turn improves shear resistance mechanisms (Park and Paulay 1975). The distribution and quantity of stirrup have an impact on the shear resistance mechanisms of RC beams (Russo and Puleri 1997; Russo et al. 2013; Pendyala and Mendis 2000). It is feasible to ensure the maximum explicable shear stress provided by the stirrups ($\rho_v f_{yv}$) and it improves the contribution of concrete (v_c) in shear resistance mechanisms with RC beams, according to Russo and Puleri (1997). Moreover, the contribution of concrete and stirrups are not independent in the shear resistance mechanisms of RC beam with transverse reinforcement, because the presence of stirrups influenced the contribution of concrete in shear resistance mechanisms (Russo et al. 2013). In this context, v_{si} term is introduced in order to account for the increment in shear strength of a transversely reinforced beam with respect to the RC beam without having stirrups (Russo and Puleri 1997; Russo et al. 2013). The v_{si} term accounts for both the inconsistency of the truss action as well

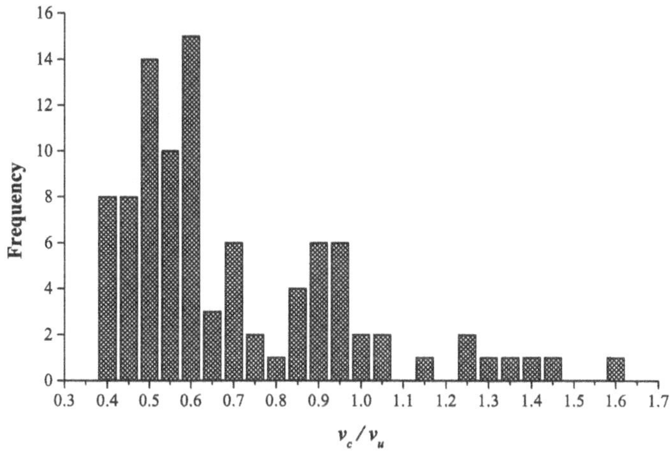

Fig. 9.26 Frequency distribution of v_c/v_u for RAC beams (Pradhan et al. 2018b)

as the increment in v_c due to beam action. Further, v_{si} ensures the interaction of the contributions by concrete and stirrup in shear resistance mechanisms. Consequently, the ultimate shear stress experienced in an RC beam with stirrups can be represented by Eq. 9.6.9 (Russo and Puleri 1997; Russo et al. 2013).

$$v_u = v_c + v_{si} \tag{9.6.9}$$

The increase in shear strength of RAC beams as a result of the addition of stirrups was evaluated using the flexure database (Appendix C). The value of v_c as shown in Eq. 9.6.2 is deemed to represent the contribution of concrete to the shear resistance mechanisms of RAC beams with stirrups. For 50% of the tested RAC beams, the contribution of v_c was found to be within 5% of v_u, and the frequency distribution of v_c/v_u is shown in Fig. 9.26. Therefore, it is anticipated that for RAC beams with transverse reinforcement, v_c contributes 55% of v_u and v_{si} contributes the remaining 45% of v_u. As a result, v_{si} can be written as 9/11 times v_c (Eq. 9.6.10).

$$v_{si} = 1.3r^{-0.1} f_c^{0.6} \rho^{0.45} \left(\frac{d_a}{d}\right)^{0.48} \left(\frac{a}{d}\right)^{-0.91} \tag{9.6.10}$$

9.7 Closure

The behaviour and performance of RAC beams without transverse reinforcement as well as with transverse reinforcement were discussed and compared with the results of NAC beams.

- Both NAC and RAC beams without transverse reinforcement had a very similar overall fracture pattern and failure mode (diagonal tension). The number of cracks and growth rate were higher in the RAC beams, though.
- The initial cracking load, ultimate failure load, and related mid-span deflection were observed to be lower for RAC beams without transverse reinforcement even though they had the same longitudinal reinforcement ratio and shear span-to-depth ratio.
- A reduction of about 14% was recorded in ultimate shear strength for RAC beams.
- For RAC beams with transverse reinforcement, a reduced cracking load was noted. However, RAC beams showed a very slight difference between yield load and ultimate load.
- The RAC beams encountered the unusual diagonal tension mode of failure at greater longitudinal reinforcement ratios ($\rho = 1.31$ and 1.61%). This shows that the transverse reinforcement was insufficient to cause the intended flexure tension mode of failure at greater longitudinal reinforcement ratios.
- In order to forecast v_c of RAC beams using the shear strength data of RAC beams without shear reinforcement, a formula was put out by Pradhan et al. (2018b), which is essentially a function of RCA replacement ratio. Comparing the proposed equation to the existing ones, the correlation is better and it is more pertinent.
- In order to determine the ultimate load carrying capacity of RAC beams with stirrups, an expression that takes into account the RCA content, longitudinal reinforcement content, mechanical properties of RAC, size of the beam, and transverse reinforcement content was proposed by Pradhan et al. (2018a).
- Using the results of RAC beam specimens with and without shear reinforcement, the contribution of stirrups in shear resistance mechanisms was investigated.

References

ACI 318 (2008) ACI committee 318. Building code requirements for structural concrete and commentary. Farmington Hills (MI). American Concrete Institute

Ajdukiewicz AB, Kliszczewicz AT (2007) Comparative tests of beams and columns made of recycled aggregate concrete and natural aggregate concrete. J Adv Concr Technol 5(2):259–273

Arezoumandi M, Smith A, Volz JS, Khayat KH (2014) An experimental study on shear strength of reinforced concrete beams with 100% recycled concrete aggregate. Constr Build Mater 53:612–620

Arezoumandi M, Drury J, Volz JS, Khayat KH (2015a) Effect of recycled concrete aggregate replacement level on shear strength of reinforced concrete beams. ACI Mater J 112(4):559–568

Arezoumandi M, Smith A, Volz JS, Khayat KH (2015b) An experimental study on flexural strength of reinforced concrete beams with 100% recycled concrete aggregate. Eng Struct 88:154–162

Arslan G (2008) Shear strength of reinforced concrete beams with stirrups. Mater Struct 41(1):113–122

Bai WH, Sun BX (2010) Experimental study on flexural behavior of recycled coarse aggregate concrete beam. Appl Mech Mater 29–32:543–548

Bažant ZP, Yu Q (2005a) Designing against size effect on shear strength of reinforced concrete beams without stirrups: I. Formulation. J Struct Eng 131(12):1877–1885

Bažant ZP, Yu Q (2005b) Designing against size effect on shear strength of reinforced concrete beams without stirrups: II. Verification and calibration. J Struct Eng 131(12):1886–1897
Bažant ZP, Kim J-K (1984) Size effect in shear failure of longitudinally reinforced beams. ACI J 81(38):456–468
Bažant ZP, Sun H-H (1987) Size effect in diagonal shear failure: influence of aggregate size and stirrups. ACI Mater J 84(4):259–272
BS 8110-1 (1997) Structural use of concrete - part 1: code of practice for design and construction
Butler LJ, West JS, Tighe SL (2014) Bond of reinforcement in concrete incorporating recycled concrete aggregates. J Struct Eng 141(3)
Butler LJ, West JS, Tighe SL (2011) The effect of recycled concrete aggregate properties on the bond strength between RCA concrete and steel reinforcement. Cem Concr Res 41(10):1037–1049
Choi HB, Yi CK, Cho HH, Kang KI (2010) Experimental study on the shear strength of recycled aggregate concrete beams. Mag Concr Res 62(2):103–114
Choi W-C, Kim S-W, Yun H-D (2012) Flexural performance of reinforced recycled aggregate concrete beams. Mag Concr Res 64(9):837–848
Etxeberria M (2004) Experimental study on microstructure and structural behaviour of recycled aggregate concrete. PhD thesis, Universitat Politècnica de Catalunya
Etxeberria M, Marí AR, Vázquez E (2007) Recycled aggregate concrete as structural material. Mater Struct 40(5):529–541
Eurocode 2 (2004) Eurocode 2: design of concrete structures: part 1-1: general rules and rules for buildings. British Standards Institution
Fathifazl G, Razaqpur AG, Isgor OB, Abbas A, Fournier B, Foo S (2010) Shear strength of reinforced recycled concrete beams with stirrups. Mag Concr Res 62(10):685–699
Fathifazl G, Razaqpur AG, Burkan Isgor O, Abbas A, Fournier B, Foo S (2011) Shear capacity evaluation of steel reinforced recycled concrete (RRC) beams. Eng Struct 33(3):1025–1033
Fathifazl G, Abbas A, Razaqpur A, Isgor O, Fournier B, Foo S (2009a) Shear strength of reinforced recycled concrete beams without stirrups. Mag Concr Res 61(7):477–490
Fathifazl G, Razaqpur AG, Isgor OB, Abbas A, Fournier B, Foo S (2009b) Flexural performance of steel-reinforced recycled concrete beams. ACI Struct J 106(6):858–867
fib MC (2010) fib model code for concrete structures. Wilhelm Ernst & Sohn, Berlin
Gastebled O, May I (2001) Fracture mechanics model applied to shear failure of reinforced concrete beams without stirrups. ACI Struct J 98(2):184–190
González-Fonteboa B, Martínez-Abella F (2007) Shear strength of recycled concrete beams. Constr Build Mater 21(4):887–893
González-Fonteboa B, Martínez-Abella F, Martínez-Lage I, Eiras-López J (2009) Structural shear behaviour of recycled concrete with silica fume. Constr Build Mater 23(11):3406–3410
González-Fonteboa B, Martínez-Abella F (2004) Shear strength of concrete with recycled aggregates. In: International RILEM conference on the use of recycled materials in buildings and structures, Barcelona, Spain, pp 619–628
Ignjatović IS, Marinković SB, Mišković ZM, Savić AR (2012) Flexural behavior of reinforced recycled aggregate concrete beams under short-term loading. Mater Struct 46(6):1045–1059
Ignjatović IS, Marinković S, Tošić N (2017) Shear behaviour of recycled aggregate concrete beams with and without shear reinforcement. Eng Struct 141:386–401
IS: 456 (2000) Plain and reinforced concrete - code of practice. Bureau of Indian Standards, New Delhi, India
Ji SK, Lee WS, DoYun H (2008) Shear strength of reinforced concrete beams with recycled aggregates. Tailor made concrete structures, pp 1089–1092. Taylor & Francis Group, London
JSCE (2007) Standard specifications for concrete structures - 2007 "Design". Number 15. Japan Society of Civil Engineers
Kang THK, Kim W, Kwak Y-K, Hong S-G (2014) Flexural testing of reinforced concrete beams with recycled concrete aggregates. ACI Struct J 111(3):607
Katkhuda H, Shatarat N (2016) Shear behavior of reinforced concrete beams using treated recycled concrete aggregate. Constr Build Mater 125:63–71

Kim J-K, Park Y-D (1996) Prediction of shear strength of reinforced concrete beams without web reinforcement. ACI Mater J 93(3):213–222
Kim S-W, Jeong C-Y, Lee J-S, Kim K-H (2013) Size effect in shear failure of reinforced concrete beams with recycled aggregate. J Asian Archit Build Eng 12(2):323–330
Knaack AM, Kurama YC (2013) Service-load deflection behavior of reinforced concrete beams with recycled concrete aggregate. In: Structures congress 2013: bridging your passion with your profession, pp 2705–2716
Knaack AM, Kurama YC (2014) Behavior of reinforced concrete beams with recycled concrete coarse aggregates. J Struct Eng 141(3)
Knaack AM (2013) Sustainable concrete structures using recycled concrete aggregate: short-term and long-term behavior considering material variability. PhD thesis, University of Notre Dame
Knaack AM, Kurama YC (2015) Creep and shrinkage of normal-strength concrete with recycled concrete aggregates. ACI Mater J 112(3):451–462
Maruyama I, Sogo M, Sogabe T, Sato R, Kawai K (2004) Flexural properties of reinforced recycled concrete beams. In: International RILEM conference on the use of recycled materials in buildings and structures, vol 1, pp 526–535
Michaud KS (2015) Evaluation of the environmental, material, and structural performance of recycled aggregate concrete. PhD thesis, Queen's University, Kingston, Ontario, Canada
New Zealand code (1984) New Zealand standard code of practice for general structural design and design loadings for buildings. Standards Association of New Zealand
Niwa J, Yamada K, Yokozawa K, Okamura H (1986) Revaluation of the equation for shear strength of reinforced concrete beams without web reinforcement. Doboku Gakkai Ronbunshu 1986(372):167–176
Park R, Paulay T (1975) Reinforced concrete structures. Wiley
Pendyala RS, Mendis P (2000) Experimental study on shear strength of high-strength concrete beams. Struct J 97(4):564–571
Pradhan S, Kumar S, Barai SV (2017) Recycled aggregate concrete: particle packing method (PPM) of mix design approach. Constr Build Mater 152:269–284
Pradhan S, Kumar S, Barai SV (2018a) Performance of reinforced recycled aggregate concrete beams in flexure: experimental and critical comparative analysis. Mater Struct 51(58):1–17
Pradhan S, Kumar S, Barai SV (2018b) Shear performance of recycled aggregate concrete beams: an insight for design aspects. Constr Build Mater 178:593–611
Rahal KN, Alrefaei YT (2017) Shear strength of longitudinally reinforced recycled aggregate concrete beams. Eng Struct 145:273–282
Rebeiz KS (1999) Shear strength prediction for concrete members. J Struct Eng 125(3):301–308
Russo G, Puleri G (1997) Stirrup effectiveness in reinforced concrete beams under flexure and shear. ACI Struct J 94(3):227–238
Russo G, Somma G, Mitri D (2005) Shear strength analysis and prediction for reinforced concrete beams without stirrups. J Struct Eng 131(1):66–74
Russo G, Mitri D, Pauletta M (2013) Shear strength design formula for RC beams with stirrups. Eng Struct 51:226–235
Sadati S, Arezoumandi M, Khayat KH, Volz JS (2016) Shear performance of reinforced concrete beams incorporating recycled concrete aggregate and high-volume fly ash. J Cleaner Prod 115:284–293
Sato R, Maruyama I, Sogabe T, Sogo M (2007) Flexural behavior of reinforced recycled concrete beams. J Adv Concr Technol 5(1):43–61
Sogo M, Sogabe T, Maruyama I, Sato R, Kawai K (2004) Shear behavior of reinforced recycled concrete beams. In: Proceedings of the international RILEM conference on the use of recycled materials in building and structures, pp 610–618
Tošić N, Marinković S, Ignjatović I (2016) A database on flexural and shear strength of reinforced recycled aggregate concrete beams and comparison to Eurocode 2 predictions. Constr Build Mater 127:932–944

Xiao J, Falkner H (2007) Bond behaviour between recycled aggregate concrete and steel rebars. Constr Build Mater 21(2):395–401
Zsutty TC (1968) Beam shear strength prediction by analysis of existing data. J Proc 65:943–951
Zsutty TC (1971) Shear strength prediction for separate categories of simple beam tests. ACI J Proc 68(2):138–143

Chapter 10
Structural Applications: Column

10.1 Introduction

The basic characterization of RAC prepared by the combined approach of PPM mix design and TSMA confirmed better performance in comparison to the conventional mix design method. This motivated to incorporate the PPM mix-designed concrete in the reinforced structural members to examine their behaviour. The experimental investigation on reinforced RAC columns subjected to an axial load and the effect of transverse reinforcement spacing on the axial strength and ductility of RAC columns are both covered in this chapter. Additionally, the effectiveness of the current design principles for creating RAC columns with full RCA utilization is confirmed.

10.2 Performance of RAC Column in Axial Loading

The behaviour of RAC columns subjected to the axial load was experimentally investigated by Ajdukiewicz and Kliszczewicz (2007). Only recycled coarse aggregate (RCA) and recycled fine aggregate (RFA) were used in the preparation of the RAC. The axial load sustained by RAC columns that only had RCA was comparable to that of a typical concrete column, whereas columns that had both RCA and RFA showed a maximum reduction of 14% in axial load. In comparison to standard concrete columns, the RAC columns' lateral and longitudinal strains were higher. Due to the slower rate of failure, RAC columns showed a better degree of ductility.

The level-2 reliability analysis of the reinforced eccentrically loaded RAC columns was carried out by Breccolotti and Materazzi (2010). RAC results for compressive strength served as the study's foundation. Without altering the current design criteria, the RCA recovered from high-strength concrete can be totally replaced to prepare structural concrete. Additionally, increasing the partial factor of safety for the RAC will help to offset the detrimental effects of the increase in the RCA replacement

S. Pradhan et al., *Particle Packing Method for Recycled Aggregate Concrete*,
https://doi.org/10.1007/978-981-99-7516-7_10

ratio. To reduce the variability in the mechanical properties of RAC due to the various replacement ratios of RCA and to provide comparable reliability performance with conventional concrete members, a thorough review of the current design approach is necessary.

Choi and Yun (2012) used three distinct RCA replacement ratios (30%, 60%, and 100%) and five different RFA replacement ratios (15%, 30%, 45%, 60%, and 100%) to conduct an experimental examination on the compressive behaviour of RAC columns subjected to axial load. As the RCA replacement ratio rises, RAC columns' axial strength has been reported to decrease. On the axial behaviour of RAC columns, the RFA was shown to have little effect. The presence of RCA or both RCA and RFA has no impact on the ductility of the RAC column. It has also been suggested to use the ACI 318 (2008) code for the design of RAC columns without making any additional changes.

To examine the behaviour of reinforced RAC columns, Xu et al. (2018) used SeismoStruct software to perform a fibre section-based nonlinear finite element analysis. First, the simulated findings were verified against the experimental data reported in Choi and Yun (2012) and Wang et al. (2011). The study demonstrated that ACI 318 (2008), Eurocode 2 (2004), NZS 3101 (2006), and equilibrium equation underestimate the axial strength of reinforced RAC columns as determined by simulations and testing. The model was then used to investigate how the RCA contents, yield strength and longitudinal reinforcement content, and yield strength and transverse reinforcement spacing affected the axial strength of reinforced RAC columns. The results of the grey correlation analysis revealed that the compressive strength of concrete, the yield strength of the longitudinal reinforcement, and the yield strength of the transverse reinforcement, respectively, have a primary, secondary, and tertiary correlation with the axial load carrying capacity of reinforced RAC columns. The authors emphasize that when simulating the behaviour of reinforced RAC columns, the sensitivity of CRA content cannot be disregarded. Additionally, the necessity of future experimental research was emphasized.

The impact of RCA contents (0%, 50%, and 100%) and eccentricity in loading (0 mm, 30 mm, 82 mm, and 100 mm) on the bearing capacity of reinforced RAC columns was investigated by Xiao (2018). As the eccentricity of the loading increased, the bearing capacity decreased. It was found that RAC columns had a lower bearing capacity than NAC columns. The compressive strength of concrete was considered to be 0.76 times the cube strength for estimating the column carrying capacity. The experimental and predicted bearing capacities of columns showed good concordance with the exception of a 100 mm eccentricity. The interaction curves were found to be independent of the RCA content.

By employing 20%, 40%, 60%, 80%, and 100% of RCA and recycled asphalt pavement (RAP), the experimental examination on the axial compressive strength of reinforced RAC columns by Shatarat et al. (2019) was carried out. According to the study, the axial compressive strength of both RCA- and RAP-incorporated columns gradually decreased; at 100% replacement level, a maximum drop of 28.27% and 31.06% was observed for RCA- and RAP-incorporated columns in comparison to NAC columns. The axial strength of RAC columns was enhanced by the mixture of

80% RCA and 20% RAP compared to NAC columns, but this behaviour was not seen for mixtures of RCA and RAP in the ratios of 60:40, 40:60, and 80:20. Shatarat et al. (2019) advised using reference ACI 318 (2008) and Japanese codes for the design of RAC columns by adding RCA, RAP, and RCA-RAP mixture based on the trial findings.

The purpose of the experimental program by Pradhan et al. (2023), which used a 100% RCA and PPM mix design technique, was to examine the effects of various tie spacings on the axial compressive strength and ductility of reinforced RAC columns. Utilizing the test results, it was therefore possible to confirm the effectiveness of the current design guidelines for RAC preparation columns. In addition, by simulating finite element models, the axial compressive strength of RAC columns with various transverse reinforcement contents (varying the spacing) was examined.

10.3 Impact of PPM on Compressive Strength of RAC Column

10.3.1 Experimental Program

Through a complete substitution of RCA for NCA, the experimental program Pradhan et al. (2023) sought to examine the behaviour of RAC columns. On eight square columns, the monotonic axial compression test was performed in this regard. RAC was used to prepare four columns, while NAC was used to prepare the other four. Additionally, utilizing two various volumetric ratios of confinement reinforcement, the impact of confinement reinforcement was examined. The experimental program is described in depth in the section that follows.

In order to evaluate the performance and behaviour of RAC columns under uniaxial compression, the experimental design of this research effort contains eight reinforced concrete columns (four NAC and four RAC). The columns were positioned horizontally when they were cast. The tested columns had a 1500 mm height and a cross-sectional area of 200 mm square. The columns can be categorized as long or slender because their slenderness ratio (λ) is around 26. A 25 mm clear cover was provided for the specimens. The longitudinal reinforcement content (rho_l) for each specimen was 2%, however the transverse reinforcement ratios varied between 0.14% (NAC (200)–C1, NAC (200)–C2, RAC (200)–C1, and RAC (200)–C2) and 0.18% (NAC (155)–C3, NAC (155)–C4, RAC (155)–C3, and RAC (155)–C4) were used to fabricate the specimens (Fig. 10.1). For transverse reinforcement content (ρ_t) of 0.14% and 0.18%, respectively, the spacing of the transverse reinforcement was 200 mm and 155 mm. The lateral tie spacing complies with the minimum standard for transverse reinforcement outlined in IS: 456 (2000). The compaction was done using a needle vibrator, and the columns were cast in a horizontal orientation. Wet jute bags were used to cover the specimens during the curing process. The samples were examined following a 28-day curing period.

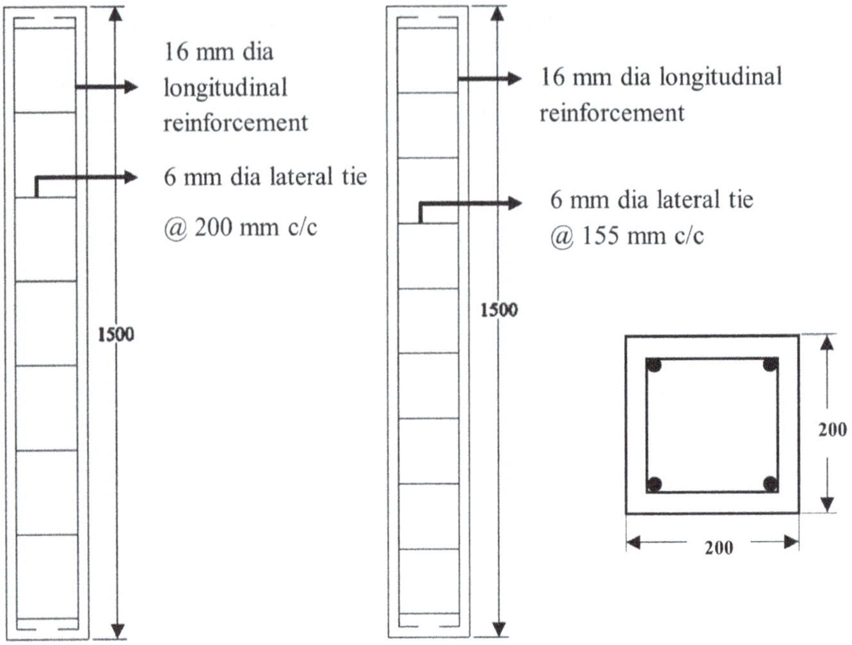

Fig. 10.1 Dimensions and reinforcement arrangement of the tested specimens (Pradhan et al. 2023)

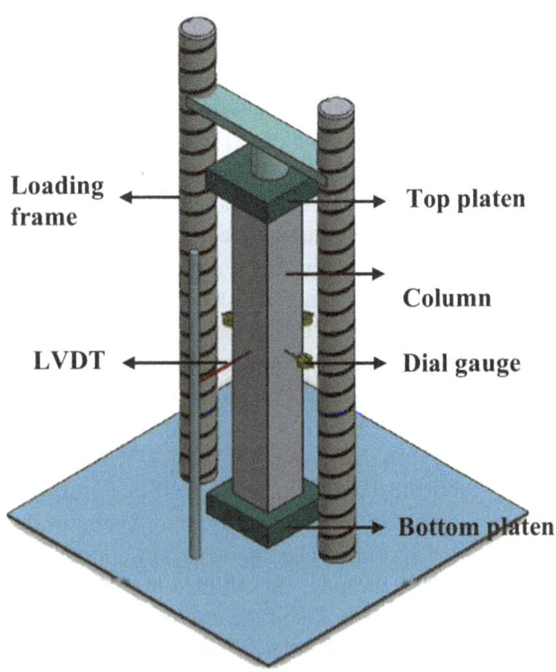

Fig. 10.2 Test setup and instrumentation details of the tested columns (Pradhan et al. 2023)

10.3.2 Test Setup and Test Procedure

A compression testing machine with a 3000 kN capacity was used to perform the monotonic incremental axial compression test on each specimen. Displacement was controlled at a rate of 0.5 mm/min as the load was applied until failure. Using a system displacement encoder, the column's axial displacement was recorded. Three dial gauges and one linear variable displacement transducer (LVDT) were mounted to measure the lateral displacement. To measure the corresponding strains during the test, electrical strain gauges were fitted in both the longitudinal and transverse reinforcements. Electrical strain gauges were mounted to the concrete's surface in both a longitudinal and transverse orientation at the midpoint of the specimens. The test setup's specifics are shown in Fig. 10.2.

10.4 Results and Discussion

10.4.1 Crack Pattern and Failure Mode

At the beginning of the loading, the concrete column's core and cover displayed comparable behaviour. This behaviour is ascribed to concrete's negligible longitudinal strain and Poisson's ratio's minimal impact. The crack was started by an increase in axial stress at the loading platen's contact surface, and it spread vertically and at an angle to the direction of longitudinal reinforcement. The longitudinal reinforcement eventually yielded, leading to the spalling of the concrete cover, which was then followed by the longitudinal reinforcement buckling and the opening of the lateral ties. In the case of RAC columns, it was noted that cracks were forming and the concrete cover was spalling earlier than respective NAC columns. Additionally, sizable concrete portions spalled off from the centre of RAC columns. The weaker bond strength of RAC may be partially responsible for this. In earlier investigations (Choi and Yun 2012; Shatarat et al. 2019), a similar failure pattern was seen for both NAC columns and RAC columns. In Fig. 10.3, the typical failure pattern of the tested columns is displayed.

10.4.2 Load–Displacement Relationship

In Fig. 10.4, the load–displacement relationships of the tested columns are depicted. Due to the settlement of the column's two ends, the axial displacement was found to be higher up to 200 kN of compressive force. The load–displacement curve, however, almost follows a linear path as the axial compressive stress rises before the longitudinal reinforcement gives way. In general, the load–displacement relationship showed a higher ascending branch for NAC columns. In Table 10.1, the maximum axial load

Fig. 10.3 Failure pattern of tested columns (Pradhan et al. 2023)

each specimen could withstand is listed. In comparison to the NAC columns with the same confinement reinforcement content, the RAC columns withstood axial loads with a resistance of about 18% less. Both Choi and Yun (2012) and Ajdukiewicz and Kliszczewicz (2007) reported a maximum reduction of 18% and 16%, respectively, in maximum axial load at 100% application of RCA. The maximum axial load was increased by 2.7% and 2.8% for NAC and RAC columns, respectively, with the increase in transverse reinforcement content. When RAC columns were compared to equivalent NAC columns, it was found that the axial displacement at maximum axial load was 1–3.5% higher for RAC columns. The axial displacement, which corresponds to the maximal axial load of NAC columns and RAC columns, increased by roughly 7.5% and 10%, respectively, when the transverse reinforcement content increased from 0.14% to 0.18%.

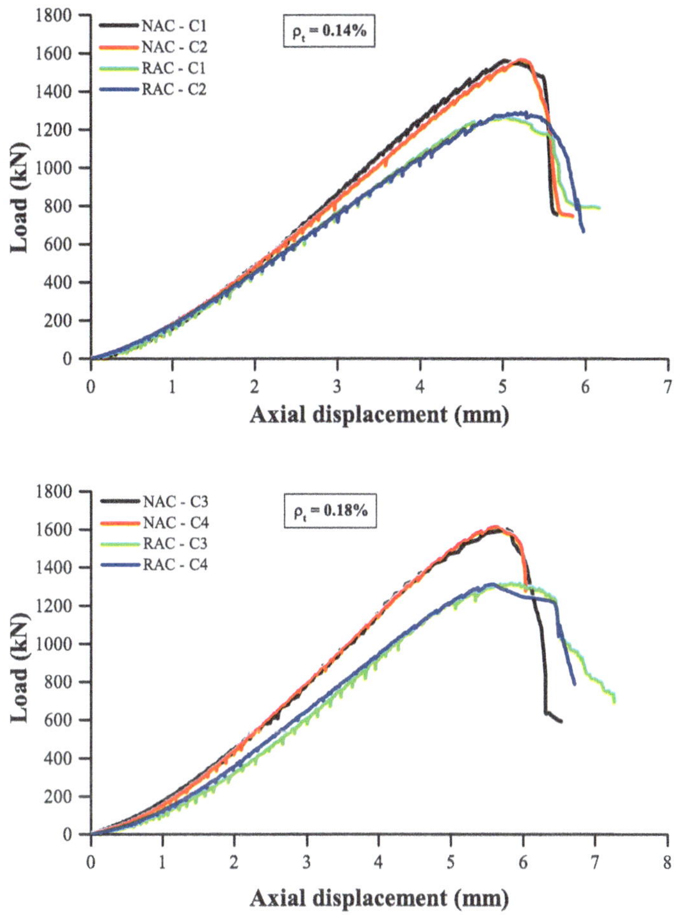

Fig. 10.4 Load vs. axial displacement of columns (Pradhan et al. 2023)

Table 10.1 Summary of the test results of column specimens

Column identifier	ρ_l (%)	ρ_t (%)	P_{max} (kN)	$0.85\,P_{max}$ (kN)	δ_y (mm)	δ_{85} (mm)	μ
NAC–C1	2	0.14	1564.31	1329.66	4.81	5.53	1.15
NAC–C2			1575.25	1338.96	5.09	5.49	1.08
RAC–C1			1274.33	1083.18	4.69	5.60	1.19
RAC–C2			1298.79	1103.97	4.83	5.72	1.18
NAC–C3		0.18	1602.27	1361.93	5.21	6.06	1.16
NAC–C4			1621.62	1378.38	5.33	6.02	1.13
RAC–C3			1327.20	1128.12	5.27	6.46	1.23
RAC–C4			1318.39	1120.63	5.25	6.46	1.23

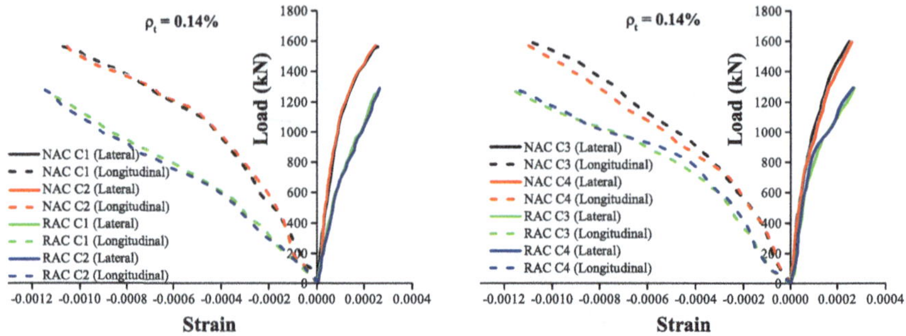

Fig. 10.5 Measured strains in concrete of the column specimens (Pradhan et al. 2023)

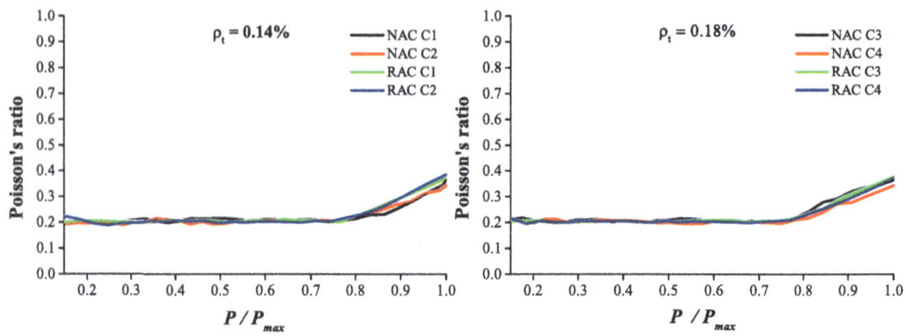

Fig. 10.6 Poisson's ratio for columns with different lateral tie spacing (Pradhan et al. 2023)

10.4.3 Strain in Concrete and Poisson's Ratio

Electrical strain gauges were used to measure the specimens' longitudinal and transverse strains at the mid-height. In comparison to NAC columns, the measured longitudinal and lateral strains of RAC columns are much higher (Fig. 10.5). The same behaviour of RAC columns was seen by Ajdukiewicz and Kliszczewicz (2007) regardless of the extent of RCA replacement. The earlier spalling of cover concrete in the case of RAC columns is due to the rapid development in concrete strain for RAC. In order to resist the axial load, restricted concrete and RAC column reinforcement are crucial.

The measured longitudinal and transverse strains were used to estimate the specimens' Poisson's ratios. At the beginning of loading, Poisson's ratio has relatively little of an impact. Therefore, variations in Poisson's ratios during different loading stages are presented beyond $P/P_{max} = 0.15$ (Fig. 10.6). The average Poisson's ratio of the columns was found to be about 0.21, regardless of the type of aggregate. Choi and Yun (2012) also observed a similar behaviour and reported a Poisson's ratio of 0.20.

10.4.4 Effect of Confinement

The distance between the lateral ties controls the confined volume of concrete and, as a result, controls how well a column resists axial load and lateral pressure. The axial load carrying capacity and ductility are both increased by the transverse reinforcement that is placed closely together. For both NAC and RAC columns, the axial strength of the columns with a tie spacing of 155 mm was marginally higher than that of the columns with a tie spacing of 200 mm. Additionally, the impact of tie spacing on the displacement ductility (μ) of NAC and RAC columns was investigated. According to Pessiki and Pieroni (1997), the ratio of displacement in the descending portion of the load–displacement curve corresponds to the axial load of $0.85P_{max}$ to the displacement that corresponds to the limit of elastic behaviour (δ_y) is the displacement ductility of reinforced concrete columns under axial compression load. A best fit line is obtained by applying linear regression analysis to the linear component of the load–displacement curve. The displacement δ_y is determined by the intersection of the best fit line with the maximum axial force that the column can resist.

$$\mu = \frac{\delta_{85}}{\delta_y} \tag{10.4.1}$$

Using Eq. 10.4.1, the displacement ductility of each column was determined, and the results are displayed in Table 10.4. For both NAC and RAC columns with ρ_t of 0.18% compared to the columns with ρ_t of 0.14%, an increase of roughly 3% was observed. Comparing RAC columns to NAC columns with the same ρ_t, the μ in RAC columns is about 6% greater. The findings of Choi and Yun (2012) are in-congruent with this outcome, though. In order to provide a more definitive statement about the ductility of RAC columns, additional research on RAC columns with various configurations of transverse reinforcement and tie spacing is necessary.

10.4.5 Assessment of the Test Results with Design Specifications

Axial Load

The results of the experimental examination suggested that the presence of RCA had a negative impact on the ability to carry axial loads. As a result, it was confirmed that the current design guidelines for RAC columns were applicable. The test results are corroborated with the design axial load specified in IS: 456 (2000), ACI 318 (2008), BS 8110-1 (1997), CSA A23.3-04 (2004), and NZS 3101 (2006) (Table 10.2). The nominal axial load calculated by the addition law equation overestimates the capacity of the RAC columns. It was observed that, RAC columns could achieve only 80% of the predicted capacity. The expression suggested by ACI 318 (2008) and NZS 3101 (2006) considering the strength reduction factor also overestimates the axial load

Table 10.2 Existing expression for axial load capacity of reinforced concrete column

Codes	Expressions
Nominal axial load	$P = 0.85 f_c^{'}(A_g - A_{st}) + f_y A_{st}$
IS: 456 (2000)	$P = 0.4 f_{ck}(A_g - A_{st}) + 0.67 f_y A_{st}$
ACI 318 (2008)	$P = 0.8\left[0.85 f_c^{'}(A_g - A_{st}) + f_y A_{st}\right]$
BS 8110-1 (1997)	$P = 0.4 f_{cu}(A_g - A_{st}) + 0.75 f_y A_{st}$
CSA A23.3-04 (2004)	$P = \alpha_1 \phi_c f_c^{'}(A_g - A_{st}) + \phi_s f_y A_{st}$
	where, $\alpha_1 = 0.85 - 0.0015 f_c^{'}$ (should be more than 0.67)
NZS 3101 (2006)	$P = \alpha_1 f_c^{'}(A_g - A_{st}) + \phi_c f_y A_{st}$
	where $\alpha_1 = 0.85$ (for $f_c^{'} \leq 55$ MPa)
	where $\alpha_1 = 0.85 - 0.008(f_c^{'} - 30)$ (for $f_c^{'} > 55$ MPa)

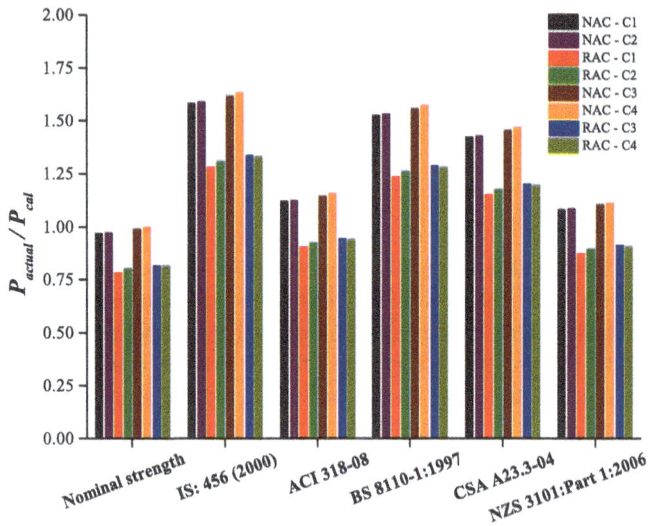

Fig. 10.7 Comparison of experimental and predicted axial load of the tested columns (Pradhan et al. 2023)

carrying capacity of RAC columns (Fig. 10.7). However, the expressions suggested in IS: 456 (2000), BS 8110-1 (1997), and CSA A23.3-04 (2004) underestimate the axial load of RAC columns (Fig. 10.7). The strength reduction factor or factor of safety used in the aforementioned expressions offers the requisite conservative efficacy for RAC column design.

where $f_c^{'}$ = compressive strength of concrete cylinder (MPa), f_{ck} and f_{cu} = compressive strength of concrete cube (MPa), A_g = gross area of column, A_{st} = total area of longitudinal reinforcement, f_y = yield strength of longitudinal reinforcement (MPa), ϕ_c = resistance factor for concrete (0.65), and ϕ_s = resistance factor for longitudinal reinforcement (0.85).

Table 10.3 Specifications for the spacing of lateral ties

Codes	Specifications
IS: 456 (2000)	$s \leq$ { Least lateral dimension of the column; 16 times the smallest diameter of longitudinal reinforcement; 300 mm
ACI 318 (2008)	$s \leq$ { 16 times the diameter of longitudinal reinforcement; 48 tie bar; Least dimension of the compression member

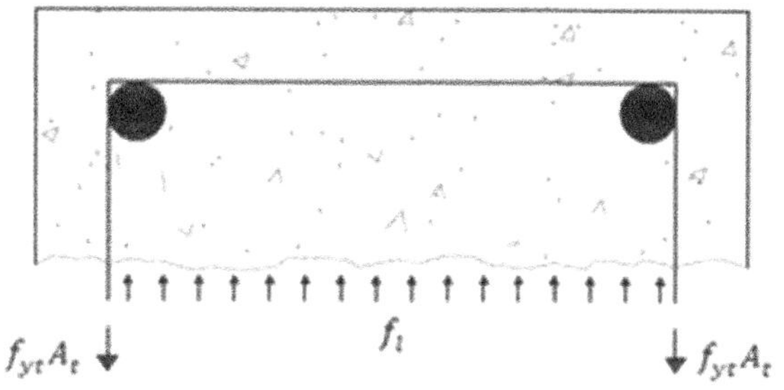

Fig. 10.8 Confining stresses by the transverse reinforcement (Pradhan et al. 2023)

Tie Spacing

The axial strength and ductile behaviour of the reinforced concrete column are controlled by the spacing of the lateral ties. Although the use of closely spaced rectangular ties increases ductility without significantly increasing the axial strength of the column (Park and Paulay 1975). Table 10.3 lists the minimal requirements for transverse reinforcement as defined by several codes. According to References IS: 456 (2000) and ACI 318 (2008), 200 mm is the maximum spacing of transverse reinforcement that can be offered. The observed axial load of the RAC columns was less than the nominal axial load capability for both the 200 mm and 155 mm tie spacings. The confined concrete experiences more compressive stress due to the early spalling of the concrete cover in the RAC column. The RAC columns' ultimate axial load was decreased as a result. Reducing the lateral ties spacing will strengthen the column's concrete core. In order to increase the ductility and axial strength of RAC columns, transverse reinforcement is therefore required and must be validated.

This research proposes a conservative approach to estimate the spacing of the lateral ties. The nominal axial strength of the column as a result of confinement is employed in this case to calculate the tie spacing. According to Richart et al. (1929), the compressive strength of a confined specimen (f_c'') under an axial load can be calculated by adding the compressive strength of an unconfined specimen (f_c') and the lateral confining pressure (f_l) (Eq. 10.4.2). Figure 10.8 displays a free body diagram of the stresses acting on a rectangular confined specimen. Equation 10.4.3 can be used to measure the maximum lateral confining pressure experienced by the core concrete.

$$f_c'' = f_c' + 4.1 f_l \tag{10.4.2}$$

$$f_l = \frac{2 f_{yt} A_t}{b_c s} \tag{10.4.3}$$

By taking into account the contributions of confined concrete, transverse reinforcement, and longitudinal reinforcement, the ultimate axial load of a confined specimen can be expressed (Eq. 10.4.4). The nominal axial load of a confined specimen can be represented by Eq. 10.4.5 by incorporating Eqs. 10.4.2 and 10.4.3 into 10.4.4 and taking into account the reduction in compressive strength of concrete caused by the size and shape effect as well as vertical casting disadvantages (sedimentation of aggregate and gain of water at the top). The compressive strength of the concrete and the yield strength of the longitudinal reinforcement are taken into account in the addition law equation for the maximum load of an axially loaded column, which is given by Eq. 10.4.6. In order to determine the minimum tie spacing in a reinforced concrete column, Eqs. 10.4.5 and 10.4.6 are operated, and the result is shown in Eq. 10.4.7. Using Eq. 10.4.7, the RAC column's minimum tie spacing of 70 mm is determined. The suggested method for measuring the minimum spacing of transverse reinforcement is conservative in nature. As a result, it is suggested that additional experimental research be done for reinforced RAC columns with various tie spacing and tie arrangement.

$$P_u = f_c'' A_{cc} + f_y A_{st} \tag{10.4.4}$$

$$P_u = \left(0.85 f_c' + \frac{8.2 f_{yt} A_t}{b_c s}\right) A_{cc} + f_y A_{st} \tag{10.4.5}$$

$$P_u = 0.85 f_c' \left(A_g - A_{st}\right) + f_y A_{st} \tag{10.4.6}$$

$$s = \frac{8.2 f_{yt} A_t \left(A_t - A_{st}\right)}{0.85 f_c' \left(A_g - A_c\right) b_c} \tag{10.4.7}$$

10.5 Closure

The behaviour of reinforced columns under axial load was covered in this chapter. Additionally, it was examined how the spacing of the transverse reinforcement affected the axial load and ductility of the RAC columns.

- The failure pattern was the same for both NAC and RAC columns, regardless of the type of aggregate and transverse reinforcement content.
- The presence of RCA had a negative impact on the compressive strength of RAC columns under axial loading. Both NAC and RAC columns' axial loads were moderately enhanced by the closely spaced lateral ties. RAC columns are comparable to NAC columns in terms of Poisson's ratio and displacement ductility.
- The RAC columns were unable to reach the design strength requirements outlined in ACI 318 (2008) and NZS 3101 (2006). For the spacing of transverse reinforce-

ment in this situation, Pradhan et al. (2023) recommended a conservative approach in order to effectively confine the core concrete and increase the axial load of RAC columns.

- For a better understanding of the impact of varied tie spacing and arrangement on the compressive strength of RAC columns under axial stress, additional experimental research is needed.

References

ACI 318 (2008) ACI committee 318. Building code requirements for structural concrete and commentary, American Concrete Institute, Farmington Hills (MI)

Ajdukiewicz AB, Kliszczewicz AT (2007) Comparative tests of beams and columns made of recycled aggregate concrete and natural aggregate concrete. J Adv Concr Technol 5(2):259–273

Breccolotti M, Materazzi AL (2010) Structural reliability of eccentrically-loaded sections in RC columns made of recycled aggregate concrete. Eng Struct 32(11):3704–3712

BS 8110-1 (1997) Structural use of concrete–part 1: code of practice for design and construction

Choi W-C, Yun H-D (2012) Compressive behavior of reinforced concrete columns with recycled aggregate under uniaxial loading. Eng Struct 41:285–293

CSA A23.3-04 (2004) Design of concrete structures. Canadian Standards Association, Mississauga, Ont

Eurocode 2 (2004) Eurocode 2: design of concrete structures: part 1-1: general rules and rules for buildings. British Standards Institution

IS: 456 (2000) Plain and reinforced concrete-code of practice. Bureau of Indian Standards, New Delhi, India

NZS 3101 (2006) Concrete structures standard, part 1–the design of concrete structures

Park R, Paulay T (1975) Reinforced concrete structures. John Wiley & Sons

Pessiki S, Pieroni A (1997) Axial load behavior of large-scale spirally-reinforced high-strength concrete columns. ACI Struct J 94(3):304–314

Pradhan S, Nayak TK, Kumar S, Barai SV (2023) Experimental and numerical study of recycled aggregate concrete column. Struct Concr

Richart FE, Brandtzaeg A, Brown RL (1929) The failure of plain and spirally bound concrete in compression. Univ Illinois Eng Exp Station Bull 26(190):1–74

Shatarat N, Alhaq AA, Katkhuda H et al (2019) Investigation of axial compressive behavior of reinforced concrete columns using recycled coarse aggregate and recycled asphalt pavement aggregate. Constr Build Mater 217:384–393

Wang Y, Chen J, Zong B, Geng Y (2011) Mechanical behavior of axially loaded recycled aggregate concrete filled steel tubular stubs and reinforced recycled aggregate concrete stubs. J Build Struct 32(12):170–177

Xiao J (2018) Recycled aggregate concrete structures. Springer-Verlag, Heidelberger, Berlin, Germany

Xu J-J, Chen Z-P, Ozbakkaloglu T, Zhao X-Y, Demartino C (2018) A critical assessment of the compressive behavior of reinforced recycled aggregate concrete columns. Eng Struct 161(2017):161–175

Chapter 11
Structural Applications: Slab

11.1 Introduction

The performance of RAC in both the fresh and hardened stages is adversely affected by the inherent inferior material qualities of RCA caused by the existence of micro-cracks and attached mortar layer. PPM mix design approach is used in conjunction with the TSMA to reduce the impact of RCA and enhance the quality of RAC, as detailed in chapters "Particle Packing Method of Mix Proportioning and Modified Mixing Approach" and "Macro-level Performance Assessment of Concrete: Conventional Approach". Results from the experimental research on the punching shear performance of RAC slabs with 100% RCA use were inconsistent. Reis et al. (2015) and Francesconi et al. (2016) observed insignificant difference in punching shear strength of RAC slabs and NAC slabs, whereas Rao et al. (2012) and Mahmoud et al. (2017) reported the adverse effect of RCA on punching shear capacity. Rao et al. (2012), Reis et al. (2015), Mahmoud et al. (2017), and Francesconi et al. (2016) used longitudinal reinforcement content of 0.85%, 0.93%, 0.95%, and 0.56%, respectively. As a result, Pradhan (2019) examined the punching shear capability of RAC slab by inserting 100% RCA and a lower flexural reinforcement ratio (0.35%). The ability of current formulations to calculate the punching shear capacity of RAC slab is discussed in this chapter. Additionally, a rigorous analysis is done on the impact of various design factors on the punching shear capacity of RAC slab.

11.2 Performance of Reinforced RAC Slab

When reinforced concrete is exposed to concentrated loads of great magnitude, the slab suddenly ruptures. Evaluation of the punching shear strength of the RAC slab is crucial because there were no visible impressions previous to the punching shear failure, which makes the situation even more critical. Initially, Rao et al. (2012) conducted the experimental investigation essentially varying the RCA content. Replacing

S. Pradhan et al., *Particle Packing Method for Recycled Aggregate Concrete*,
https://doi.org/10.1007/978-981-99-7516-7_11

20%, 40%, 60%, 80%, and 100% of NCA by RCA, the punching shear performance of RAC slabs of size 1100 × 1100 × 50 mm and with all four edges in simply supported condition was studied. The study reported a reduction in punching shear strength as well as ultimate deflection (δ_u) of RAC slab with the increment in RCA content. In addition, the estimated stiffness and strain energy absorption of RAC slabs also decreased proportionately as the RCA content increased. However, up to 40% use of RCA the punching shear strength of RAC slabs was marginally influenced. Later, Reis et al. (2015) substituted 20%, 50%, and 100% of NCA by RCA and performed the punching shear test on 1100 mm × 1100 mm × 90-mm-sized specimens. Moreover, eight specimens were tested on symmetrically placed simply supported bases. The supports were in a circular pattern at a radial distance of 0.22L (L is the span of the slab), as suggested in the earlier studies (Guandalini et al. 2009; Ruiz et al. 2014; Inácio et al. 2015). The punching shear strength of tested specimens exhibited only marginal variation with the variation in RCA content. However, the pre-cracked stiffness and initial cracking load of RAC slabs were lower than the conventional concrete slabs and these parameters further reduced as the RCA content increased. The experimental results were further corroborated analytically using three-dimensional nonlinear finite element model. They have observed satisfactory predictions of RAC slabs for Fib MC (2010) with level II, level III, and level IV approximations, whereas Fib MC (2010) with level I of approximation, ACI 318 (2008) and Eurocode 2 (2004) exhibited conservative predictions.

Francesconi et al. (2016) conducted punching shear test on 1100 mm × 1100 mm × 50-mm-sized slab specimens with all four edges in simply supported condition. RAC slabs were prepared using four different RCA replacement percentages of 30%, 50%, 80%, and 100%. Similar performance of RAC slabs was recorded with respect to the conventional concrete slabs. The experimental results were compared with the predicted values obtained from ACI 318 (2008), Eurocode 2 (2004), and Fib MC (2010) formulations. In ACI 318 (2008) formulation, it was recommended to adopt the value of multiplying factor related to the aggregate type as 0.85. ACI 318 (2008) and Eurocode 2 (2004) conservatively predicted the experimental results, whereas Fib MC (2010) expression overestimated the punching shear strength of RAC slabs. Mahmoud et al. (2017) studied the influence of RCA content as well as maximum size of aggregate on punching shear behaviour of RAC slabs. In this context, RCA with a maximum size of 12.5 mm and 25 mm was incorporated in three different percentages (30%, 60%, and 100%) to prepare RAC. The slab specimens were of size 1200 mm × 1200 mm × 50 mm and tested with all four edges in simply supported condition. The reported results indicate the adverse effect of RCA on initial cracking load, ultimate punching load, stiffness, and strain energy absorption, and the reduction of these parameters was more significant with the increment in RCA content. Furthermore, the specimens with maximum size of aggregate 25 mm exhibited higher punching shear strength, stiffness, and strain energy absorption in comparison to the specimens with maximum aggregate size of 12.5 mm. Xiao et al. (2019) studied the failure pattern, energy consumption, punching shear capacity, and load–deflection relationship of RAC slabs with different RCA content (0%, 50%, and 100%) and steel fibre volumetric ratio (0%, 0.5%, and 1%). The mechanism of failure in RAC slabs without steel fibre happened in cable-stayed damage. However, RAC slabs with steel fibre induced the punching shear failure by reaching the maximum strain limit of concrete in compression-shear area. The failure pattern transformed from shear failure to bending-shear failure by the inclusion of steel fibre. It also improved the ductility, energy consumption, and deformation of RAC slabs. Although the punching shear capacity of RAC slabs improved by 7–15% due to the incorporation of steel fibre, it was still lower than NAC slabs. Sahoo and Singh (2020, 2021) conducted punching shear test of 1200 mm × 1200 mm × 100-mm-sized reinforced RAC slabs, which were supported by eight roller support in a circular pattern of diameter 1100 mm. RACs used in the slabs preparation were consisting of different RCA contents (0%, 50%, and 100%) and compressive strength. Irrespective of these variability all the slabs failed through punching shear crack. Punching strength of RAC slabs was independent of RCA content in RAC. Eurocode 2 (2004) and strut-and-tie model predicted the punching shear strength of RAC slabs conservatively. Pradhan (2019) conducted the punching shear test of reinforced RAC slabs, which were prepared using 100% RCA and PPM mix design approach. The influence of PPM mix design approach on its performance was evaluated and discussed in detail in the subsequent sections.

11.3 Impact of PPM on Punching Shear Strength of RAC Slab

11.3.1 Experimental Program

The experimental investigation by Pradhan (2019) was performed on four numbers of slab specimens. In this context, two numbers of NAC slab specimens (NAC-S1 and NAC-S2) and RAC specimens (RAC-S1 and RAC-S2) were subjected to punching shear test. The dimensions of all the specimens were 1000 mm × 1000 mm × 100 mm (Fig. 11.1). The slabs were supported on four sides and the distance between the supports was 950 mm in both the directions. The specimens were longitudinally reinforced with 8 mm diameter bar of yield strength 500 MPa and the longitudinal reinforcement content was kept to be 0.35%. The yield strength (f_y) and modulus of elasticity (E_s) are observed to be 580 MPa and 200 GPa, respectively. Reinforcement meshes of 190 mm × 190 mm were prepared using 8 mm diameter bar for the flexural reinforcement. The flexural reinforcement was provided at the bottom of the slab with a clear cover of 20 mm. The specimens were cured for 28 days using jute bags and each specimen was tested within 1 week after the curing period.

11.3.2 Test Setup and Test Procedure

The slab specimens were tested in the upside down condition with respect to the usual state of slab and column interaction. For this, at the bottom side, the flexural reinforcement of the slab was placed and the column was simulated at the top surface.

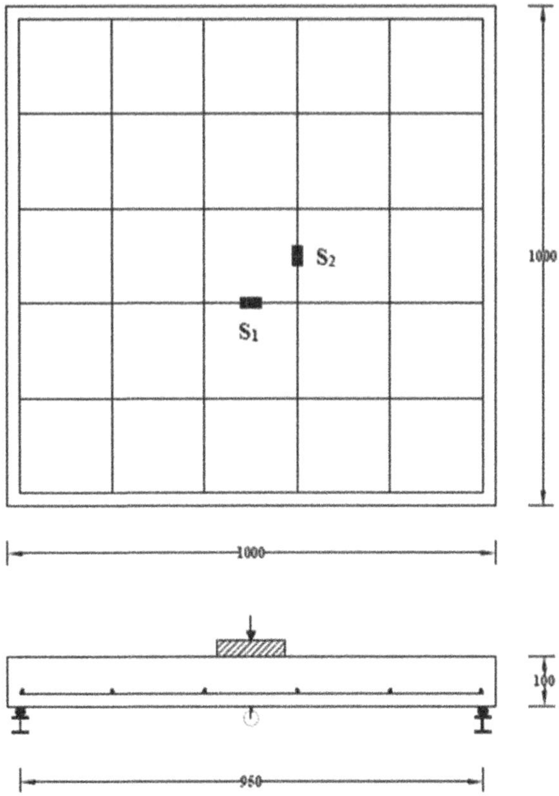

Fig. 11.1 Layout of the longitudinal reinforcement (Pradhan 2019)

The specimens were placed in the simply supported condition over the loading platform. A steel rod of 16 mm was welded to the top of the flange of the loading platform in order to allow the free rotation of the specimen along the edges (Fig. 11.1). A concentrated circular patch load was applied on the geometric centre of each slab. In this regard, a rigid circular bearing plate of diameter 140 mm and thickness 75 mm was placed at the centre of the slab to simulate a column. The concentrated incremental load was introduced using a hydraulic jack of capacity 300 kN, which was mounted on a steel reaction frame. A dial gauge of least count 0.01 mm was installed at the geometric centre of the slab to measure the vertical deflection of the slab (Fig. 11.1). The electrical strain gauges (S_1 and S_2) were mounted on the orthogonally placed longitudinal reinforcement close to the centre of the slab to measure the strain in steel (Fig. 11.1). The applied load and corresponding displacement of the centre of the slab and strain gauge readings were recorded at an interval of 5 kN (Fig. 11.2).

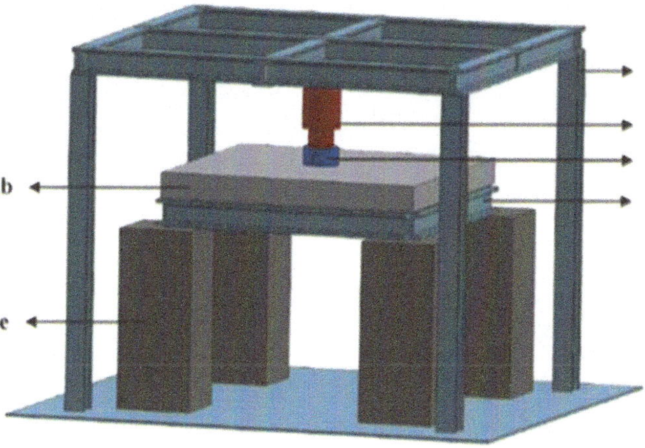

Fig. 11.2 Test setup for punching shear test of the slabs (Pradhan 2019)

11.4 Results and Discussion

11.4.1 Crack Pattern and Failure Mode

The progression of the cracks was observed to be similar in NAC and RAC slabs. The tangential cracks were initially witnessed close to the periphery of the column as the tensile strength of concrete exceeded at the bottom surface. These tangential cracks were confined only to the vicinity of the column. Furthermore, a radial crack was observed more prominently parallel to the flexural reinforcement. Upon a further increase in the applied load, the radial cracks were propagated approximately from the axis of the column to the centre of the edges as well as the corners of the slab. Hence, the slabs were segmented into a number of panels by the radial flexural cracks at the time of collapse. Figure 11.3 suggests that the crack pattern of the tested specimens is similar to the typical crack pattern depicted as per the yield line theory for a reinforced concrete slab with all edges in simply supported condition. Consequently, the punching shear failure of both NAC and RAC slabs with large plastic deformation was triggered by the formation of diagonal yield line mechanism.

11.4.2 Load–Deflection Relationship

The load–deflection relationship was measured by means of a dial gauge at the geometric centre of the soffit of the slabs, which are depicted in Fig. 11.4. Initially a linear elastic branch was observed for each specimen. Subsequently, due to the formation of tangential and radial cracks, a significant reduction in the stiffness of the slabs was noticed. Moreover, for both NAC and RAC slabs, the plastic plateau

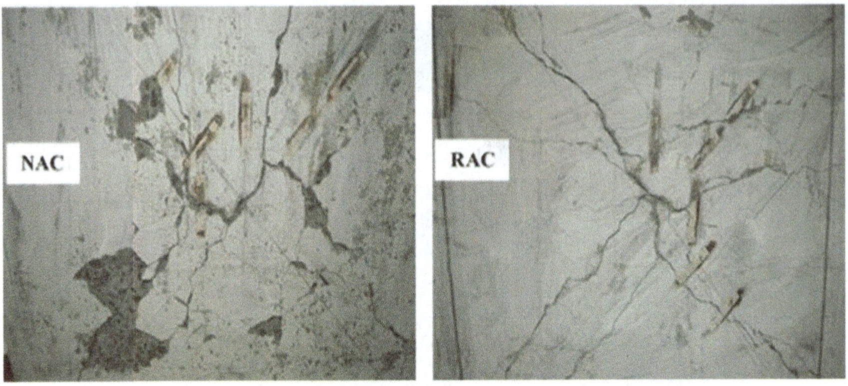

Fig. 11.3 Failure pattern and final condition of the tested slabs (Pradhan 2019)

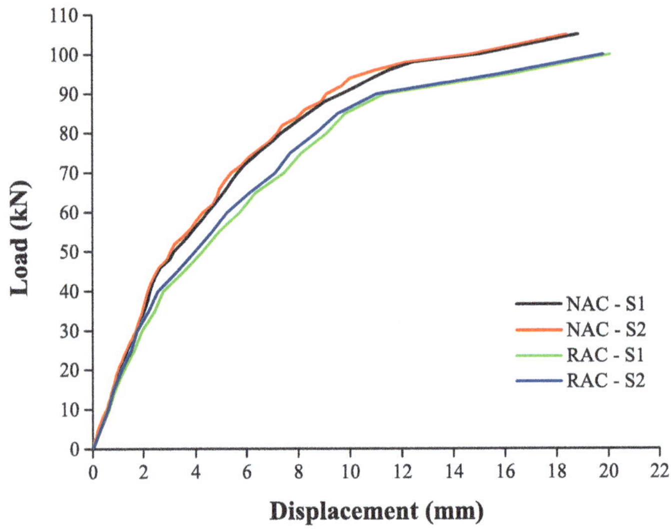

Fig. 11.4 Load–deflection curves of the tested specimens (Pradhan 2019)

can be distinguished from the load–deflection relationship (Fig. 11.4). Owing to the smaller slab thickness and lower longitudinal reinforcement content, the tested specimens exhibited a plastic plateau (Guandalini et al. 2009).

Figure 11.4 shows that the initial cracking load ($V_{R,cr}$) as well as the ultimate punching load ($V_{R,u}$) of RAC slabs is lower than the conventional concrete slabs. A reduction of 12.5% and 4.8% is observed in $V_{R,cr}$ and $V_{R,u}$, respectively. Using 100% RCA, Rao et al. (2012), Reis et al. (2015), Mahmoud et al. (2017), Francesconi et al. (2016), and Sahoo and Singh (2020, 2021) observed a maximum reduction of 14%, 2%, 24%, 10%, and 19% in punching shear strength of RAC slabs with respect to the NAC slabs. This indicates that the PPM mix design approach in combination with the

Table 11.1 Summary of experimental investigation

Slab identifier	$V_{R,cr}$ (kN)	$V_{R,u}$ (kN)	δ_{cr} (mm)	δ_u (mm)	Strain energy (kNm)
NAC-S1	40	105	2.16	18.39	1.518
NAC-S2	40	105	2.26	19.32	1.442
RAC-S1	35	100	2.43	19.90	1.447
RAC-S2	35	100	2.20	19.51	1.433

TSMA positively influenced the punching shear strength of RAC slabs. Moreover, RAC slabs exhibited a maximum of 8.2% higher ultimate deflection (δ_u). Reis et al. (2015) and Francesconi et al. (2016) also reported an increment in δ_u, whereas Rao et al. (2012) and Mahmoud et al. (2017) reported a reduction in δ_u of RAC slabs with respect to the NAC slabs.

The strain energy absorbed by the specimens is estimated as the area under the load–deflection relationship (Fig. 11.4) and represented in Table 11.1. The strain energy absorbed by the RAC specimens is marginally lower (maximum 6%) than the NAC specimens. Rao et al. (2012) and Mahmoud et al. (2017) reported a reduction in absorbed strain energy for RAC specimens with 100% use of RCA. On the contrary, Francesconi et al. (2016) and Sahoo and Singh (2021) observed an incremental trend in absorbed strain energy for RAC slabs as the replacement percentage of RCA increases. This behaviour may be ascribed to the higher δ_u observed for RAC slabs with respect to the conventional concrete slabs.

11.4.3 Strain in Longitudinal Reinforcement and Concrete

The strain in longitudinal reinforcement was monitored by fixing the electrical strain gauges in the orthogonal direction. The measured strains are depicted in Fig. 11.5. For each specimen, the yielding of longitudinal reinforcement can be observed from Fig. 11.5. The corresponding strain in reinforcement at failure was observed to be higher for RAC beams. This may be attributed to the higher δ_u recorded for RAC slabs.

11.5 Comparison with Design Codes

11.5.1 IS: 456 (2000)

According to IS: 456 (2000), the punching shear strength ($V_{Rd,c}$) can be estimated by using Eq. 11.5.1, which is a function of the compressive strength of concrete and side dimensions of the column.

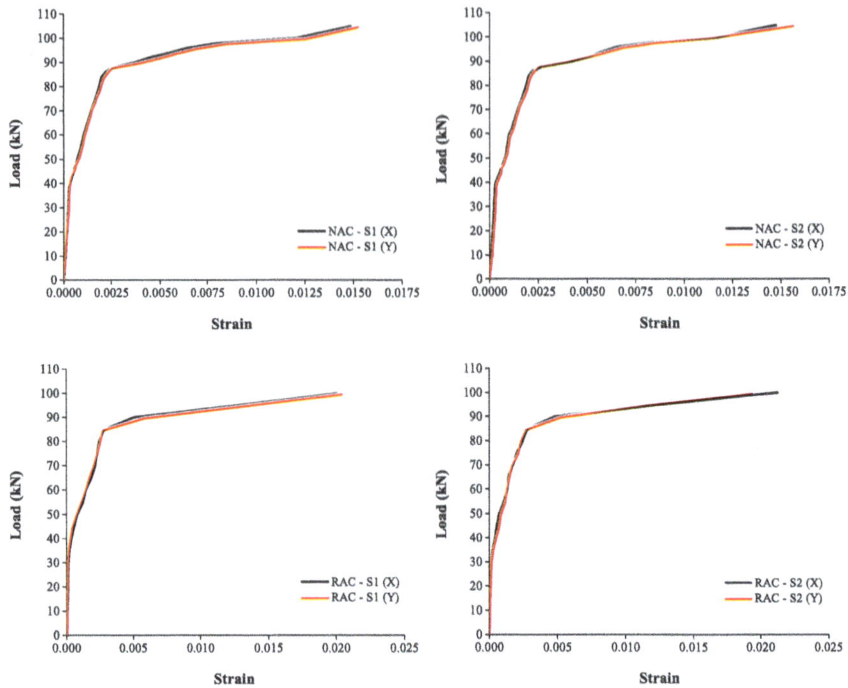

Fig. 11.5 Measured strain in longitudinal reinforcement of the slabs (Pradhan 2019)

$$V_{Rd,c} = 0.25(0.5 + \beta_c)\sqrt{f_{ck}}b_0 d \leq 0.375\sqrt{f_{ck}}b_0 d \tag{11.5.1}$$

where b_0 is assumed to be at a distance $d/2$ from the side of load area.

11.5.2 ACI 318 (2008)

The $V_{Rd,c}$ of reinforced concrete slab can be predicted using Eq. 11.5.2. As suggested by ACI 318 (2008), $V_{Rd,c}$ is the minimum of the following relations. The expression accounts the influence of the cross-sectional dimension of the column, location of the column, and compressive strength of concrete.

$$V_{Rd,c} \leq \begin{cases} 0.18\left(1 + \frac{2}{\beta}\right)\lambda\sqrt{f_c}b_0 d \\ 0.083\left(\frac{\alpha_s d}{b_0} + 2\right)\lambda\sqrt{f_c}b_0 d \\ 0.33\lambda\sqrt{f_c}b_0 d \end{cases} \tag{11.5.2}$$

where α_s = 40, 30, and 20 for the interior, edge, and corner column, respectively; $\lambda = 1$ and 0.75 for normal weight concrete and light weight concrete, respectively. b_0 is located at a distance $d/2$ from the edges of the column. For $\lambda = 1$, the $V_{Rd,c}$ of RAC slabs is overestimated by the expression of ACI 318 (2008) (Francesconi et al. 2016). Hence, for RAC, Francesconi et al. (2016) adopted λ value as 0.85, which yields a conservative prediction of the $V_{Rd,c}$ of RAC slabs. In the present study also λ is adopted as 0.85.

11.5.3 BS 8110-2 (1985)

The expression (Eq. 11.5.3) suggested by BS 8110-2 (1985) to estimate the $V_{Rd,c}$ of reinforced concrete slab accounts the influence of longitudinal reinforcement content, compressive strength of concrete, and size of the slab. The control perimeter is considered to be at a distance of 1.5d from the column face.

$$V_{Rd,c} = 0.79\left(\frac{100A_s}{b_0 d}\right)^{\frac{1}{3}}\left(\frac{f_{ck}}{25}\right)^{\frac{1}{3}}\frac{\left(\frac{400}{d}\right)^{\frac{1}{4}}}{\gamma_m}b_0 d \tag{11.5.3}$$

where γ_m = partial factor of safety (1.25).

11.5.4 Eurocode 2 (2004)

The design value of $V_{Rd,c}$ of reinforced concrete slab as suggested by Eurocode 2 (2004) is represented in Eq. 11.5.4. The influence of the size of the slab, longitudinal reinforcement content, and the compressive strength of concrete is considered in Eurocode 2 (2004) expression. The control perimeter (u_1) is located at a distance of 2d from the face of the column instead of $d/2$ and 1.5d as suggested by ACI 318 (2008) and BS 8110-2 (1985).

$$V_{Rd,c} = C_{Rd,c}k\left(100\rho f_{ck}\right)^{\frac{1}{3}}u_1 d_{eff} \geq v_{min}u_1 d_{eff} \tag{11.5.4}$$

where $C_{Rd,c} = 0.18$ and k is the size parameter, which can be defined by the following expression (Eq. 11.5.5):

$$k = 1 + \sqrt{\frac{200}{d_{eff}}} \leq 2 \tag{11.5.5}$$

where $d_{eff} = \frac{d_x+d_y}{2}$. d_x and d_y are the depth of extreme compression fibre of slab from the centroid of the tensile reinforcement in the X and Y orthogonal directions, respectively.

$$\rho_l = \sqrt{\rho_{lx}\rho_{ly}} \leq 0.02 \tag{11.5.6}$$

where ρ_l = longitudinal reinforcement ratio, $\rho_{lx} = \frac{A_{lx}}{b_x d_{eff}}$ and $\rho_{lx} = \frac{A_{ly}}{b_y d_{eff}}$, $b_x = a_x + 6d_{eff}$ and $b_y = a_y + 6d_{eff}$, and a_x and a_y are the sides of the column in the X and Y orthogonal directions, respectively. The u_1 is represented by the following expression (Eq. 11.5.7):

$$u_1 = 2\left(a_x + a_y\right) + 4\pi d_{eff} \tag{11.5.7}$$

The lower limit value of the punching shear stress (ν_{min}) is represented in Eq. 11.5.8.

$$\nu_{min} = 0.035k^{3/2} f_{ck}^{1/2} \tag{11.5.8}$$

11.5.5 Fib MC (2010)

Fib MC (2010) accounts the rotation (ψ) of the slab and maximum size of the aggregate in addition to the compressive strength, control perimeter (b_1), and effective depth of the slab while assessing the $V_{Rd,c}$. b_1 is considered at a distance of $d/2$ from the column face.

$$V_{Rd,c} = k_{\psi} \frac{\sqrt{f_c}}{\gamma_c} b_1 d \tag{11.5.9}$$

where γ_c = concrete partial factor of safety. The k_{ψ} parameter depends on the rotation of the slab and can be represented by Eq. 11.5.10.

$$k_{\psi} = \frac{1}{1.5 + 0.9k_{dg}\psi d} \leq 0.6 \tag{11.5.10}$$

The k_{dg} parameter accounts the influence of aggregate size on $V_{Rd,c}$ (Eq. 11.5.11).

$$V_{Rd,c} \leq \begin{cases} \frac{32}{16+d_g} \geq 0.75(for, d_g \leq 16mm) \\ 1(for, d_g \geq 16mm) \end{cases} \tag{11.5.11}$$

The calculation of ψ of the slab around the supported area is characterized by various levels of approximation. For level I of approximation, ψ can be calculated by Eq. 11.5.12.

$$\psi = 1.5 \frac{r_s}{d} \frac{f_{yd}}{E_s} \tag{11.5.12}$$

where r_s = distance from the axis of the column to the position where the radial bending moment is zero. r_s can be approximated as $0.22L_x$ or $0.22L_y$ in X- and Y-directions, respectively. L_x and L_y are the span of the slab in X- and Y-directions, respectively.

11.5.6 NZS 3101 (2006)

The expression suggested by NZS 3101 (2006) to predict $V_{Rd,c}$ of reinforced concrete slab is given in Eq. 11.5.13. The influence of the size of the slab is accounted by k_{ds} parameter. Apart from this, the influence of the location of the column and compressive strength of concrete on $V_{Rd,c}$ are considered.

$$V_{Rd,c} \leq \begin{cases} \frac{1}{6}k_{ds}\left(1+\frac{2}{\beta_c}\right)\sqrt{f_c}b_0 d \\ \frac{1}{6}k_{ds}\left(\frac{\alpha_s d}{b_0}+2\right)\sqrt{f_c}b_0 d \\ \frac{1}{3}k_{ds}\sqrt{f_c}b_0 d \end{cases} \quad (11.5.13)$$

where α_s = 20, 15, and 10 for the interior, edge, and corner column, respectively, $k_{ds} = \sqrt{\frac{200}{d}}$ $(1 \leq k_{ds} \leq 0.5)$. b_0 is located at a distance $d/2$ from the edges of the column.

11.5.7 JSCE (2007)

According to JSCE (2007), the design value of $V_{Rd,c}$ is given by Eq. 11.5.14. The influence of the size of the slab, flexural reinforcement content, and compressive strength of concrete is accounted in the following expression:

$$V_{Rd,c} = \beta_d \beta_p \beta_r \beta_{pcd} u_p \frac{d}{\gamma_b} \quad (11.5.14)$$

where $\beta_d = \left(\frac{1000}{d}\right)^{1/4} \leq 1.5$, $\beta_p = (100\rho)^{1/3} \leq 1.5$, $\beta_r = 1 + \frac{1}{0.25\frac{u}{d}}$, $\beta_{pcd} = 0.2\sqrt{f_c} \leq 1.2MPa$, and γ_b = member factor = 1.3. The critical perimeter is represented by Eq. 11.5.15, which is considered at a distance $d/2$ from the column face.

$$u_p = u + \pi d \quad (11.5.15)$$

where u = perimeter of the loading pad = $4c$ and c is the sides of the column cross section.

11.5.8 Critical Analysis

The experimental results of reinforced RAC slabs are collected from the available literature (Rao et al. 2012; Reis et al. 2015; Mahmoud et al. 2017; Francesconi et al. 2016; Xiao et al. 2019; Pradhan 2019; Sahoo and Singh 2020, 2021) and reported in

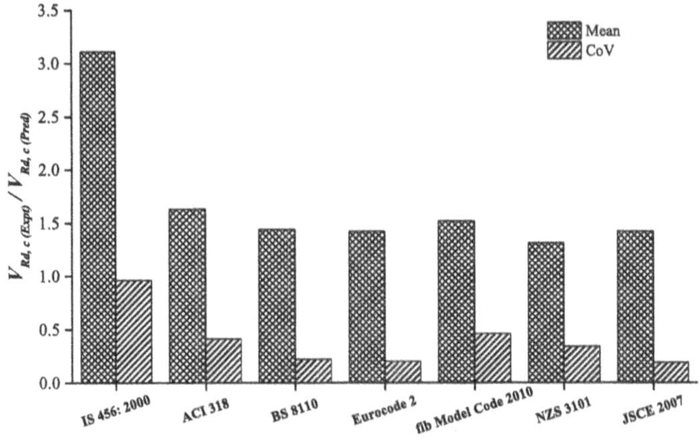

Fig. 11.6 Comparison of experimental and predicted punching shear strength of reinforced RAC slabs (Pradhan 2019)

Appendix D. The database consists of 46 numbers of RAC slab results. The accuracy of the discussed design standards (IS: 456 2000; ACI 318 2008; BS 8110-2 1985; Eurocode 2 2004; Fib MC 2010; NZS 3101 2006; JSCE 2007) in estimating the punching shear capacity of reinforced RAC slab is verified. The mean and coefficient of variation (CoV) of the ratio of the experimental punching shear strength to the predicted values ($V_{Rd,c(expt)}/V_{Rd,c(pred)}$) are depicted in Fig. 11.6. The mean value is the index of the accuracy in prediction of the considered expression and CoV is the index of uniformity in prediction. The mean value closer to 1 and the lower CoV value confirm the better prediction capability of the considered formula.

The mean value of 3.11 (CoV = 0.96) indicates that, IS: 456 (2000) underestimates the punching shear strength of RAC slabs. It signifies the conservative nature of IS: 456 (2000). The mean values of $V_{Rd,c(expt)}/V_{Rd,c(pred)}$ for ACI 318 (2008), Fib MC (2010), and NZS 3101 (2006) are 1.63, 1.52, and 1.31, respectively. However, the CoV values are slightly on the higher side and observed to be 0.41, 0.46, and 0.34, respectively. The mean (CoV) values of $V_{Rd,c(expt)}/V_{Rd,c(pred)}$ for BS 8110-2 (1985), Eurocode 2 (2004), and JSCE (2007) are observed to be 1.44 (0.22), 1.42 (0.20), and 1.42 (0.19), respectively. Considering both mean and CoV values, BS 8110-2 (1985), Eurocode 2 (2004), NZS 3101 (2006), and JSCE (2007) are more suitable in predicting the punching shear strength of RAC slab.

It can be highlighted that the control perimeters assumed in BS 8110-2 (1985), Eurocode 2 (2004), and NZS 3101 (2006) are $1.5d$, $2d$, and $d/2$, respectively. Although there is a significant difference in the assumption of the control perimeter of slab in BS 8110-2 (1985), Eurocode 2 (2004), NZS 3101 (2006), and JSCE (2007), the accuracy in prediction of punching shear strength by these formulations exhibits marginal variation. Hence, in addition to the control perimeter, the governing design parameters and their coefficients also determine the accuracy of the formulation in estimating the punching shear strength of the slab.

The accuracy of aforementioned formulations in estimating the punching shear capacity of reinforced RAC slabs is thus verified with respect to the variation in different design parameters. In this context, the prepared database (Appendix D) is operated to analyze the influence of the variation in compressive strength (f_c), effective depth of the slab (d_{eff}), longitudinal reinforcement content (ρ), RCA replacement ratio, yield strength of the longitudinal reinforcement ratio (f_y), and the maximum size of the aggregate (d_g) on the accuracy of the prediction of punching shear strength. For this, the variation in $V_{Rd,c(expt)}/V_{Rd,c(pred)}$ for the discussed expressions is analysed with respect to the aforementioned design parameters and depicted in Fig. 11.7.

Compressive Strength of Concrete: Figure 11.7 suggests that, for compressive strength of concrete up to 40 MPa the predicted values of the punching shear strength $V_{Rd,c(pred)}$ of RAC slab as per above-discussed expressions do not vary significantly. However, for compressive strength of concrete beyond 40 MPa, $V_{Rd,c(expt)}/V_{Rd,c(pred)}$ value decreases for IS: 456 (2000), ACI 318 (2008), Fib MC (2010), and NZS 3101 (2006). Moreover, for compressive strength beyond 50 MPa Fib MC (2010) and NZS 3101 (2006) overestimate the punching shear strength of RAC slabs. In this regard, Eurocode 2 (2004), BS 8110-2 (1985), and JSCE (2007) exhibit uniformity in prediction irrespective of the variation in compressive strength of concrete. For compressive strength of concrete above 50 MPa, $V_{Rd,c(expt)}/V_{Rd,c(pred)}$ marginally decreases as the compressive strength of concrete increases. This behaviour indicates that the punching shear strength of RAC slab decreases as the compressive strength of concrete increases above 50 MPa. Muttoni (2009) observed that the expression suggested by Eurocode 2 (2004) predicts the punching shear strength of the slab more consistently than ACI 318 (2008). In Eurocode 2 (2004) and BS 8110-2 (1985) formulations for punching shear strength, the contribution of compressive strength of concrete is accounted in $f_c^{1/3}$ term, whereas in IS: 456 (2000), ACI 318 (2008), Fib MC (2010), NZS 3101 (2006), and JSCE (2007) expression $f_c^{1/2}$ is considered. Sacramento et al. (2012) also observed a consistent prediction by considering $f_c^{1/3}$ term as a function of punching shear strength.

Effective Thickness of Slab: The formulations discussed above conservatively estimate the punching shear strength of RAC slabs for higher slab thickness. However, for lower effective depth of slabs, Fib MC (2010) and NZS 3101 (2006) overestimate punching shear capacity.

Flexural Reinforcement Ratio: $V_{Rd,c(expt)}/V_{Rd,c(pred)}$ increases as the longitudinal reinforcement content (ρ) increases (Fig. 11.7). Moreover, for higher longitudinal reinforcement content, the discussed expressions conservatively estimate the punching shear capacity of RAC slabs. This can be attributed to the increment in punching shear capacity of slabs with the increase in ρ. The uniformity in prediction is not observed for IS: 456 (2000), ACI 318 (2008), Fib MC (2010), and NZS 3101 (2006), as these expressions do not account the influence of ρ on punching shear strength. However, in BS 8110-2 (1985), Eurocode 2 (2004), and JSCE (2007) formulations, the $\rho^{1/3}$ is a function of punching shear strength. Sacramento et al. (2012) also witnessed similar relationship for ρ and punching shear strength.

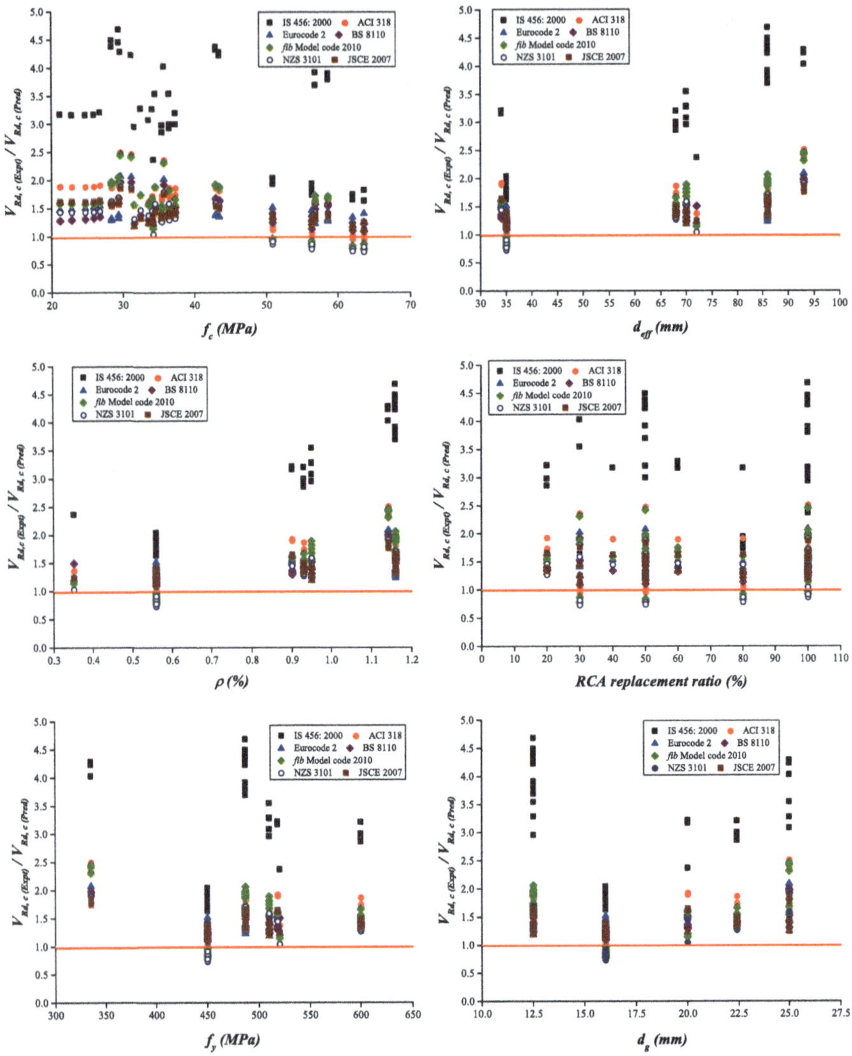

Fig. 11.7 Influence of different parameters on the predicted punching shear strength of reinforced RAC slabs (Pradhan 2019)

RCA Replacement Ratio: The accuracy of prediction of the existing expressions with respect to the RCA replacement ratio is studied. Apart from Fib MC (2010) and NZS 3101 (2006), the other discussed expressions conservatively predict the punching shear strength of RAC slab and the uniformity in prediction of the expressions is observed irrespective of the RCA content.

Yield Strength of Reinforcement: It can be evidenced from Fig. 11.7 that, Fib MC (2010) and NZS 3101 (2006) overestimate the punching shear strength of RAC slab for the yield strength of 450 MPa of the reinforcement. However, for yield strength of reinforcement beyond 500 MPa, all the expressions conservatively estimate the punching shear strength of RAC slab.

Maximum Size of Aggregate: In the available literature on the experimental investigation of RAC slabs, variation in the maximum size of RCA is observed. In this context, the accuracy of the existing expressions in estimating the punching shear strength of RAC slab is verified with respect to different size of RCA. For the maximum RCA size of 16 mm, Fib MC (2010) and NZS 3101 (2006) overestimate the punching shear strength of RAC slab, whereas the other discussed expressions conservatively predict the same. However, for the maximum size of RCA above 16 mm, the conservative prediction is observed for all the expressions discussed in the present study. The $V_{Rd,c(expt)}/V_{Rd,c(pred)}$ values increase marginally as the maximum size of RCA increases beyond 16 mm. This indicates that the punching shear strength of RAC slab increases as the maximum size of the RCA increases above 16 mm.

11.6 Yield Line Theory

The yield load (P_{yield}) of the reinforced concrete slab supported along all four edges and free to lift at the corners can be predicted by using Eqs. 11.6.1 and 11.6.2 of the yield line theory (Park and Gamble 2000).

$$P_{yield} = 8m\left(\frac{L}{s-c}\right) \tag{11.6.1}$$

$$m = A_s f_y\left[d - 0.59A_s\left(\frac{f_y}{f_c}\right)\right] \tag{11.6.2}$$

where m = ultimate flexural moment per unit width, L = side dimension of the square slab (mm), s = side dimension between slab support (mm), c = side dimension of the column (mm).

11.6.1 Critical Analysis

The failure mode of the slab can be determined by calculating the ratio of the measured failure load (P_u) to the estimated P_{yield} (Hognestad 1953). The mode of failure is shear if $\frac{P_u}{P_{yield}} \leq 1$ and for $\frac{P_u}{P_{yield}} > 1$, the failure mode of the reinforced concrete slab is flexure. Accordingly, the modes of failure of the RAC tested slabs (Rao et al. 2012; Reis et al. 2015; Mahmoud et al. 2017; Francesconi et al. 2016; Pradhan 2019; Xiao et al. 2019; Sahoo and Singh 2020, 2021) are determined and presented in Appendix D. The tested slabs by Pradhan (2019) and Francesconi et al. (2016) exhibited the flexure mode of failure. Such behaviour of RAC slabs may be due to the lower longitudinal reinforcement content. However, in the other studies (Rao et al. 2012; Reis et al. 2015; Mahmoud et al. 2017; Xiao et al. 2019; Sahoo and Singh 2020, 2021), the RAC slabs failed in shear mode irrespective of the RCA content.

11.7 Critical Shear Crack Theory

Muttoni and Schwartz (1991) observed a reduction in shear strength of reinforced concrete slab owing to the formation of critical shear crack, which propagates along the inclined compression strut and responsible for the transfer of shear force to the column. The experimental investigation shows that the concentric cracks are formed in the vicinity of the column and in the other parts of the slab only radial cracks are developed. In the tangential direction, the shear force is not transferred. Hence, the curvature of the slab is concentrated in the radial direction close to the support. The width of the critical shear crack is proportional to the product of critical rotation and depth of the slab (ψd). The punching shear capacity of slab decreases as the rotation of the slab increases. The transfer of shear force across the critical shear crack in concrete is related to its roughness, which is a function of the maximum size of the aggregate (Walraven 1981; Vecchio and Collins 1986). On the basis of these fundamental concepts, Muttoni (2003) proposed a formulation for the failure criterion of reinforced concrete slab (Eq. 11.7.1), which is known as critical shear crack theory (CSCT).

$$\frac{V_{Rd,c}}{b_0 d\sqrt{f_c}} = \frac{3/4}{1 + 15\frac{\psi d}{d_{g0}+d_g}} \tag{11.7.1}$$

where ψ = rotation of the slab outside the column region, which can be estimated by using the following expression (Eq. 11.7.2):

$$\psi = 1.5\frac{r_s}{d}\frac{f_y}{E_s}\left(\frac{V}{V_{flex}}\right)^{\frac{3}{2}} \tag{11.7.2}$$

where r_s = plastic radius around the column $\approx B/2$, B = side dimension of the slab, V = shear force, and V_{flex} = flexural shear strength of slab. The flexural shear strength of slab (V_{flex}) can be estimated using Eq. 11.7.3 or 11.7.4.

$$V_{flex} = \frac{4m_R}{r_q\left(\cos\frac{\pi}{8} + \sin\frac{\pi}{8}\right) - c}\,\frac{B^2 - Bc - \frac{c^2}{4}}{B - c} \tag{11.7.3}$$

$$V_{flex} = 2\pi m_R \frac{r_s}{r_q - r_c} \tag{11.7.4}$$

where c = size of the column; r_q = radius of load introduction at perimeter; r_c = radius of the circular column; and m_R = nominal moment capacity per unit width, which can be calculated from the following expression (Eq. 11.7.5):

$$m_R = \rho f_y d^2 \left(1 - \frac{\rho}{2}\frac{f_y}{f_{cp}}\right) \tag{11.7.5}$$

11.7.1 Critical Analysis

The experimental results of RAC slabs as well as companion NAC slabs reported in Appendix D are compared with the predicted values obtained by the formulation (Eq. 11.7.1) of CSCT (Muttoni 2003). Test results include the slabs in which the punching shear failure was witnessed prior to the attainment of the flexural strength (V_{flex}) as well as after reaching V_{flex}. The rotation (ψ) of the slab is estimated by using Eq. 11.7.2 proposed by Muttoni (2009). In order to nullify the effects of the depth of the slab and aggregate size $d/(d_{g0} + d_g)$ factor is multiplied with ψ (Muttoni 2009). It is observed from Fig. 11.8 that, for lower $\psi d/(d_{g0} + d_g)$ values, the CSCT (Muttoni 2003) overestimates the punching shear capacity of slender RAC slabs, as most of the test results are below the failure criterion obtained using CSCT (Muttoni 2003) formulations. However, for RAC slabs with higher values of $\psi d/(d_{g0} + d_g)$ conservative prediction is witnessed for CSCT (Muttoni 2003).

The experimental results are also compared with the lower limits of ACI 318 (2008) and IS: 456 (2000) (Fig. 11.8). The value of λ is assumed to be 0.85 for RAC while determining the limit using ACI 318 (2008) formulation. It is observed that, for lower values of $\psi d/(d_{g0} + d_g)$, both ACI 318 (2008) and IS: 456 (2000) conservatively estimate the punching shear strength of RAC slabs. However, for higher $\psi d/(d_{g0} + d_g)$ values, ACI 318 (2008) overestimates the punching shear strength of RAC slabs, whereas IS: 456 (2000) provides reasonable conservative predictions. For $\psi d/(d_{g0} + d_g)$ values higher than 0.2, the ability of prediction of ACI 318 (2008) and IS: 456 (2000) is unknown. Hence, further experimental investigation on RAC slab with larger slab thickness is necessitated. The experimental results obtained by Pradhan (2019) are conservatively estimated by CSCT (Muttoni 2003), ACI 318 (2008) and IS: 456 (2000). This may be attributed to the positive effect of PPM mix design in improving the punching shear strength of RAC slab. PPM mix-designed concrete possibly helped in improving the aggregate interlocking mechanism and propagation of cracks in a tortuous path. Consequently, an improvement in the shear transfer mechanism might have happened, which in turn increases the punching shear strength of the slab.

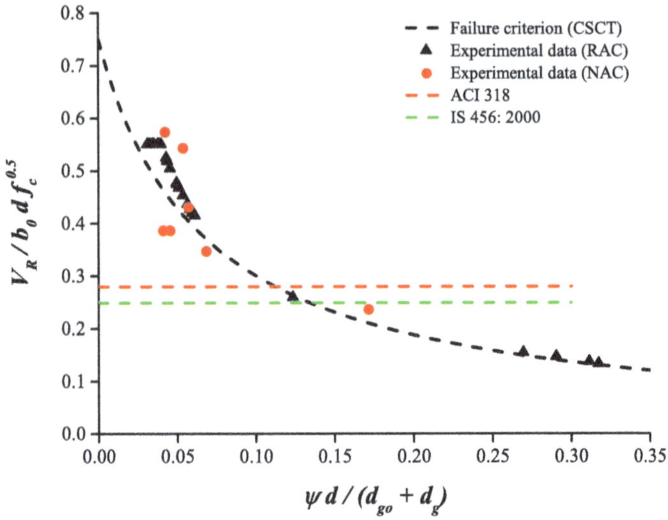

Fig. 11.8 Comparison of the experimental punching shear strength of both RAC and NAC slabs as a function of the width of critical shear crack (Pradhan 2019)

11.8 Parametric Study

The influence of the RCA replacement ratio, longitudinal reinforcement ratio (ρ), effective thickness of the slab (d_{eff}), size of the column with respect to the depth of the slab (b_0/d), compressive strength of concrete (f_c), yield stress of the tensile reinforcement (f_y), span-to-depth ratio of the slab (r_s/d), and maximum size of the aggregate (d_g) on the punching shear strength of RAC slab are investigated and demonstrated in Fig. 11.9.

RCA Replacement Ratio: Rao et al. (2012), Mahmoud et al. (2017), and Xiao et al. (2019) reported a reduction in punching shear strength of RAC slab as the RCA replacement ratio increases, whereas Reis et al. (2015), Francesconi et al. (2016), and Sahoo and Singh (2020, 2021) observed no significant variation for the same reason. It is observed from Fig. 11.9 that the RCA replacement ratio has no significant influence on the punching shear strength of RAC slabs. The estimated lower values of $V_R/b_0 d\sqrt{f_c}$ for the results of Francesconi et al. (2016) may be attributed to the lower longitudinal reinforcement content.

Flexural Reinforcement Ratio: The punching shear capacity of the RAC slab increases as the longitudinal reinforcement content increases (Fig. 11.9). This is similar to the behaviour of NAC slab as reported by Kinnunen and Nylander (1960) and Muttoni (2009). The influence of flexural reinforcement content on the punching shear strength of slab is accounted in BS 8110-2 (1985), Eurocode 2 (2004), JSCE (2007), and CSCT (Muttoni 2003). However, IS: 456 (2000), ACI 318 (2008), Fib

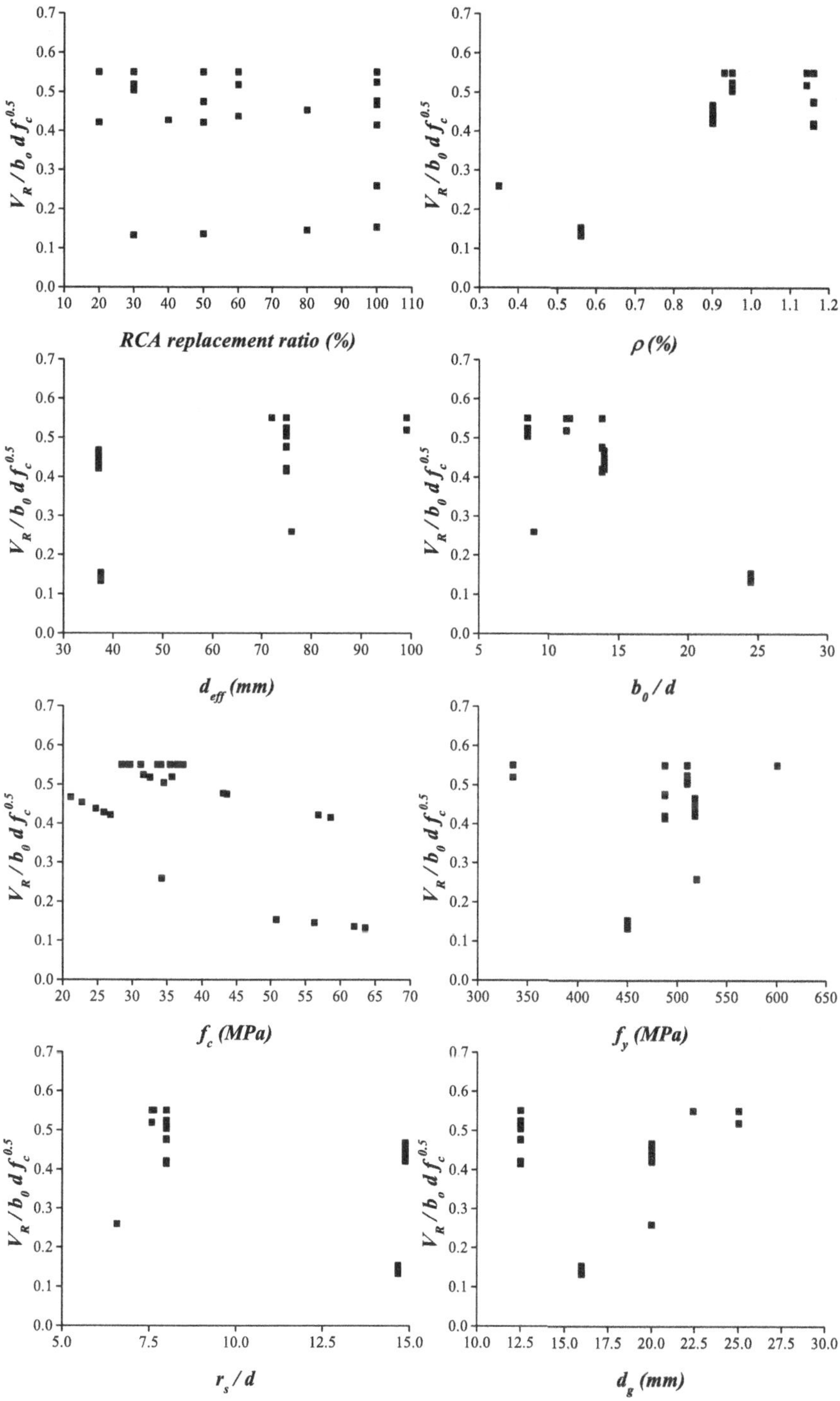

Fig. 11.9 Influence of different parameters on punching shear strength of RAC slab (Pradhan 2019)

MC (2010), and NZS 3101 (2006) do not consider the effect of longitudinal reinforcement content.

Effective Thickness of Slab: The punching shear strength of RAC slabs shows marginal improvement as the effective thickness of the slab increases (Fig. 11.9). It is contrary to the observations of Guandalini and Muttoni (Guandalini et al. 2009; Guandalini and Muttoni 2004). The expressions of BS 8110-2 (1985), Eurocode 2 (2004), NZS 3101 (2006), JSCE (2007), and CSCT (Muttoni 2003) also suggest a reduction in punching shear strength of slab as the effective thickness of slab increases. This behaviour of RAC slab advocates for further study by varying the effective thickness of the RAC slab.

Column Size-to-Slab Thickness Ratio: Concerning the effect of the size of the column with respect to the slab thickness (b_0/d) on the punching shear strength of slab, the control perimeter is addressed in the available expression. As observed from Fig. 11.9, the punching shear capacity of RAC slab decreases as b_0/d value increases. This is similar to the behaviour of conventional concrete slab.

Compressive Strength of Concrete: The influence of compressive strength of concrete on the punching shear strength of RAC slab is similar to that of conventional concrete slab. The punching shear strength of RAC slab decreases as the compressive strength of concrete increases (Fig. 11.9). This may be attributed to the increase in brittleness of concrete as the compressive strength increases. The effect of compressive strength is accounted in all the expressions discussed in section "Comparison with Design Codes".

Yield Strength of Reinforcement: The effect of the yield stress of longitudinal reinforcement in punching shear capacity is not very pronounced (Fig. 11.9). A similar observation is reported by Muttoni (2009). The punching shear strength of RAC slab increases as the yield stress of reinforcement increases beyond 450 MPa. However, for the increment in yield stress above 500 MPa, the increase in punching shear strength is not observed. Moreover, further study is necessitated in this regard.

Span-to-Depth Ratio: For an isolated slab element, the span-to-depth ratio (r_s/d) also has influence on the punching shear capacity (Muttoni 2009). Similar to the conventional concrete slab, the punching shear strength of RAC slab decreases as r_s/d increases (Fig. 11.9). This influence of r_s/d is accounted only in CSCT (Muttoni 2003) and Fib MC (2010). Further investigation on the RAC slab with significant thickness is substantiated in order to understand the punching shear strength of RAC slab at lower r_s/d.

Maximum Size of Aggregate: The effect of maximum size of the RCA on the punching shear capacity of RAC slab is studied. Apart from the maximum RCA size of 12.5 mm, the punching shear strength of RAC slab increases as the size of the RCA increases (Fig. 11.9). The influence of the maximum size of the aggregate is accounted in CSCT (Muttoni 2003) as well as Fib MC (2010). However, the other formulations discussed in section "Comparison with Design Codes" do not consider the effect of the maximum size of aggregate.

11.9 Closure

The present chapter discussed the experimental investigation on the punching shear behaviour of slabs.

- The punching shear capacity of slabs marginally reduced by the inclusion of 100% RCA, whereas RAC slabs exhibited higher ultimate deflection at failure.
- The expressions in BS 8110-2 (1985), Eurocode 2 (2004), NZS 3101 (2006), and JSCE (2007) are recommended to predict the punching shear capacity of RAC slabs.
- The influence of different design parameters was analysed critically. Further experimental investigation is necessitated on RAC slabs with higher slab thickness, different longitudinal reinforcement content and dimensions of column.

References

ACI 318 (2008) ACI committee 318. Building code requirements for structural concrete and commentary, American Concrete Institute, Farmington Hills (MI)

BS 8110-2 (1985) Structural use of concrete–part 2: code of practice for special circumstances

Eurocode 2 (2004) Eurocode 2: design of concrete structures: part 1-1: general rules and rules for buildings. British Standards Institution

Fib MC (2010) Fib model code for concrete structures. Wilhelm Ernst & Sohn, Berlin

Francesconi L, Pani L, Stochino F (2016) Punching shear strength of reinforced recycled concrete slabs. Constr Build Mater 127:248–263

Guandalini S, Burdet OL, Muttoni A (2009) Punching tests of slabs with low reinforcement ratios. ACI Struct J 106(1):87–95

Guandalini S, Muttoni A (2004) Symmetric punching tests on reinforced concrete slabs without shear reinforcement. Test report, EPFL, Lausanne, Switzerland

Hognestad E (1953) Shearing strength of reinforced concrete column footings. J Proceed 50:189–208

Inácio MMG, Almeida AFO, Faria DMV, Lúcio VJG, Ramos AP (2015) Punching of high strength concrete flat slabs without shear reinforcement. Eng Struct 103:275–284

IS: 456 (2000) Plain and reinforced concrete-code of practice. Bureau of Indian Standards, New Delhi, India

JSCE (2007) Standard specifications for concrete structures–2007 design, 15th edn. Japan Society of Civil Engineers

Kinnunen S, Nylander H (1960) Punching of concrete slabs without shear reinforcement. Elander

Mahmoud ZI, El tony EtM, Saeed KS (2017) Punching shear behavior of recycled aggregate reinforced concrete slabs. Alex Eng J 4–12

Muttoni A (2003) Shear and punching strength of slabs without shear reinforcement. Beton- und Stahlbetonbau 98(2):74–84

Muttoni A (2009) Punching shear strength of reinforced concrete slabs without transverse reinforcement. ACI Struct J 105(4):440–450

Muttoni A, Schwartz J (1991) Behavior of beams and punching in slabs without shear reinforcement. IABSE Colloq 62:703–708

NZS 3101 (2006) Concrete structures standard, part 1–the design of concrete structures

Park R, Gamble WL (2000) Reinforced concrete slabs. John Wiley & Sons

Pradhan S (2019) Performance of recycled aggregate concrete and structural members: particle packing method of mix design approach. PhD Thesis, Indian Institute of Technology Kharagpur

Rao HS, Reddy VSK, Ghorpade VG (2012) Influence of recycled coarse aggregate on punching behaviour of recycled coarse aggregate concrete slabs. Int J Modern Eng Res 2(4):2815–2820

Reis N, de Brito J, Correia JR, Arruda MR (2015) Punching behaviour of concrete slabs incorporating coarse recycled concrete aggregates. Eng Struct 100:238–248

Ruiz F, Mirzaei Y, Muttoni A (2014) Post-punching behavior of flat slabs. ACI Struct J 110(5):801–812

Sacramento PVP, Ferreira MP, Oliveira DRC, Melo GSSA (2012) Punching strength of reinforced concrete flat slabs without shear reinforcement. Ibracon Struct Mater J 5(5):659–674

Sahoo S, Singh B (2020) Recycled aggregate concrete slab punching shear capacity. Structures 24(2):426–443

Sahoo S, Singh B (2021) Punching shear capacity of recycled-aggregate concrete slab-column connections. J Build Eng 41(3):102430

Vecchio FJ, Collins MP (1986) The modified compression-field theory for reinforced concrete elements subjected to shear. ACI J 83(2):219–231

Walraven JC (1981) Fundamental analysis of aggregate interlock. J Struct Divis 107(11):2245–2270

Xiao J, Wang W, Zhou Z, Tawana MM (2019) Punching shear behavior of recycled aggregate concrete slabs with and without steel fibres. Front Struct Civil Eng 13(3):725–740

Appendix A
Primary Data Regarding NCA and RCA Production

A.1 Natural Coarse Aggregate Production

- **Capacity:** $\approx$ 180 t/h
- **Water consumption for dust control:** 5000 lt/day

A.2 Recycled Coarse Aggregate Production

- **Capacity:** $\approx$ 40 t/h
- **Water consumption for dust control:** 1500 lt/day

The production capacity of NCA crushing plant is 180 t/h and in this process it consumes 700.7 kWh energy. So, the energy required for the production of 1 t of NCA is 3.89 kWh. However, in accordance to Coelho and de Brito (2013) and Braga et al. (2017) the stoppages because of the loading, unloading, and maintenance activities, 70% of the entire production period is effectively utilized. Hence, for the production of 1 t of NCA, effectively 2.72 kWh energy is consumed. Similarly, for the production of RCA 3.07 kWh/t energy is estimated. However, by accounting the effective running period of the equipment during production, 2.15 kWh/t energy is consumed (Pradhan et al. 2019).

S. Pradhan et al., *Particle Packing Method for Recycled Aggregate Concrete*,
https://doi.org/10.1007/978-981-99-7516-7

Table A.1 Basalt extraction

Equipment	Quantity	Fuel consumption
Drill rig with compressor	1	5 lt/h

Table A.2 Transportation of extracted basalt

Equipment	Quantity	Fuel consumption
Loader (excavator)	2	26.5 lt/h

Table A.3 NCA processing

Equipment	Quantity	Total power (kW)
Vibrating feeder	6	26.4
Conveyor belt	5	39.6
Jaw crusher	1	110
Conveyor belt	3	29.1
Cone crusher	1	160
Conveyor belt	2	15
Vibrating screen	1	37
Conveyor belt	2	8
Conveyor belt	2	10
Vertical shaft impact crusher	2	260
Conveyor belt	2	5.6
Total		700.7

Table A.4 Transfer of prepared NCA to open air pile

Equipment	Quantity	Fuel consumption
Excavator (for loading operation to truck)	3	12 lt/h

Table A.5 Transfer of waste concrete to recycling process

Equipment	Quantity	Fuel consumption
Excavator (for feeding of waste concrete and collection of prepared aggregate)	2	4–5 lt/h

Table A.6 RCA processing

Equipment	Quantity	Total power (kW)
Conveyor belt	1	4
	1	5
	4	8.8
	1	2.8
Rubble master 60 (impact crusher)	1	75
Jaw crusher	1	30 (Runs 20 minutes in 1 hour period)
Suction bag	6	0.36
Feeder to RM 60	2	1.9
Vibrating screens	1	10
Water sprinkler	1	5
Total		122.86

Table A.7 Transfer of prepared RCA to open air pile

Equipment	Quantity	Fuel consumption
Excavator (for loading operation to truck)	1	12 lt/h

Table A.8 Direct burdens for production of 1 t of aggregate

Type of aggregate	Process	Explosive (g)	Diesel (MJ)	Lubricating oil (kg)	Electricity (kWh)	Water (kg)	Distance (tkm)
NCA	Extraction of basalt	150	1.67	0.001	–	–	–
	Truck loading	–	4.24	0.001	–	–	–
	Transportation	–	–	–	–	–	5
	Crushing and screening	–	–	–	2.72	3	–
	Storing in open air pile	–	2.88	0.001	–	–	–
RCA	Transportation	–	–	–	–	–	35
	Feeding of waste concrete	–	3.6	0.005	–	–	–
	Crushing and screening	–	–	–	2.13	5	–
	Storing in open air pile	–	4.32	0.003	–	–	–

Appendix B
Database of RAC Beams Without Shear Reinforcement

S. Pradhan et al., *Particle Packing Method for Recycled Aggregate Concrete*,
https://doi.org/10.1007/978-981-99-7516-7

Authors	r (%)	b (mm)	d (mm)	D (mm)	d_a (mm)	ρ (%)	f_y (MPa)	f_c (MPa)	L (mm)	l (mm)	a (mm)	a/d	V_u (kN)
González-Fonteboa and Martínez-Abella (2004, 2007)	100	200	303	350	25	2.98	571	39.65	3050	2450	1000	3.3	90.64
Etxeberria et al. (2007)	25	200	303	350	25	2.98	500	42.38	3050	2600	1000	3.3	104
	50							41.34					89
	100							39.75					84
Sato et al. (2007)	100	150	160	200	19	1.06	331	46.5	2800	2200	700	4.4	21
								32.9					21.7
								46.6					21.4
						0.59		30.4					12.1
								28.4					12.6
								34.5					13.2
								31.8					13.5
						1.06		30.4					19.7
								28.4					20
								34.5					20
								31.8					21.4
						1.65	342	30.4					27.3
								28.4					27.7
								34.5					28.3
								31.8					31.1
Fathifazl et al. (2011)	63.5	200	300	–	19	1	420	41.6	1900	–	–	1.5	186.7
			300			1.5			2200			2	169.5
			309			1.62			2600			2.59	103.9
			201			1.99			2080			2.69	89.3
			305			2.46			3400			3.93	83.2
			381			1.83			3180			2.73	99.5

Authors	r (%)	b (mm)	d (mm)	D (mm)	d_a (mm)	ρ (%)	f_y (MPa)	f_c (MPa)	L (mm)	l (mm)	a (mm)	a/d	V_u (kN)
Choi et al. (2010)	30	200	360	400	25	1.61	456	24.56	1840	1440	540	1.5	161.7
						1.61	456		2560	2160	900	2.5	81.34
						1.61	456		3100	2700	1170	3.25	56.7
						0.53	522		2560	2160	900	2.5	56.7
						0.83	486		2560	2160	900	2.5	78.4
	50					1.61	456	24.15	1840	1440	540	1.5	152.88
						1.61	456		2560	2160	900	2.5	87.9
						1.61	456		3100	2700	1170	3.25	71.54
						0.53	522		2560	2160	900	2.5	57.82
						0.83	486		2560	2160	900	2.5	67.13
	100					1.61	522	22.56	1840	1440	540	1.5	107.8
						1.61	486		2560	2160	900	2.5	84.77
						1.61	456		3100	2700	1170	3.25	57.77
						0.53	456		2560	2160	900	2.5	59.78
						0.83			2560	2160	900	2.5	70.07
Kim et al. (2013)	50	200	300	350	25	2.85	651	32.6	2400	2100	750	2.5	60.6
		200	450	530		2.85	610		3150	2850	1125		108.9
		200	600	680		2.85	651		3900	3600	1500		126.1
		300	450	530		3.02	600		3150	2850	1125		154.2
		400	600	680		2.85	651		3900	3600	1500		261.5
	100	200	300	350		2.85	651	34.9	2400	2100	750		72.9
		200	450	530		2.85	610		3150	2850	1125		96.4
		200	600	680		2.85	651		3900	3600	1500		125.1
		300	450	530		3.02	600		3150	2850	1125		159.8
		400	600	680		2.85	651		3900	3600	1500		256.6

Authors	r (%)	b (mm)	d (mm)	D (mm)	d_a (mm)	ρ (%)	f_y (MPa)	f_c (MPa)	L (mm)	l (mm)	a (mm)	a/d	V_u (kN)
Arezoumandi et al. (2014, 2015)	50	300	400	460	25	1.27	449	32.1	4300	3600	1200	3	117.5
			375			2.03		32.1					151.3
			375			2.71		32.1					171.8
			400			1.27		35.5					111.7
			375			2.03		35.5					148.6
			375			2.71		35.5					168.7
	100		400			1.27		30					114.8
			375			2.03		30					143.2
			375			2.71		30					131.4
			400			1.27		34.1					113.0
			375			2.03		34.1					124.1
			375			2.71		34.1					140.3
Knaack and Kurama (2014)	50	150	200	230	19	1.34	572	41.8	1980	1680	765	3.875	44.0
								41.8					39.1
								37.4					43.7
								37.4					41.2
	100							39.1					36.4
								39.1					38.0
								39.2					39.9
								39.2					39.9
Katkhuda and Shatarat (2016)	50	206	260	306	19	1.9	457	25.2	2440	2000	520	2	58.94
											800	3	49.07
	100							23.2			520	2	55.04
											800	3	46.45

Authors	r (%)	b (mm)	d (mm)	D (mm)	d_a (mm)	ρ (%)	f_y (MPa)	f_c (MPa)	L (mm)	l (mm)	a (mm)	a/d	V_u (kN)
Ignjatović et al. (2017)	50	200	235	300	31.5	4.09	555	33.44	3500	3000	1000	4.2	91.75
	100							34.48					104.75
Rahal and Alrefaei (2017)	10	150	388	420	19	0.79	534	36.6	2900	2599	1162	3	44.5
	20							35					40.05
	20							35.3					48.9
	35							35.3					45.05
	50							38.1					46.95
	75							36.6					47.4
	100							35.8					42.5
	5							37.4					56
	10							34.8					52.5
	16							35.4					54.2
	23							34					47.25
	35							35.1					42.5
Pradhan et al. (2018)	100	200	270	300		1.31	650	46.67	2400	2100	700	2.6	92.28
						0.75	590	46.75					81.29
								46.48					81.1

Appendix C
Database of RAC Beams with Shear Reinforcement

S. Pradhan et al., *Particle Packing Method for Recycled Aggregate Concrete*,
https://doi.org/10.1007/978-981-99-7516-7

Authors	r (%)	b (mm)	d (%)	D (%)	d_g (mm)	ρ (%)	f_y (MPa)	ρ_w (%)	f_{yw} (MPa)	f_c (MPa)	L (mm)	l (mm)	θ (°)	a/d	V_{cr} (kN)	V_y (kN)	V_u (kN)
González-Fonteboa and Martínez-Abella (2004, 2007)	50	200	303	350	25	2.98	571	0.12	500	39.3	3050	2600	25	3.3	–	–	164.3
								0.17		41.5			35				177.0
								0.22		40.5			21				233.6
Etxeberria (2004), Etxeberria et al. (2007)	25	200	304	350	25	2.97	500	0.12	544	42.38	3050	2600	26	3.3	109.0	–	238.0
								0.17					30		100.5	140.5	169.0
								0.22					32		86.0	146.0	186.5
	50							0.12		41.34			28		91.0	–	164.0
								0.17					22		93.5	163.5	176.0
								0.22					21		92.5	176.0	220.0
	100							0.12		39.75			22		80.0	–	168.0
								0.17					24		90.0	152.5	163.0
								0.22					29		86.0	165.0	189.5
Ajdukiewicz and Kliszczewicz (2007)	100	200	250	300	16	0.87	483	0.28	234.1	34.6	2600	2400	26	3.2	20.0	58.5	64.0
										56.4			26		15.0	63.0	78.0
										40.1			26		15.0	67.5	81.5
										60.2			26		20.0	63.5	68.0
										85.3			26		20.0	65.0	79.0
										35.3			26		20.0	66.0	75.0
										57.6			26		20.0	62.0	71.5
										105.3			26		25.0	70.0	83.0
						1.55				36.6			26		15.0	107.0	118.0
										58.3			26		20.0	107.0	118.5*

Authors	r (%)	b (mm)	d (%)	D (%)	d_g (mm)	ρ (%)	f_y (MPa)	ρ_w (%)	f_{yw} (MPa)	f_c (MPa)	L (mm)	l (mm)	θ (°)	a/d	V_{cr} (kN)	V_y (kN)	V_u (kN)
										39.3			26		15.0	100.0	116.5*
										59.6			26		15.0	109.0	118.5*
										89.1			26		20.0	97.0	121.0*
										35.8			26		22.5	107.0	120.5
										59.6			26		30.0	106.0	119.0*
										107.8			26		25.0	110.0	130.5*
Fathifazl et al. (2009, 2010)	63.5	200	304	350	19	0.49	420	0.39	450	41.6	–	2200	–	2.6	16.25	48.00	57.50
						1.99		0.88	450				30	2.7	17.25	154.25	184.50
						3.26		1.77	450				–	2.7	26.38	263.25	279.70
						3.31		1.01	530				–	2.7	30.88	241.25	305.50
			309	375		2.46		0.25	530				–	2.7	–	–	172.00
			302	385		3.20		0.5	530				29	2.6	–	–	308.00
			301	385		4.00	431	0.5	450				–	2.6	–	–	341.00*
	74.3		304	350		0.49	420	0.39	450	49.1			–	2.6	20.25	49.88	58.40
						1.99		0.88	450				–	2.7	19.00	170.75	185.70
						3.26		1.77	450				–	2.7	–	–	283.80
						3.31		1.01	530				–	2.7	36.38	275.25	305.00
			309	375		2.46	418	0.25	530				27	2.6	–	–	235.00
			302	385		3.20	407	0.5	530				36	2.6	–	–	308.00
			301	385		4.00	431	0.5	530				–	2.7	–	–	327.00*
Bai and Sun (2010)	50	150	300	340	20	0.68	420	0.25	250	42.3	2100	1800	–	2.0	16.67	46.67	56.67
	70					0.68				43.7					18.67	46.67	59.17
	100					0.68				43.5					16.67	44.17	56.67
						0.89									17.67	54.17	67.33
						1.13									17.67	66.67	80.67

Authors	r (%)	b (mm)	d (%)	D (%)	d_g (mm)	ρ (%)	f_y (MPa)	ρ_w (%)	f_{yw} (MPa)	f_c (MPa)	L (mm)	l (mm)	θ (°)	a/d	V_{cr} (kN)	V_y (kN)	V_u (kN)
Choi et al. (2012)	100	400	525	600	25	2.34	420	0.4	483	26.9	6400	6000	29	5.14	21.88	284.08	302.82
						2.10				31.3			29		13.25	275.69	316.04
Ignjatović et al. (2012, 2017)	50	200	238	300	31.5	0.28	640	0.335	555	35.60	3500	3000	26	4.2	10.00	22.45	27.00
			263			1.46	550	1.05	555				26		20.00	105.00	110.55
			244			2.54	550	1.31	555				26		30.00	–	160.35*
			235			4.09	640	0.19	300	33.44			26		–	–	156.90*
	100		268			0.28	640	0.335	555	34.00			30		10.00	23.35	26.80
			263			1.46	550	1.05	555				30		20.00	101.00	105.40
			244			2.54	550	1.31	555				30		15.00	–	142.60*
			235			4.09	640	0.19	300	33.44			30		–	–	163.40*
Knaack (2013), Knaack and Kurama (2014, 2015)	50	150	200	230	19	1.34	572	1.1	420	36.7	1980	1680	–	3.83	11.50	54.00	54.64
										36.7					11.24	55.16	56.34
										37.6					10.46	53.60	54.00
										37.6					10.46	53.60	54.00
	100									42.0					9.28	52.94	54.51
										42.0					10.85	52.94	54.51
										37.5					12.42	54.64	57.65
										37.5					11.11	52.30	57.56
Kang et al. (2014)	15	135	230	270	25	0.5	377	1.16	400	59.4	3030	2700	–	3.9	6.40	14.40	20.20
						1	408								7.35	24.15	32.55
						1.5	389								8.80	33.60	42.90
						1.8	410								10.10	50.25	64.55

Authors	r (%)	b (mm)	d (%)	D (%)	d_g (mm)	ρ (%)	f_y (MPa)	ρ_w (%)	f_{yw} (MPa)	f_c (MPa)	L (mm)	l (mm)	θ (°)	a/d	V_{cr} (kN)	V_y (kN)	V_u (kN)
	30					0.5	377			48.8					5.85	12.40	18.80
						1	408								6.90	24.35	31.95
						1.5	389								7.80	34.50	42.65
						1.8	410								8.80	52.10	54.45
	15					0.5	377			32.7					7.35	12.10	16.40
						1	408								8.85	25.10	30.25
						1.5	389								9.70	34.00	39.75
						1.8	410								13.75	52.60	57.35
	30					0.5	377			31.7					6.35	13.90	16.35
						1	408								8.85	26.10	29.20
						1.5	389								9.30	36.05	39.25
						1.8	410								13.75	48.90	55.80
	50					0.5	377			29.0					7.35	12.35	15.15
						1	408								8.85	25.10	27.10
						1.5	389								9.80	35.85	36.45
						1.8	410								12.25	54.20	56.10
Arezoumandi et al. (2015)	100	300	400	460	25	0.47	568	0.13	494	30.5	–	2700	33	2.24	38.44	132.56	166.22
										31.3			33	2.24	39.89	139.33	172.11
						0.63	517			30.5			30	2.24	46.67	163.78	191.78
										31.3			30	2.24	47.44	164.67	188.11

Authors	r (%)	b (mm)	d (%)	D (%)	d_g (mm)	ρ (%)	f_y (MPa)	ρ_w (%)	f_{yw} (MPa)	f_c (MPa)	L (mm)	l (mm)	θ (°)	a/d	V_{cr} (kN)	V_y (kN)	V_u (kN)
Pradhan et al. (2018)	100	200	265	300	20	1.61	508	0.323	352	41.52	2400	2100	34	2.6	23.91	–	161.97*
						1.61				45.47			34		23.82	–	162.12*
						1.31				46.67			28		31.78	159.30	165.04
						0.75	590	0.283		46.04			36		26.52	101.24	114.75
						0.75				46.54			36		26.76	104.43	114.40
						0.42	543			46.35			–		15.37	54.60	65.05
						0.42				45.99			–		16.11	55.43	65.89

Appendix D
Database of Punching Shear Test of RAC Slabs

S. Pradhan et al., *Particle Packing Method for Recycled Aggregate Concrete*,
https://doi.org/10.1007/978-981-99-7516-7

Authors	r (%)	B (mm)	d (mm)	d_g (mm)	ρ (%)	f_y (MPa)	f_c (MPa)	f_{ct} (MPa)	E_c (MPa)	r_c (mm)	b_0 (mm)	P (kN)	$\frac{V_R}{b_0 d \sqrt{f_c}}$	Mode of failure
Rao et al. (2012)	20	1100	37	20	0.85	518	33.50	3.25	–	63.66	516.2	47.2	0.43	Shear
	40						32.37	3.16				45.8	0.42	
	60						30.86	3.11				44.6	0.42	
	80						28.40	2.88				42.8	0.42	
	100						26.42	2.78				41.4	0.42	
Reis et al. (2015)	20	1100	72	22.4	0.93	600	44.30	2.90	32800	95.50	826.2	158.6	0.40	Shear
	20						44.30	2.90	32800			152.2	0.38	
	50						46.60	2.94	32700			163.6	0.40	
	50						46.60	2.94	32700			174.8	0.43	
	100						45.60	2.86	31500			161.8	0.40	
	100						45.60	2.86	31500			158.3	0.39	
Mahmoud et al. (2017)	30	1200	37.5	12.5	0.95	510	34.50	3.71	–	63.66	635.6	153.0	0.55	Shear
	60						32.50	3.58				137.5	0.51	
	100						31.60	3.50				122.0	0.46	
	30			25			36.40	3.73				157.0	0.55	
	60						34.10	3.67				140.5	0.50	
	100						33.60	3.63				131.0	0.47	
Francesconi et al. (2016)	30	1100	37.5	16	0.56	450	63.60	4.40	40381	127.32	917.8	64.9	0.24	Flexure
	30											72.5	0.26	
	30											72.5	0.26	
	50						62.00	3.94	37980			64.9	0.24	
	50											68.7	0.25	
	50											64.9	0.24	

Authors	r (%)	B (mm)	d (mm)	d_g (mm)	ρ (%)	f_y (MPa)	f_c (MPa)	f_{ct} (MPa)	E_c (MPa)	r_c (mm)	b_0 (mm)	P (kN)	$\frac{V_R}{b_0 d\sqrt{f_c}}$	Mode of failure
	80						56.30	3.83	28818			68.7	0.27	
	80											64.9	0.25	
	80											72.5	0.28	
	100						50.80	3.65	31390			68.7	0.28	
	100											68.7	0.28	
	100											72.5	0.30	
Pradhan (2019)	100	1000	76	20	0.35	580	42.82	2.88	35062	70.00	678.6	100.0	0.30	Flexure
	100											100.0	0.30	
Xiao et al. (2019)	30	1500	99	25	1.142	335	44.65	–	35000	127.3	1111.0	313.4	0.48	Shear-compression
	50						38.95		29600			307.1	0.50	
	100						37.05		27400			303.4	051	
Sahoo and Singh (2020, 2021)	50	1200	75	12.5	1.16	487	28.4	–	–	95.5	835.7	211.5	0.63	Shear
												217.0	0.65	
	50						43.5					252.6	0.61	
												255.8	0.62	
	50						56.8					252.4	0.53	
												267.5	0.56	
	100						29.4					218.9	0.64	
												230.0	0.68	
	100						43.0					257.1	0.63	
												260.1	0.63	
	100						58.6					263.4	0.55	
												270.4	0.56	

References

Ajdukiewicz AB, Kliszczewicz AT (2007) Comparative tests of beams and columns made of recycled aggregate concrete and natural aggregate concrete. J Adv Concr Technol 5(2):259–273

Arezoumandi M, Smith A, Volz JS, Khayat KH (2014) An experimental study on shear strength of reinforced concrete beams with 100% recycled concrete aggregate. Constr Build Mater 53:612–620

Arezoumandi M, Drury J, Volz JS, Khayat KH (2015) Effect of recycled concrete aggregate replacement level on shear strength of reinforced concrete beams. ACI Mater J 112(4):559–568

Arezoumandi M, Smith A, Volz JS, Khayat KH (2015) An experimental study on flexural strength of reinforced concrete beams with 100% recycled concrete aggregate. Eng Struct 88:154–162

Bai WH, Sun BX (2010) Experimental study on flexural behavior of recycled coarse aggregate concrete beam. Appl Mech Mater 29–32:543–548

Braga AM, Silvestre JD, de Brito J (2017) Compared environmental and economic impact from cradle to gate of concrete with natural and recycled coarse aggregates. J Clean Prod 162:529–543

Choi HB, Yi CK, Cho HH, Kang KI (2010) Experimental study on the shear strength of recycled aggregate concrete beams. Mag Concr Res 62(2):103–114

Choi W-C, Kim S-W, Yun H-D (2012) Flexural performance of reinforced recycled aggregate concrete beams. Mag Concr Res 64(9):837–848

Coelho A, de Brito J (2013) Environmental analysis of a construction and demolition waste recycling plant in Portugal - Part I: Energy consumption and CO2emissions. Waste Manag 33(5):1258–1267

Etxeberria M (2004) Experimental study on microstructure and structural behaviour of recycled aggregate concrete. PhD thesis, Universitat Politècnica de Catalunya

Etxeberria M, Marí AR, Vázquez E (2007) Recycled aggregate concrete as structural material. Mater Struct 40(5):529–541

Fathifazl G, Razaqpur AG, Isgor OB, Abbas A, Fournier B, Foo S (2009) Flexural performance of steel-reinforced recycled concrete beams. ACI Struct J 106(6):858–867

Fathifazl G, Razaqpur AG, Isgor OB, Abbas A, Fournier B, Foo S (2010) Shear strength of reinforced recycled concrete beams with stirrups. Mag Concr Res 62(10):685–699

Fathifazl G, Razaqpur AG, Burkan Isgor O, Abbas A, Fournier B, Foo S (2011) Shear capacity evaluation of steel reinforced recycled concrete (RRC) beams. Eng Struct 33(3):1025–1033

Francesconi L, Pani L, Stochino F (2016) Punching shear strength of reinforced recycled concrete slabs. Constr Build Mater 127:248–263

González-Fonteboa B, Martínez-Abella F (2004) Shear strength of concrete with recycled aggregates. Int. RILEM Conf. Use Recycl. Mater. Build. Struct, Barcelona, Spain, pp 619–628

González-Fonteboa B, Martínez-Abella F (2007) Shear strength of recycled concrete beams. Constr Build Mater 21(4):887–893

Ignjatović IS, Marinković SB, Mišković ZM, Savić AR (2012) Flexural behavior of reinforced recycled aggregate concrete beams under short-term loading. Mater Struct 46(6):1045–1059

Ignjatović IS, Marinković S, Tošić N (2017) Shear behaviour of recycled aggregate concrete beams with and without shear reinforcement. Eng Struct 141:386–401

Kang THK, Kim W, Kwak Y-K, Hong S-G (2014) Flexural testing of reinforced concrete beams with recycled concrete aggregates. ACI Struct J 111(3):607

Katkhuda H, Shatarat N (2016) Shear behavior of reinforced concrete beams using treated recycled concrete aggregate. Constr Build Mater 125:63–71

Kim S-W, Jeong C-Y, Lee J-S, Kim K-H (2013) Size effect in shear failure of reinforced concrete beams with recycled aggregate. J Asian Architect Build Eng 12(2):323–330

Knaack AM, Kurama YC (2014) Behavior of reinforced concrete beams with recycled concrete coarse aggregates. J Struct Eng 141(3)

Knaack AM (2013) Sustainable concrete structures using recycled concrete aggregate: short-term and long-term behavior considering material variability. PhD thesis, University of Notre Dame

Knaack AM, Kurama YC (2015) Creep and shrinkage of normal-strength concrete with recycled concrete aggregates. ACI Mater J 112(3):451–462

Mahmoud ZI, El tony ETM, Saeed KS (2017) Punching shear behavior of recycled aggregate reinforced concrete slabs. Alex Eng J 4–12

Pradhan S (2019) Performance of recycled aggregate concrete and structural members: particle packing method of mix design approach. PhD thesis, Indian Institute of Technology Kharagpur

Pradhan S, Kumar S, Barai SV (2018) Performance of reinforced recycled aggregate concrete beams in flexure?: experimental and critical comparative analysis. Mater Struct 51(58):1–17

Pradhan S, Kumar S, Barai SV (2018) Shear performance of recycled aggregate concrete beams: an insight for design aspects. Constr Build Mater 178:593–611

Pradhan S, Tiwari BR, Kumar S, Barai SV (2019) Comparative LCA of recycled and natural aggregate concrete using Particle Packing Method and conventional method of design mix. J Clean Prod 228:679–691

Rahal KN, Alrefaei YT (2017) Shear strength of longitudinally reinforced recycled aggregate concrete beams. Eng Struct 145:273–282

Rao HS, Reddy VSK, Ghorpade VG (2012) Influence of recycled coarse aggregate on punching behaviour of recycled coarse aggregate concrete slabs. Int J Mod Eng Res 2(4):2815–2820

Reis N, de Brito J, Correia JR, Arruda MR (2015) Punching behaviour of concrete slabs incorporating coarse recycled concrete aggregates. Eng Struct 100:238–248

Sahoo S, Singh B (2020) Recycled aggregate concrete slab punching shear capacity. Structures 24(2):426–443

Sahoo S, Singh B (2021) Punching shear capacity of recycled-aggregate concrete slab-column connections. J Build Eng 41(3):102430

Sato R, Maruyama I, Sogabe T, Sogo M (2007) Flexural behavior of reinforced recycled concrete beams. J Adv Concr Technol 5(1):43–61

Xiao J, Wang W, Zhou Z, Tawana MM (2019) Punching shear behavior of recycled aggregate concrete slabs with and without steel fibres. Front Struct Civil Eng 13(3):725–740

The manufacturer's authorised representative in the EU is Springer Nature Customer Service Centre GmbH, Europaplatz 3, 69115 Heidelberg, Germany. If you have any concerns regarding our products, please contact ProductSafety@springernature.com

Printed and bound by CPI Group (UK) Ltd, Croydon, CR0 4YY
30/09/2026
02225292-0002